DISSIPATIVE PHASE TRANSITIONS

Series on Advances in Mathematics for Applied Sciences – Vol. 71

DISSIPATIVE PHASE TRANSITIONS

Editors

Pierluigi Colli
University of Pavia, Italy

Nobuyuki Kenmochi
Chiba University, Japan

Jürgen Sprekels
WIAS – Berlin, Germany

World Scientific

NEW JERSEY • LONDON • SINGAPORE • BEIJING • SHANGHAI • HONG KONG • TAIPEI • CHENNAI

Published by

World Scientific Publishing Co. Pte. Ltd.

5 Toh Tuck Link, Singapore 596224

USA office: 27 Warren Street, Suite 401-402, Hackensack, NJ 07601

UK office: 57 Shelton Street, Covent Garden, London WC2H 9HE

Library of Congress Cataloging-in-Publication Data

Dissipative phase transitions / editors, Pierluigi Colli, Nobuyuki Kenmochi & Jürgen Sprekels.
p. cm. -- (Series on advances in mathematics for applied sciences ; v. 71)
Includes bibliographical references.
ISBN 981-256-650-3 (alk. paper)
1. Phase transformations (Statistical physics). 2. Phase transformations (Statistical physics)--Mathematical models. 3. Energy dissipation. I. Colli, P. (Pierluigi), 1958– II. Kenmochi, Nobuyuki. III. Sprekels, J. IV. Series.

QC175.16.P5 D57 2006
530.4'74--dc22

2005046719

British Library Cataloguing-in-Publication Data

A catalogue record for this book is available from the British Library.

Printed in Singapore by B & JO Enterprise

PREFACE

Phase transition phenomena arise in a variety of relevant real world situations like e.g. melting and freezing in a solid-liquid system, evaporation, solid-solid phase transitions in shape memory alloys, combustion, crystal growth, damage in elastic materials, glass formation, phase transitions in polymers, plasticity, and so on.

The practical interest of such phenomena is evident and has deeply influenced the technological development of our society, indeed stimulating an intense mathematical research. The crucial issue of describing some relevant phase change situations adequately requires the joint contribution of several different knowledges, ranging from continuum thermomechanics, to the mathematical analysis of PDE's, to approximation and computation.

The possibly dissipative character of phase transition phenomena is often a guideline in all of the aforementioned steps of the study. In fact, the concept of dissipation occurs in the analysis of the thermodynamic consistency of the models. Furthermore, it is related to the expected regularity of solutions and the possible stability of suitable discretization schemes.

Let us discuss some particular examples in which dissipation plays a key role. As a matter of fact, the well-known and recently developed phase field models have been originally introduced in order to include phase dissipation effects in the dynamics of moving interfaces, arising in thermally induced phase transitions. Actually, in many cases the moving front is spread and forms a thin layer, which looks like a neighborhood of a moving surface. For instance, to represent such a moving surface the Penrose-Fife model yields a thermodynamically consistent description in terms of the inverse temperature.

Damage of materials can be modelled by damage quantities within the framework of continuum mechanics. The constitutive equations are derived by a free energy functional and a pseudo-potential of dissipation, and they are coupled with balance laws, resulting from the principle of virtual power to provide predictive theories. In this order of ideas, an example of mathematical model is the one proposed by Frémond, consisting of differential equations for the elastic displacement and the damage field, which is treated as an internal quantity. When compared to experimental results, the resulting model can predict the behaviour of concrete structures and the structural size effect in civil engineering.

Another example of phase transformation models is given by the recently investigated mathematical systems for the thermo-mechanical evolution of shape memory alloys. In this framework as well, dissipation is very important in the phenomenon dynamics. Indeed, it sheds light on this surprising effect of recovering the former shape by thermal influence. Moreover, dissipation may lead to the appearance of hysteretic behaviour.

This contributed book follows the above guidelines and aims to extend the field of investigation and applications. In fact, it collects some efforts in the direction of describing, approximating, and especially analysing models and related PDE's problems, which involve phase transitions in different contexts. Of course, the occurrence of dissipation effects is a leading theme in most of the articles. The editors are very grateful to the authors for their valuable work and to the WS staff for their accompanying support. Finally, special thanks are due to Riccarda Rossi for her competent and significant help with finalizing this book.

Pierluigi Colli
Department of Mathematics
University of Pavia
Pavia, Italy

Nobuyuki Kenmochi
Department of Mathematics
Faculty of Education
Chiba University, Japan

Jürgen Sprekels
WIAS – Weierstrass Institute
for Applied Analysis and Stochastics
Berlin, Germany

CONTENTS

DISSIPATIVE PHASE TRANSITIONS

MATHEMATICAL MODELS INCLUDING A HYSTERESIS OPERATOR

Toyohiko Aiki

Department of Mathematics
Faculty of Education, Gifu University
Yanagido 1-1, Gifu, 501-1193, Japan
E-mail: aiki@cc.gifu-u.ac.jp

The purpose of the present paper is to summarize our recent works dealing with hysteresis. Although there are several approaches to study hysteresis operators, here we focus on the characterization of hysteresis operators by using ordinary differential equations. Applying this method, we have made some progress in the fields related to shape memory alloys, solid-liquid phase transitions, magnetic materials and biological models. In this paper, we introduce our mathematical models consisting of partial differential equations and ordinary differential equations, and show theorems concerned with the well-posedness of the models.

1. Introduction

In this paper, we address the mathematical treatment of hysteresis operators. We can see hysteresis effects when we deal with some nonlinear phenomena such as, for example, shape memory alloys, magnetic materials, solid-liquid phase transitions, biological models, saturation problems and so on. All models treated here are analyzed by applying the differential equation characterization method to the related hysteresis operator. In view of this characterization, the input-output relationship of the hysteresis operator is expressed in terms of an ordinary differential equation, including the subdifferential of the indicator function of a closed interval, where the interval depends on the input function.

Let us state the advantages of this method. First, it is easy to approximate the hysteresis operator. When we analyze a mathematical model for a phenomena in which hysteresis effects appear, the model often consists of nonlinear partial differential equations. Hence, in some cases it is necessary to consider an approximate problem of the original one, since it may be hard to get mathematical results because of the nonlinearity of the original problem. The differential equation characterization method provides more possibilities than the direct treatment to approximate the hysteresis operator, such as, for example, Yosida approximation, to add a laplacian term to the ordinary different equation.

The second advantage of this method is that it enables to describe the dependence of the hysteresis operator on some parameter. To deal with hysteresis phenomena, we may regard the shape of the hysteresis loop as a function of some parameter. If we assume that this dependence is Lipschitz continuous, then, by applying the differential equation characterization method it is sufficient to suppose that the functions defining the interval depend on the parameter. Thus, this method is useful in order to investigate a system including several parameters.

The plan of this paper is as follows. In Section 2 we shall introduce the differential equation characterization method by using examples of hysteresis operators. The subject of Section 3 is shape memory alloys, which is a typical example of phenomenon displaying hysteresis effects. In the dynamics of shape memory alloys the relationship between strain, stress and temperature is hysteresis. In this section, we shall illustrate the mathematical modelling for 1-D shape memory alloys and the well-posedness of our problem. Also, we mention recent works on the three dimensional problems. In the last section, we give a brief review about results on some mathematical models for phase transition phenomena, magnetization processes, and biological problems.

2. Mathematical treatment for hysteresis operator

The content of this section can be traced back to the books by Krasnoselskii and Pokrovskii [23], Visintin [38], and Brokate and Sprekels [9]. Here, we consider only the differential equation characterization method and omit the precise definition of hysteresis operator (see the above references).

2.1. *Play operator*

First, we introduce a play operator as a typical example of hysteresis operator. It is easy to see that $u(t)$ is decided by the play operator defined by Figure 1, with input function $w(t)$, if and only if u is a solution of the

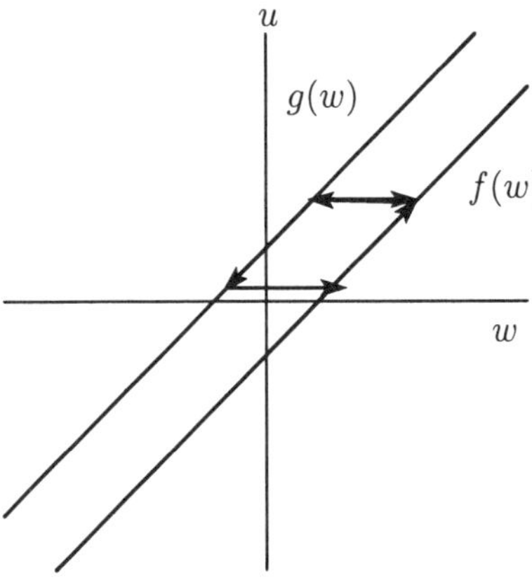

Fig. 1.1. Play operator

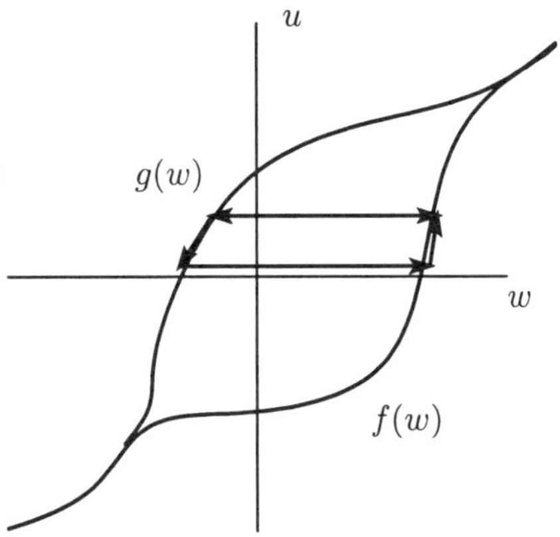

Fig. 1.2. Generalized play operator

following ordinary differential equation:

$$\frac{d}{dt}u + \partial I(w; u) \ni 0, \tag{1}$$

where $I(w; \cdot)$ is the indicator function of the closed interval of $[f(w), g(w)]$, ∂I is the subdifferential of I, and $f(w) = w - 1$, $g(w) = w + 1$ are the lower and the upper lines in Figure 1.1, respectively.

Next, we consider the generalized play operator defined by Figure 1.2. Also, this operator can be represented by equation (1), where f and g are the given lower and upper curves in Figure 1.2, respectively. Let us now show some results concerned with a system including the generalized play operator. In [21], the following system with some boundary conditions and initial conditions was treated:

$$u_t - \Delta u + w = 0 \quad \text{in } (0, T) \times \Omega,$$
$$w_t - \Delta w + \partial I(u; w) \ni 0 \quad \text{in } (0, T) \times \Omega,$$

where $T > 0$, Ω is a bounded domain in R^3, and $I(u; \cdot)$ is the indicator function of the closed interval $[f(u), g(u)]$, for given functions f and g with $f \leq g$. The second equation is an approximation of the generalized play operator, while the first equation is a heat equation. In this system, we can regard the play operator as a mathematical description of a thermostat device. Existence and uniqueness of the solution were obtained.

Also, Colli, Kenmochi and Kubo [13] studied the following system:

$$u_t + w_t - \Delta u = 0 \quad \text{in } (0,T) \times \Omega,$$
$$w_t - \Delta w + \partial I(u; w) \ni 0 \quad \text{in } (0,T) \times \Omega.$$

In this case, the system is one of the mathematical models of phase transition phenomena, and only the existence of a solution was proved.

2.2. *Stop operator*

By the stop operator we can describe the stress-strain relation in a one-dimensional elastic-plastic element, depicted in Figure 1.3. This type of

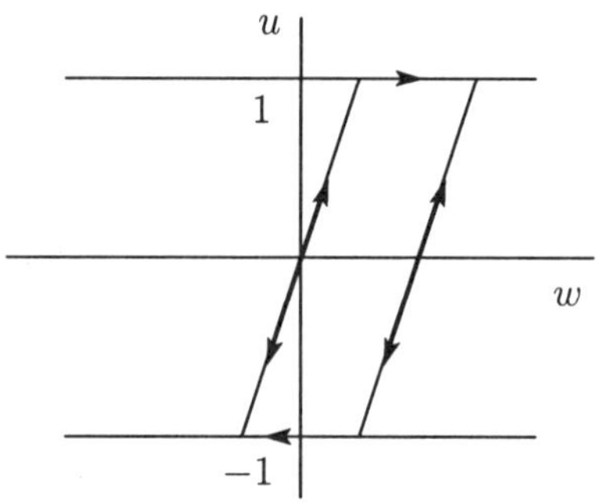

Fig. 1.3. Stop operator

hysteresis operator can be characterized by ordinary differential equations in the following manner:

$$u_t + \partial I(w; u) \ni c w_t,$$

where I is the indicator function of the closed interval $[-1, 1]$, and the positive constant c corresponds to the slope of the line in the hysteresis loop. This fact was already pointed out by Krejčí [24]. As in the case of the play operator, we can consider a generalized stop operator. Minchev, Okazaki and Kenmochi [29] treated the system:

$$c w_t + u_t - \Delta u = h(u, w) \quad \text{in } (0,T) \times \Omega,$$
$$a w_t - \Delta w + b u_t + \partial I(u; w) \ni f(w, u) \quad \text{in } (0,T) \times \Omega,$$

where $h : R^2 \to R$ and $f : R^2 \to R$ are given continuous functions, a and c are positive constants, and b is a real constant (it may be negative), whereas

$I(u;\cdot)$ is the indicator function of the closed interval $[f_*(u), f^*(u)]$, f^* and f_* are corresponding to the upper and the lower curves of the hysteresis loop, respectively. The second equation provides an approximation of the generalized stop operator.

2.3. *The Duhem model*

A magnetization process is a famous example of hysteresis phenomenon, and it has more complicated features than the play and stop operators. The complexity is due to the fact that in the hysteresis loop the increasing and the decreasing ways are different. Several mathematical models for magnetization have been proposed. In Chapter 9 of the book by Visintin [39] the literature on early works for ferro-magnetism is reviewed. Also, Visintin [40, 41, 42] has proposed some mathematical models for ferromagnetism by using some types of hysteresis operators. As for the other approach to micro-magnetics, we shall show results obtained by Roubíček, Kružík and Prohl [35, 25].

In our work [4], we have adopted the following Duhem model as a mathematical description for the magnetization process:

$$\frac{dm}{dt} = g_1(m,h)[\frac{dh}{dt}]^+ - g_2(m,h)[\frac{dh}{dt}]^-,$$

where m and h denote the magnetization and the applied field, respectively, g_1 and g_2 are given functions on R^2. Basic properties of the Duhem model are discussed in Chapter 5 of the book by Visintin [38]. In the paper [4], we have considered a generalization of the Duhem model, i.e.

$$\frac{dm}{dt} + \partial I(h;m) \ni g_1(h,m)[\frac{dh}{dt}]^+ - g_2(h,m)[\frac{dh}{dt}]^-, \tag{2}$$

where $I(h;\cdot)$ is the indicator function of the closed interval $[f_*(h), f^*(h)]$ and f_* and f^* are continuous functions corresponding to the lower and the upper curves in the hysteresis loop, respectively, which can be determined by experimental data. Here, we show the graph, Figure 1.4, obtained by the numerical simulation for (2), with the input function $h(t) = t\sin t$.

From Figure 1.4 we have chosen the Duhem model as the mathematical description of magnetization process. There is another reason for adopting the Duhem model for the mathematical description of the magnetization process. Indeed, some magnetic materials have a piezomagneticity. This means that the shape of the hysteresis loop and the behavior of m in the interior of the loop may change according to the stress σ and the magnetization m. Hence, we must assume that the functions f^*, f_*, g_1 and

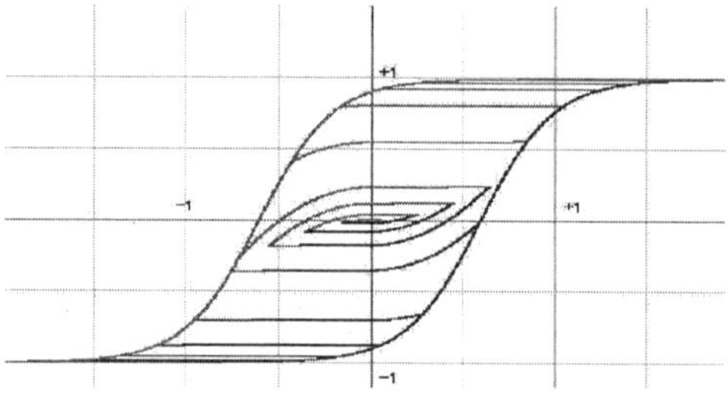

Fig. 1.4. Numerical result for the Duhem model

g_2 in (2) depend on h, σ and m. In our formulation, it is easy to handle mathematically such a nonlinear effect. Thus, we have proposed the following differential equation as a mathematical model for magnetization under piezomagneticity:

$$\frac{dm}{dt} + \partial J(h, \sigma_e, m; m) \ni G(h, h_t, m, \sigma_e),$$

where σ_e is the elastic stress and $G(h, h_t, m, \sigma_e) = (g_1(m,h) + \phi(m, h, \sigma_e))[\frac{dh}{dt}]^+ - (g_2(m,h) + \phi_2(m, h, \sigma_e))[\frac{dh}{dt}]^-$, ϕ being a given function on R^3 with $\phi(m, h, 0) = 0$, and $J(h, \sigma_e, m; \cdot)$ being the indicator function of the closed interval $[f_*(h) + \phi(m, h, \sigma_e), f^*(h) + \phi(m, h, \sigma_e)]$. We note that the function ϕ is defined from the piezomagneticity of ferromagnet materials. The mathematical result concerning the Duhem model is shown in the last section of this paper.

3. Shape memory alloys

Throughout this section, let σ be the stress, ε be the strain, u be the deformation, and θ be the temperature field. We always assume that the deformation is very small so that $u_x = \varepsilon$. By some experiments we have already obtained the following load-deformation curves (Figure 1.5), where θ_c is some critical temperature.

There are several ways of describing this relationship mathematically. The Falk model proposes a meaningful approach. Now, we give a brief explanation of the Falk model (see Brokate and Sprekels [9] in detail). I. Müller and his co-workers have developed a mathematical model for one dimensional shape memory alloy materials. Let the Helmholtz free energy function Ψ be defined by:

$$\Psi(\theta, \varepsilon) := -\theta \log \theta + \theta + \kappa_1(\theta - \theta_c)\varepsilon^2 - \kappa_2\varepsilon^4 + \kappa_3\varepsilon^6,$$

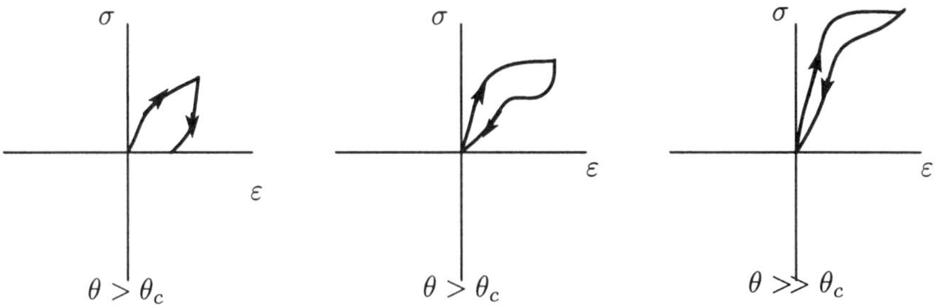

Fig. 1.5. The relationship between stress, strain and temperature

where κ_1, κ_2 and κ_3 are positive constants. Then, by standard free energy theory we obtain

$$\sigma = \frac{\partial \Psi}{\partial \varepsilon} = 2\kappa_1(\theta - \theta_c)\varepsilon - 4\kappa_2\varepsilon^3 + 6\kappa_3\varepsilon^5.$$

So, the relationship between the strain and the stress is approximated by means of a polynomial function of ε.

To include non-local effects like interfacial energies, Falk [14] added a quadratic strain gradient term to Ψ, so that the total free energy density is

$$F(\theta, \varepsilon) = \Psi(\theta, \varepsilon) + \frac{\gamma}{2}\varepsilon_x^2,$$

where γ is a positive constant. By using this total free energy function, we obtain the following system (called the Falk model):

$$\begin{aligned}
&u_{tt} + \gamma u_{xxxx} - (f_1(u_x)\theta + f_2(u_x))_x = 0 \quad \text{in } Q(T) := (0,T) \times (0,1),\\
&\theta_t - k\theta_{xx} - f_1(u_x)\theta u_{xt} = 0 \quad \text{in } Q(T),\\
&u(t,0) = u(t,1) = u_{xx}(t,0) = u_{xx}(t,1) = 0 \quad \text{for } 0 \le t \le T,\\
&\theta_x(t,0) = \theta_x(t,1) = 0 \quad \text{for } 0 \le t \le T,\\
&u(0) = u_0, u_t(0) = v_0, \theta(0) = \theta_0 \quad \text{on } (0,1),
\end{aligned}$$

where $T > 0$, f_1 and f_2 are given continuous functions whose typical examples are $f_1(u_x) = 2\kappa_1 u_x$ and $f_2(u_x) = -4\kappa_2 u_x^3 + 6\kappa_3 u_x^5$.

Sprekels and Zheng [36] proved the existence and uniqueness of a classical solution of the above problem. Bubner, Sokolowski and Sprekels [12, 11] gave existence results for weak solutions, under the assumptions $f_1(\varepsilon) = \kappa_1\varepsilon$, $f_2(\varepsilon) = C_1 - \kappa_2'\varepsilon^3 + \kappa_3\varepsilon^5$, and $u_0 \in H^3(0,1)$, $v_0 \in H_0^1(0,1)$ and $\theta_0 \in H^1(0,1)$, where κ_2', κ_3' and C_1 are positive constants. Also, the well-posedness of the above system was shown by Aiki [1], when $f_1, f_2 \in C^2(R)$, $\hat{f}_2(r) \ge -C_2$ for $r \in R$, and $u_0 \in H^3(0,1)$, $v_0 \in H_0^1(0,1)$, $\theta_0 \in H^1(0,1)$, where $\hat{f}_2$ is a primitive of f_2 and C_2 is a positive constant. Recently, Yoshikawa [44] has obtained a more general well-posedness result, with data $u_0 \in H^2(0,1)$, $v_0 \in L^2(0,1)$ and $\theta_0 \in L^1(0,1)$.

Moreover, in order to handle this model easily, one may assume the viscosity of the stress. In this case, the shape memory alloy problem has been treated by Hoffmann and Zochowski [19].

From now on, we state some results developed by the differential equation characterization method. Precisely, in Aiki and Kenmochi [6] we suppose that

$$\sigma_t + \partial I(\theta, \varepsilon; \sigma) \ni c\varepsilon_t, \tag{3}$$

where $c > 0$, $I(\theta, \varepsilon; \cdot)$ is the indicator function of the closed interval $[f_*(\theta,\varepsilon), f^*(\theta,\varepsilon)]$, and f^* and f_* are given continuous functions. By using this mathematical description we have proposed the following problem: find a triplet of functions u, θ, σ satisfying

$$u_{tt} + \gamma u_{xxxx} - \mu u_{xxt} = \sigma_x \quad \text{in } Q(T), \tag{4}$$

$$\theta_t - \kappa\theta_{xx} = \sigma u_{xt} + \mu|u_{xt}|^2 \quad \text{in } Q(T), \tag{5}$$

$$\sigma_t - \nu\sigma_{xx} + \partial I(\theta, \varepsilon; \sigma) \ni c u_{xt} \quad \text{in } Q(T), \tag{6}$$

$$u(t,0) = u(t,1) = u_{xx}(t,0) = u_{xx}(t,1) = 0 \quad \text{for } t \in [0,T], \tag{7}$$

$$\theta_x(t,0) = \theta_x(t,1) = 0 \quad \text{for } t \in [0,T], \tag{8}$$

$$\sigma_x(t,0) = \sigma_x(t,1) = 0 \quad \text{for } t \in [0,T], \tag{9}$$

$$u(0) = u_0, u_t(0) = v_0, \theta(0) = \theta_0, \sigma(0) = \sigma_0 \quad \text{on } (0,1), \tag{10}$$

where $\gamma > 0$, $\nu > 0$, u_0, v_0, θ_0 and σ_0 are given initial functions on $[0,1]$.

Let us now briefly give a derivation for the above system. First, the following two differential equations are derived from the conservation laws of linear momentum and energy, respectively:

$$u_{tt} + \gamma u_{xxxx} = \hat{\sigma}_x \quad \text{in } Q(T),$$

$$U_t + q_x = \hat{\sigma}\varepsilon_t \quad \text{in } Q(T), \tag{11}$$

where $\hat{\sigma}$ is the total stress, U is the internal energy and q is the heat flux. Here, we refer Brokate-Sprekels [9] and Pawlow [31] for the physical background of these equations. By using the classical Fourier law and an elementary approximation, $U_t = \theta_t$, (11) reads $\theta_t - \kappa\theta_{xx} = \hat{\sigma}\varepsilon_t$ in $Q(T)$, where κ is a positive constant depending on the specific heat and the heat conductivity. For some mathematical reasons, we assume that there are interior frictions in the form of viscous stresses in the material. Then, we can apply Hooke's-like law, so that we have $\hat{\sigma} = \sigma + \mu\varepsilon_t$, where $\mu > 0$ is the viscosity constant. Moreover, we approximate the generalized stop operator (3) by adding the laplacian of σ to (3). Therefore, we obtain the system $\{(4) \sim (10)\}$.

Let us review some of the previous literature on the above system. In the first result [6], we studied the system obtained by replacing (5) with the following

$$\theta_t - \kappa\theta_{xx} = \sigma u_{xt} \quad \text{in } Q(T). \tag{12}$$

In order to overcome the mathematical difficulty caused from the term $(|u_{xt}|^2)$ in (5), we have applied the decomposition method to the fourth order partial differential equation (4) as follows. Let $\mu^2 > 4\gamma$. Then, we can take two positive numbers λ_1 and λ_2 such that $\lambda_1 + \lambda_2 = \mu$ and $\lambda_1\lambda_2 = \gamma$. Let us put $z := u_t - \lambda_1 u_{xx}$. Obviously, if u satisfies (4), then it yields that $z_t - \lambda_2 z_{xx} = \sigma_x$. Hence, it is possible to apply the classical theory for parabolic equations. This idea, which was already used by Pawlow and Zochowski [32], enabled us to deal with the full system $\{(4) \sim (10)\}$, and to obtain a well-posedness result in Aiki and Kenmochi [7] in the case $\nu > 0$. In the case $\nu = 0$, a theorem concerned with existence and uniqueness of a solution was proved by Aiki [1]. Let us note that in these two works the assumption $\mu^2 > 4\gamma$ was necessary, since the maximal regularity for parabolic equations is essential in their proofs. However, from the experimental results it is pointed out that $\mu = 0$ and γ is not so small in Bubner [10]. Recently. Aiki, Kadoya and Yoshikawa have obtained the well-posedness of the system $\{(4) \sim (10)\}$ for $\nu = 0$ and any positive numbers μ and γ by using the maximal regularity for parabolic equations with complex coefficients (Lemma 2). Indeed, by this result we can remove the condition $\mu^2 > 4\gamma$, so that it is possible to treat the more realistic problem in our formulation.

Now, let us give the precise assumption on the data and the definition of a solution. To define a solution, we introduce the following notation: $H := L^2(0,1)$, $V := H_0^1(0,1)$, V^* is the dual space of V, and $< \cdot, \cdot >$ is the duality pair on $V \times V^*$. These are the assumptions on the data.

(A1) γ, μ, κ and c are positive constants.

(A2) $f_*, f^* \in C^2(R^2) \cap W^{2,\infty}(R^2)$ and $f_* \leq f^*$ on R^2.

(A3) $u_0 \in H^4(0,1)$, with $u_0(0) = u_0(1) = u_{0xx}(0) = u_{0xx}(1) = 0$, $v_0 \in V \cap H^2(0,1)$, $\theta_0 \in H^1(0,1)$ and $\sigma_0 \in H^1(0,1)$. Moreover, $f_*(\theta_0, \varepsilon_0) \leq \sigma_0 \leq f^*(\theta_0, \varepsilon_0)$ on $(0,1)$, where $\varepsilon_0 = u_{0x}$.

We define a solution of P $= \{(4) \sim (10)\}$ in the following way:
For any $T > 0$ let $\{u, \theta, \sigma\}$ be a triplet of functions u, θ and σ on $Q(T)$. We say that $\{u, \theta, \sigma\}$ is a solution of P on $[0, T]$ if the following conditions (S1) $\sim$ (S4) are satisfied:
(S1) $u \in W^{2,\infty}(0,T;V^*) \cap W^{1,\infty}(0,T;V) \cap W^{1,2}(0,T;H^2(0,1)) \cap L^\infty(0,T;H^3(0,1))$,
$\theta \in W^{1,2}(0,T;H) \cap L^\infty(0,T;H^1(0,1))$, $\sigma \in W^{1,2}(0,T;H) \cap L^\infty(0,T;H^1(0,1))$.
(S2) $u(0,x) = u_0(x)$ and $u_t(0,x) = v_0(x)$ for $x \in (0,1)$, and
$< u_{tt}(t), \eta > -\gamma \int_0^1 u_{xxx}(t)\eta_x dx - \mu \int_0^1 u_{txx}(t)\eta dx = \int_0^1 \sigma_x(t)\eta dx$ for any $\eta \in V$ and a.e. $t \in [0,T]$.
(S3) (5) holds in the usual sense, (8) and the initial condition for θ hold.
(S4) There exists $\xi \in L^2(Q(T))$ such that $\xi(t) \in \partial I(\theta(t), \varepsilon(t); \sigma(t))$ for a.e. $t \in [0,T]$, (6) holds for a.e. $(t,x) \in Q(T)$, and $\sigma(0) = \sigma_0$ a.e. on $(0,1)$.

Theorem 1: (cf. Aiki, Kadoya and Yoshikawa [5]) *Suppose that (A1) $\sim$ (A3) hold. Then:*
(1) (Uniqueness) For any $T > 0$ let $\{u_i, \theta_i, \sigma_i\}$ be a solution of P on $[0,T]$, $i = 1,2$. If $u_{itx} \in L^\infty(Q(T))$ for $i = 1,2$, then $u_1 = u_2$, $\theta_1 = \theta_2$ and $\sigma_1 = \sigma_2$ on $Q(T)$.
(2) (Existence) For any $T > 0$ there exists a solution $\{u, \theta, \sigma\}$ of P on $[0,T]$ satisfying $u \in W^{2,\infty}(0,T;H) \cap W^{2,2}(0,T;H^1(0,1)) \cap W^{1,\infty}(0,T;H^2(0,1))$.

The main idea of the proof is to apply the maximal regularity for parabolic equations with a complex coefficient in the following way.

Lemma 2: *(cf. Hoffmann and Zochowski [19, Section 2]) Let $T > 0$, $p > 1$, $f \in L^2(0,T;H^1(0,1))$, α be a complex number and $w_0 \in W^{2-2/p,p}(0,1)$ with $w_0(0) = w_0(1) = 0$. If $Re\alpha > 0$, then there exists $w \in W^{1,2}(0,T;L^2(0,1)) \cap L^\infty(0,T;H^1(0,1))$ satisfying*

$$\begin{aligned} &w_t - \alpha w_{xx} = f_x \quad \text{in } Q(T), \\ &w(t,0) = w(t,1) = 0 \quad \text{for } 0 < t < T, \\ &w(0,x) = w_0(x) \quad \text{for } 0 \le x \le 1. \end{aligned}$$

Moreover, if $p > 1$ and $f \in L^p(Q(T))$, then there exists a positive constant K_{1p}, depending only on T, α and p, such that

$$|w_x|_{L^p(Q(T))} \le K_{1p}(|f|_{L^p(Q(T))} + |w_0|_{W^{2-2/p,p}(0,1)}).$$

This idea is already used in Hoffmann-Zochowski [19].

Shape memory alloy problems in the three dimensional space have also been investigated. Falk and Konopka [15] considered a three-dimensional Landau theory for martensitic phase transformations in shape memory alloys, and Pawlow [31] proposed a mathematical model, which is a system consisting of differential equations. For the mathematical results concerning the Falk-Konopka model, we refer to Pawlow and Zochowski [32], Zimmer [45], Pawlow and Zajaczkowski [34, 33], and Yoshikawa [43]. Also, the Falk-Konopka model with the generalized stop operator was investigated by Aiki [3].

At the end of this section, we mention about the Frémond Model, which was introduced by Frémond [16] and was investigated by many mathematicians (see Chapter 13 in the book by Frémond [17] and Chapter 5 in the book by Brokate and Sprekels [9]).

4. Examples of hysteresis operator

The aim of this section is to introduce several topics in which hysteresis effects appear.

4.1. *Solid-liquid phase transition*

In this section, we consider a material which occupies a bounded domain $\Omega \subset R^N$, $N \leq 3$, with smooth boundary Γ, and the domain consists of three parts, solid, liquid and mushy regions. Let θ be the temperature field. We denote by w the order parameter, which means that $-1 \leq w \leq 1$ and $w = 1$ in the liquid region, $w = -1$ in the solid region, and $|w| < 1$ in the mushy region. The Stefan problem is known as the classical mathematical model for phase transitions. Here, we consider the so-called enthalpy formulation of Stefan problem, defined by

$$(\theta + w)_t - \Delta\theta = 0, \quad w \in H(\theta) \quad \text{in } Q(T) := (0,T) \times \Omega, \tag{13}$$

where $T > 0$ and H is the maximal monotone graph in R^2 (Figure 1.6) defined by

$$H(\theta) = \begin{cases} \{1\} & \text{for } \theta > 0, \\ [-1,1] & \text{for } \theta = 0, \\ \{-1\} & \text{for } \theta < 0. \end{cases}$$

This model does not account for superheating supercooling phenomena, since the state is solid if and only if the temperature is negative. To describe such a phenomena mathematically, several models have been developed.

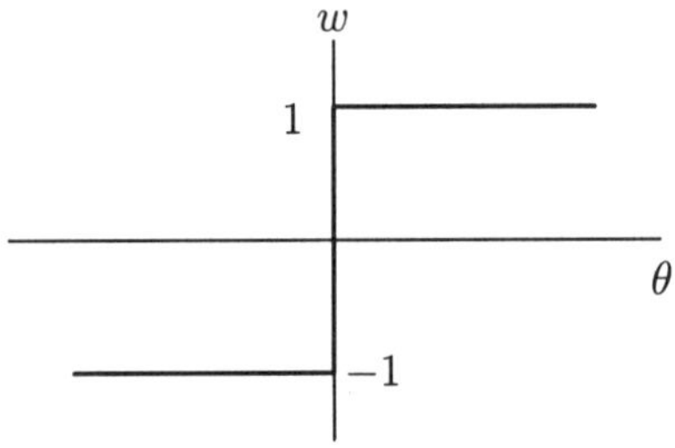

Fig. 1.6. Graph of order parameter

One of them is the following model, which includes a hysteresis operator, proposed by Visintin [37]:

$$(\theta + w)_t - \Delta\theta = 0, \quad w = \mathcal{H}(\theta) \quad \text{in } Q(T), \tag{14}$$

where $\mathcal{H}$ is the hysteresis operator (Figure 1.7) defined by

$$\begin{cases} \text{If } \theta(t) < \theta_1(\text{ respect. } \theta(t) > \theta_2), \text{ then } w(t) = -1(\text{ respect. } w(t) = 1), \\ \text{if } \theta_1 \le \theta(t) \le \theta_2, \text{ then } -1 \le w(t) \le 1, \\ \text{if } \theta_1 < \theta(t) < \theta_2, \text{ then } w(t) \text{ is constant in a neighborhood of } \theta(t), \\ \text{if } \theta(t) = \theta_1(\text{ respect. } \theta(t) = \theta_2), \text{ then } w(t) \text{ is non-increasing} \\ (\text{ respect. is non-decreasing}). \end{cases}$$

Visintin [37] proved the existence of a solution to the related PDE system, whereas Hilpert [18] proved the uniqueness. Moreover, Visintin obtained the following convergence result. Let θ_ε be a solution of (14) with $\mathcal{H}_\varepsilon$ (Figure 1.8) for any $\varepsilon > 0$. Then w_ε converges to a solution w of (14) as $\varepsilon \downarrow 0$.

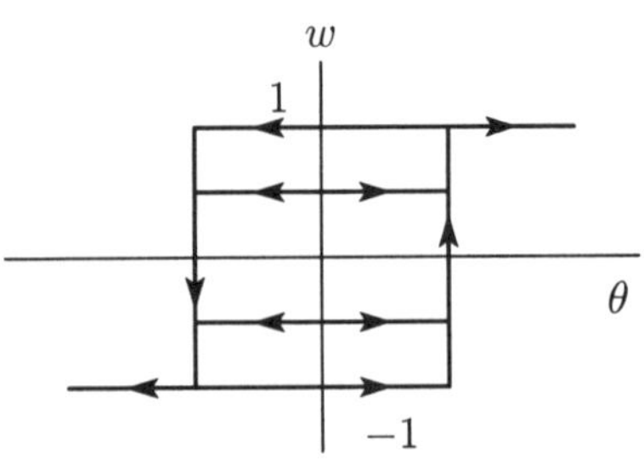

Fig. 1.7. Graph of hysteresis of order parameter

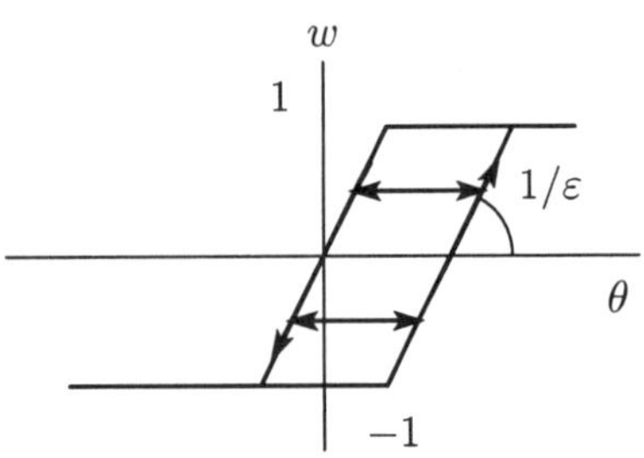

Fig. 1.8. Approximation of hysteresis

The phase-field equations provide another mathematical model for phase transitions.

$$(\theta + w)_t - \Delta\theta = 0 \quad \text{in } Q(T), \tag{15}$$

$$w_t - \kappa\Delta w + \beta(w) + g(w) \ni \theta \text{ in } Q(T), \tag{16}$$

where κ is a positive constant, β is a maximal monotone graph and g is a continuous function on R. The graph in Figure 1.9 is a typical example of $\beta + g$.

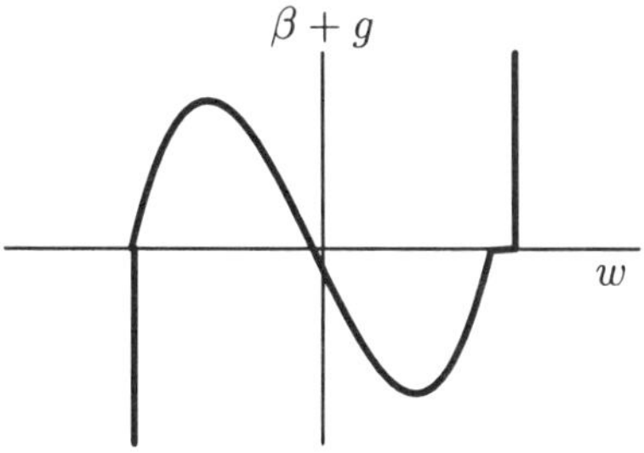

Fig. 1.9. Approximation of hysteresis

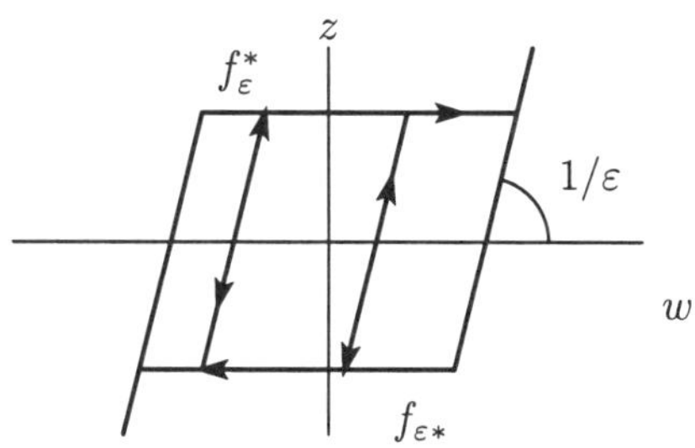

Fig. 1.10. Approximation of hysteresis

In the system $\{(15), (16)\}$ by regarding the term $\beta+g$ as the approximation of the inverse of a hysteresis operator $\mathcal{H}$, we obtain a new mathematical model for phase transitions. However, it is very difficult to deal directly with the inverse of $\mathcal{H}$. Hence, we consider the following system (P_ε), (17) $\sim$ (21), which contains the approximation of the inverse of $\mathcal{H}$ (Figure 1.10):

$$(\theta + w)_t - \Delta\theta = 0 \quad \text{in } Q(T), \tag{17}$$

$$w_t - \kappa\Delta w + z = \theta \quad \text{in } Q(T), \tag{18}$$

$$z_t + \partial I_\varepsilon(w; z) \ni \frac{1}{\varepsilon} w_t \quad \text{in } Q(T), \tag{19}$$

$$\frac{\partial\theta}{\partial\nu} + a(\theta - h) = 0 \text{ and } \frac{\partial w}{\partial\nu} = 0 \quad \text{on } (0,T) \times \Gamma, \tag{20}$$

$$\theta(0) = \theta_0, w(0) = w_0 \text{ and } z(0) = z_0 \quad \text{on } \Omega, \tag{21}$$

where a is a positive constant, ν is the outward normal unit vector on Γ, h is a given function on $(0,T) \times \Gamma$, u_0, w_0 and z_0 are initial data, $I_\varepsilon(w; \cdot)$ is the indicator function of the closed interval of $[f_{*\varepsilon}(w), f_\varepsilon^*(w)]$ (Figure 1.10), and $\partial I_\varepsilon(w; \cdot)$ is its subdifferential. The ordinary differential equation (19) is the characterization of the stop operator.

Before giving theorems on the existence of a solution to (P_ε) and the behavior of solutions as $\varepsilon \downarrow 0$, let us state the assumptions on the data and the definition of a solution.

Let κ and a be positive numbers, $H = L^2(\Omega)$, $X = H^1(\Omega)$ and $d\Gamma$ be a surface element of the boundary. Also, for $0 < \varepsilon < 1$ let us put

$$f_{\varepsilon *}(r) = \begin{cases} \frac{1}{\varepsilon}(r-1) & \text{for } r \geq 1-\varepsilon^2, \\ -\varepsilon & \text{for } -1-\varepsilon^2 < r < 1-\varepsilon^2, \\ \frac{1}{\varepsilon}(r+1) & \text{for } r \leq -1-\varepsilon^2, \end{cases}$$

$$f_\varepsilon^*(r) = \begin{cases} \frac{1}{\varepsilon}(r-1) & \text{for } r \geq 1+\varepsilon^2, \\ -\varepsilon & \text{for } -1+\varepsilon^2 < r < 1+\varepsilon^2, \\ \frac{1}{\varepsilon}(r+1) & \text{for } r \leq -1+\varepsilon^2. \end{cases}$$

Then, we define a solution of (P_ε) in the following way.

Definition 3: A triplet $\{\theta, w, z\}$ of functions θ, w and z on $Q(T)$ is a solution of (P_ε) on $[0,T]$ if and only if the following conditions hold:
$\theta \in W^{1,2}(0,T;H) \cap L^\infty(0,T;X)$, $w \in W^{1,2}(0,T;H) \cap L^\infty(0,T;X)$, and $z \in W^{1,2}(0,T;H)$;
$\int_\Omega (\theta_t(t) + w_t(t))\eta dx + \int_\Omega \nabla\theta(t) \cdot \nabla\eta dx + a\int_\Gamma u(t)\eta d\Gamma = \int_\Omega h(t)\eta d\Gamma$ for any $\eta \in X$ and a.e. $t \in [0,T]$;
$\int_\Omega w_t(t)\eta dx + \int_\Omega \nabla w(t) \cdot \nabla\eta dx + \int_\Omega z(t)\eta dx = \int_\Omega \theta(t)\eta dx$ for any $\eta \in X$ and a.e. $t \in [0,T]$;
there exists $\xi \in L^2(0,T;H)$ such that $\xi(t) \in \partial I_\varepsilon(w(t);z(t))$ and $z_t(t) + \xi(t) = \dfrac{1}{\varepsilon}w_t(t)$ in H for a.e. $t \in [0,T]$;
$\theta(0) = \theta_0$, $w(0) = w_0$ and $z(0) = z_0$.

The following theorem guarantees the well-posedness of (P_ε).

Theorem 4: (cf. Aiki [2].) If $h \in W^{1,2}(0,T;L^2(\Gamma))$, $\theta_0 \in X$, $w_0 \in X$, $z_0 \in H$ with $f_*(w_0) \leq z_0 \leq f^*(w_0)$ a.e. on Ω, then (P_ε) has a unique solution on $[0,T]$.

The following two lemmas are essential for proving uniqueness. The first lemma guarantees the continuity of the generalized stop operator (note that similar lemmas were proved in the papers [21, 6]). The second lemma yields ant L^∞-estimate, which is the classical result of the book by Ladyzenskja, Solonnikov and Ural'ceva [26].

Lemma 5: *For $\varepsilon > 0$ and $i = 1, 2$, let $w_i \in W^{1,2}(0,T;H) \cap L^\infty(Q(T))$ and z_i be a solution of*

$$z_{it} + \partial I_\varepsilon(w_i; z_i) \ni cw_{it} \quad \text{in } Q(T), \quad z_i(0) = z_0,$$

where $z_0 \in H$. If $f_{\varepsilon *}(w_i(0)) \le z_0 \le f^*_\varepsilon(w_i(0))$ a.e. on Ω for $i = 1, 2$, then

$$|z|^2_{L^\infty(0,s;L^\infty(\Omega))} \le C(|w|^2_{L^\infty(Q(s))} + e^s \int_0^s |w_\tau(\tau)|^2_H d\tau) \text{ for } 0 \le s \le T,$$

where $w = w_1 - w_2$, $z = z_1 - z_2$, and C is a positive constant.

Lemma 6: *(cf. Ladyzenskja, Solonnikov and Ural'ceva [26].) Let w be a solution of the following problem:*

$$\begin{aligned} &w_t - \Delta w = f \quad \text{in } Q(T), \\ &\frac{\partial w}{\partial \nu} = 0 \quad \text{on } (0,T) \times \Gamma, \\ &w(0) = 0 \quad \text{on } \Omega. \end{aligned}$$

If $f \in L^r(0,T;L^q(\Omega))$ with $\dfrac{1}{r} + \dfrac{3}{2q} < 1$, then there exists a positive constant C such that

$$|w|_{L^\infty(Q(t))} \le C|f|_{L^r(0,t;L^q(\Omega))} \quad \text{for } 0 \le t \le T.$$

Next, we consider the asymptotic behavior of solutions of (P_ε) as $\varepsilon \downarrow 0$. In order to give the statement, we introduce the following proper l.s.c. convex functions ψ and ϕ on H:

$$\psi(w) = \begin{cases} 0 & w \in K, \\ +\infty & \text{otherwise,} \end{cases} \quad \phi(w) = \begin{cases} \frac{\kappa}{2}|\nabla w|^2_H + \psi(w) & \text{if } w \in X \cap K, \\ +\infty & \text{otherwise,} \end{cases}$$

where $K = \{z \in H : -1 \le z \le 1 \text{ a.e. on } \Omega\}$. By using this notation, we define a solution of the system (P) consisting of (15), (16) with $g = 0$, (20), $\theta(0) = \theta_0$, and $w(0) = w_0$, as follows:

Definition 7: We say that a pair $\{\theta, w\}$ is a solution of (P) on $[0,T]$ if and only if the following conditions hold.
$\theta \in W^{1,2}(0,T;H) \cap L^\infty(0,T;X)$ and $w \in W^{1,2}(0,T;H) \cap L^\infty(0,T;X)$;
$\int_\Omega (\theta_t(t) + w_t(t))\eta dx + \int_\Omega \nabla\theta(t) \cdot \nabla\eta dx + \int_\Omega z(t)\eta dx = \int_\Omega \theta(t)\eta dx$ for any $\eta \in X$ and a.e. $t \in [0,T]$;
there exists $\xi \in L^2(Q(T))$ such that $\xi(t) \in \partial\phi(w(t))$ for a.e. $t \in [0,T]$ and $w_t + \xi = \theta$ a.e. on $Q(T)$;
$\theta(0) = \theta_0$ and $w(0) = w_0$ on Ω.

Theorem 8: (cf. Aiki [2].) Let $\theta_0 \in X$ and $w_0 \in H^2(\Omega)$, with $-1 \leq w_0 \leq 1$. For $\varepsilon \in (0,1)$, let $\{\theta_\varepsilon, w_\varepsilon, z_\varepsilon\}$ be a solution of (P_ε), with $z_0 = w_0 =: z_{0\varepsilon}$. Then we have the following convergences as $\varepsilon \downarrow 0$: $\theta_\varepsilon \to \theta$ in $C([0,T];H)$, weakly in $W^{1,2}(0,T;H)$, weakly* in $L^\infty(0,T;X)$, and $w_\varepsilon \to w$ in $C([0,T];H)$, weakly* in $W^{1,\infty}(0,T;H)$, weakly in $W^{1,2}(0,T;X)$, where $\{\theta, w\}$ is a solution of (P) on $[0,T]$.

4.2. *Biological model*

The papers [22] by Kernevez, Joly, Duban, Bunow and Thomas and [27] by Landau, Lorente, Henry and Canu indicate that hysteresis also occurs in various biological problems. However, the mathematical treatment of biological problems with hysteresis has been considered only in few papers, see Hoppensteadt, Jäger and Pöppe [20], as well as the survey paper by Macki, Nistri and Zecca [28]. So, we have studied the following problem:

$$\sigma_t - (\lambda(u))_t - \kappa\Delta\sigma + \partial I_{u,v}(\sigma) \ni F(\sigma, u, v) \quad \text{in } Q(T), \tag{22}$$

$$u_t - \Delta u = h(\sigma, u, v) \quad \text{in } Q(T), \tag{23}$$

$$v_t - \Delta v = g(\sigma, u, v) \quad \text{in } Q(T), \tag{24}$$

$$\frac{\partial \sigma}{\partial \nu} = \frac{\partial u}{\partial \nu} = \frac{\partial v}{\partial \nu} = 0 \quad \text{on } (0,T) \times \partial\Omega, \tag{25}$$

$$\sigma(0) = \sigma_0, u(0) = u_0, v(0) = v_0 \quad \text{on } \Omega, \tag{26}$$

where $T > 0$, $\Omega \subset R^N$ is a bounded domain with smooth boundary, $Q(T) = (0,T) \times \Omega$; $\kappa \geq 0$ is a constant; ν is the outward normal vector on the boundary; $\lambda : R \to R$, $F, h, g : R^3 \to R$ are given functions and σ_0, u_0 and v_0 are given initial functions; $I_{u,v}(\cdot)$ is the indicator function of the interval $[f_*(u,v), f^*(u,v)]$, $\partial I_{u,v}(\cdot)$ denotes the subdifferential of $I_{u,v}(\cdot)$, and $f_*, f^* : R^2 \to R$ are given continuous functions.

Equation (22) describes the kinetics of the density of the quantity of food σ for the prey, (23) and (24) are the evolution equations of the prey and the evolution of the predator, respectively; here u and v are the densities of the prey and predator, respectively.

Well-posedness for the above system has been discussed by Aiki, Minchev and Okazaki [8].

Definition 9: Let $\kappa \geq 0$. A triplet of functions $\{\sigma, u, v\}$ is called a solution of the system (22)-(26) if:

(i) $\sigma \in W^{1,2}(0,T;H) \cap L^\infty(0,T;X) \cap L^2(0,T;H^2(\Omega))$ if $\kappa > 0$, and $\sigma \in W^{1,2}(0,T;H)$ if $\kappa = 0$.

(ii) $u, v \in W^{1,2}(0,T;H) \cap L^\infty(0,T;X) \cap L^2(0,T;H^2(\Omega))$.
(iii) (22) $\sim$ (26) hold in the usual sense,
where $H = L^2(\Omega)$, $X = H^1(\Omega)$.

Theorem 10: (cf. Aiki, Minchev and Okazaki [8].) Assume that the following conditions (H1) $\sim$ (H4) hold.

(H1) $\kappa \geq 0$, $\lambda \in C^2(R)$, λ' and λ'' are bounded functions on R.

(H2) $f_*, f^* \in C^2(R^2)$, $0 \leq f_* \leq f^* \leq 1$ on R^2, and all partial derivatives of first and second order of f_* and f^* are bounded on R^2.

(H3) F, h and g are Lipschitz continuous functions on R^3, $h(\sigma, 0, v) = 0$ for $\sigma \in [0,1], v \in R$, and $g(\sigma, u, 0) = 0$ for $\sigma \in [0,1], u \in R$.

(H4) $\sigma_0, u_0, v_0 \in L^\infty(\Omega) \cap V$ and $u_0 \geq 0$, $v_0 \geq 0$, $f_*(u_0, v_0) \leq \sigma_0 \leq f^*(u_0, v_0)$ a.e. in Ω.

Then, there exists a constant $\kappa_0 > 0$ such that for $0 < \kappa < \kappa_0$ there exists at least one solution of the system (22) $\sim$ (26). Moreover, if $\Omega \subset R^3$, then the system admits at most one solution.

4.3. *Magnetostrictive thin film multi-layers*

In this section we show a result by Aiki, Hoffmann and Okazaki [4], dealing with thin film multi-layers combining ferromagnet with soft magnets. In order to describe the mathematical model for the thin film multi-layers we use the following notation: let $L > 0$, $\Omega = (-L, L) \times (-L, L) \times (0, 3) \subset R^3$ be a whole domain occupied by the multi-layers, $\Omega_1 = (-L, L) \times (-L, L) \times (0, 1)$ and $\Omega_3 = (-L, L) \times (-L, L) \times (2, 3)$ be soft magnetic regions, $\Omega_2 = (-L, L) \times (-L, L) \times (1, 2)$ be the ferromagnetic region, $Q(T) = (0, T) \times \Omega$, $T > 0$, $Q_q(T) = (0, T) \times \Omega_q$, $q = 1, 2, 3$, $\Sigma(T) := (0, T) \times \partial\Omega$, $\Sigma_q(T) = (0, T) \times (-L, L) \times (-L, L) \cap \{x_3 = q\}$, $q = 1, 2$, $u = (u_1, u_2, u_3)$ be the deformation vector, $m = (m_1, m_2, m_3)$ be the magnetization, $h = (h_1, h_2, h_3)$ be the applied magnetic field, and $\sigma = (\sigma_{ij})$, $\epsilon = (\epsilon_{ij})$ be the stress and the strain tensors, respectively.

The problem is to find a pair of functions u and m satisfying

$$u_{tt} + \gamma\Delta(\Delta u) - \mu\Delta u_t + \text{ div } (A_q\epsilon + B_q\lambda_q(m, h)) = 0 \quad \text{in } Q_q(T) \text{ for } q = 1, 2, 3, \tag{27}$$

$$\frac{dm_i}{dt} + \partial J(h, \sigma, m_i; m_i) \ni G_i(h, h_t, m, \sigma) \text{ in } Q_2(T) \quad \text{for } i = 1, 2, 3, \tag{28}$$

$$m_i = F_i(h) + \phi_q(m_i, h, \sigma) \quad \text{in } Q_q(T), \text{ and } q = 1, 3, \tag{29}$$

$$\sigma = A_q\epsilon(u), \ \epsilon(u) = \frac{1}{2}(\nabla u + {}^t(\nabla u)) \quad \text{in } Q_q(T), \ q = 1, 2, 3, \tag{30}$$

$$u = 0, \ \Delta u = 0 \quad \text{on } \Sigma(T), \tag{31}$$

$$= \left[\frac{\partial u_t}{\partial n}\right] = [\Delta u] = \left[\frac{\partial(\Delta u)}{\partial n}\right] = 0 \quad \text{on } \Sigma_q(T), \text{ for } q = 1, 2, \tag{32}$$

$$u(0) = u_0, \ u_t(0) = v_0 \text{ on } \Omega \ \text{ and } \ m(0) = m_{0f} \text{ on } \Omega_2, \tag{33}$$

where γ and μ are positive constants, λ_q is a continuous function and $A_q = (a^{(q)}_{ijkl})$ and $B_q = (b^{(q)}_{ijkl})$ are 4th order tensors for each $q = 1, 2, 3$, while $J(h, \sigma_e, m; \cdot)$ is the indicator function of the closed interval $[f_*(h) + \phi_2(m, h, \sigma_e), f^*(h) + \phi_2(m, h, \sigma_e)]$, f^*, f_* and ϕ are given functions, $[\cdot]$ denotes the jump across the $\Gamma_q(T)$, $q = 1, 2$, n is the outward normal vector at Ω_q, and u_0, v_0 and m_{0f} are initial functions.

Although the physical derivation of the above system (27) $\sim$ (33) is given in the paper [4], we note that (27) comes from the momentum balance law, the $\lambda_q(m, h)$ on the left-hand side in (27) depends on the magnetostrictivity, and the equation (28) is the Duhem model for hysteresis operator mentioned in section 2. We refer to [4] for an existence and uniqueness result for (27) $\sim$ (33).

References

[1] T. Aiki, Weak solutions for Falk's model of shape memory alloys. *Math. Methods Appl. Sci.* **23**, 299–319 (2000).

[2] T. Aiki, Phase field model including a hysteresis operator. To appear in: *Proceedings of the WCNA 2004*.

[3] T. Aiki, A model of 3D shape memory alloy materials. *J. Math. Soc. Japan* **57**, 903–933 (2005).

[4] T. Aiki, K.-H. Hoffmann and T. Okazaki, Wellposedness for a new mathematical model for magnetostrictive thin film multilayers. *Adv. Math. Sci. Appl.* **14**, 417–442 (2004).

[5] T. Aiki, A. Kadoya and S. Yoshikawa, One-dimensional shape memory alloy problem with small viscosity. To appear in: *Mathematical Approach to Nonlinear Phenomena; Modelling, Analysis and Simulations*, Gakkōtosho, Tokyo.

[6] T. Aiki and N. Kenmochi, Some models for shape memory alloys. In: *Mathematical aspects of modelling structure formation phenomena* (Gakkōtosho, Tokyo, 2002), p. 144.

[7] T. Aiki and N. Kenmochi, Models for shape memory alloys described by subdifferentials of indicator functions. In: *Elliptic and Parabolic Problems (Rolduc/Gaeta, 2001)* (World Sci. Publishing, River Edge NJ, 2002), p. 1.
[8] T. Aiki, E. Minchev and T. Okazaki, A prey – predator model with hysteresis effect. *SIAM J. Math. Anal.* **36**, 2020–2032 (2005).
[9] M. Brokate and J. Sprekels, *Hysteresis and Phase Transitions.* Springer, New York, 1996.
[10] N. Bubner, A mathematical model for deformation-driven experiments on shape memory alloys. *Z. Angew. Math. Mech.* **76**, 79–80 (1996).
[11] N. Bubner, J. Sokolowski and J. Sprekels, Optimal boundary control problems for shape memory alloys under state constraints for stress and temperature. *Numer. Funct. Anal. Optimization* **19**, 489–498 (1998).
[12] N. Bubner and J. Sprekels, Optimal control of martensitic phase transitions in a deformation-driven experiment on shape memory alloys. *Adv. Math. Sci. Appl.* **8**, 299–325 (1998).
[13] P. Colli, N. Kenmochi and M. Kubo, A phase-field model with temperature dependent constraints. *J. Math. Anal. Appl.* **256**, 668–685 (2001).
[14] F. Falk, Elastic phase transitions and nonconvex energy functions. In: *Free Boundary Problems: Theory and Applications. Vol. I*, Ed. K.-H. Hoffmann and J. Sprekels (Longman Sci. Tech, Harlow, 1990), p. 45.
[15] F. Falk and P. Konopka, Three-dimensional Landau theory describing the martensitic phase transformation of shape-memory alloys. *J. Phys.: Condens. Matter* **2**, 61–77 (1990).
[16] M. Freḿond, Mateŕiaux aṁeḿoire de forme. *C. R. Acad. Sci. Paris Seŕ. II* **304**, 239–244 (1987).
[17] M. Frémond, *Non-smooth thermomechanics.* Springer-Verlag, Berlin, 2002.
[18] M. Hilpert, On uniqueness for evolution problems with hysteresis. In: *Mathematical Models for Phase Change Problems*, Ed. J. F. Rodrigues (Birkhäuser, Basel, 1989), p. 377.
[19] K.-H. Hoffmann and A. Zochowski, Existence of solutions to some nonlinear thermoelastic systems with viscosity. *Math. Mech. Appl. Sci.* **15**, 187–204 (1992).
[20] F.C. Hoppensteadt, W. Jäger and C. Pöppe, A hysteresis model for bacterial growth patterns. In: *Modelling of Patterns in Space and Time*, Ed. W. Jäger and J.D. Murray (Springer-Verlag, Berlin, 1984), p. 123.
[21] N. Kenmochi, T. Koyama and G.H. Meyer, Parabolic PDEs with hysteresis and quasivariational inequalities. *Nonlinear Anal.* **34**, 665–686 (1998).
[22] J.-P. Kernevez, G. Joly, M.-C. Duban, B. Bunow and D. Thomas, Hysteresis, oscillations, and pattern formation in realistic immobilized enzyme systems. *J. Math. Biol.* **7**, 41–56 (1979).
[23] M.A. Krasnoselskii and A.V. Pokrovskii, *Systems with hysteresis.* Nauka, Moskow, 1983 (in russian). English translation: Springer, Berlin, 1989.
[24] P. Krejci, Hysteresis operators—a new approach to evolution differential inequalities. *Comment. Math. Univ. Carolin.* **30**, 525–536 (1989).
[25] M. Kružík and A. Prohl, Young measure approximation in micromagnetics. *Num. Math.* **90**, 291–307 (2001).

[26] O.A. Ladyzenskaja, V.A. Solonnikov and N.N. Ural'ceva, *Linear and Quasi-Linear Equations of Parabolic Type.* Amer. Math. Soc., Providence R.I., 1968.

[27] M. Landau, P. Lorente, J. Henry and S. Canu, Hysteresis phenomena between periodic and stationary solutions in a model of pacemaker and non-pacemaker coupled cardiac cells. *J. Math. Biol.* **25**, 491–509 (1987).

[28] J.W. Macki, P. Nistri and P. Zecca, Mathematical models for hysteresis. *SIAM Rev.* **35**, 94–123 (1993).

[29] E. Minchev, T. Okazaki and N. Kenmochi, Ordinary differential systems describing hysteresis effects and numerical simulations. *Abstr. Appl. Anal.* **7**, 563–583 (2002).

[30] M. Niezgodka, S. Zheng and J. Sprekels, Global solutions to a model of structural phase transitions in shape memory alloys. *J. Math. Anal. Appl.* **130**, 39–54 (1988).

[31] I. Pawlow, Three-dimensional model of thermomechanical evolution of shape memory materials. *Control Cybernet.* **29**, 341–365 (2000).

[32] I. Pawlow and A. Zochowski, Existence and uniqueness of solutions for a three-dimensional thermoelastic system. *Dissertationes Math. (Rozprawy Mat.)* **406**, 46 pp. (2002).

[33] I. Pawlow and W.M. Zajaczkowski, Global existence to a three-dimensional non-linear thermoelasticity system arising in shape memory materials. *Math. Methods Appl. Sci.* **28**, 407–442 (2005).

[34] I. Pawlow and W.M. Zajaczkowski, Unique global solvability in two-dimensional non-linear thermoelasticity. *Math. Methods Appl. Sci.* **28**, 551–592 (2005).

[35] T. Roubíček and M. Kružík, Microstructure evolution model in micromagnetics. *Z. Angew. Math. Phys.* **55**, 159–182 (2004).

[36] J. Sprekels and S. Zheng, Global solutions to the equations of a Ginzburg-Landau theory for structural phase transitions in shape memory alloys. *Phys. D* **39**, 59–76 (1989).

[37] A. Visintin, A phase transition problem with delay. *Control Cybernet.* **11**, 5–18 (1982).

[38] A. Visintin, *Differential Models of Hysteresis.* Springer-Verlag, Berlin, 1993.

[39] A. Visintin, *Models of Phase Transitions.* Birkhäuser, Boston, 1996.

[40] A. Visintin, Modified Landau-Lifshitz equation for ferromagnetism. *Phys. B* **223**, 365-369 (1997).

[41] A. Visintin, A Weiss-type model of ferromagnetism. *Phys. B* **275**, 87-91 (2000).

[42] A. Visintin, Vector Preisach model and Maxwell's equations. *Phys. B* **306**, 21-25 (2001).

[43] S. Yoshikawa, A three-dimensional thermoelastic system of shape memory alloys. *Adv. Math. Sci. Appl.* (to appear).

[44] S. Yoshikawa, Weak solutions for the Falk model system of shape memory alloys in energy class. *Math. Methods Appl. Sci.* **28**, 1423–1443 (2005).

[45] J. Zimmer, Global existence for a nonlinear system in thermoviscoelasticity with non-convex energy. *J. Math. Anal. Appl.* **292**, 589-604 (2004).

MODELLING PHASE TRANSITIONS VIA AN ENTROPY EQUATION: LONG-TIME BEHAVIOUR OF THE SOLUTIONS

Elena Bonetti

Dipartimento di Matematica "F. Casorati"
Università di Pavia
Via Ferrata 1, 27100 Pavia, Italy
E-mail: elena.bonetti unipv.it

The paper deals with a model of phase transitions with the possibility of thermal memory described via an entropy equation, and a generalized version of the principle of virtual power including microscopic movements, which are responsible for the thermomechanical process. We first recall an existence result for the general situation. Hence, we focus our attention on a simplified version of the problem, in which neither the thermal memory nor the diffusive effects in the phase transition are accounted for. After recalling an existence and uniqueness result for the corresponding initial and boundary value problem, we investigate the long-time behaviour of the solution. In particular, we characterize the ω-limit set.

1. Introduction

This paper is devoted to the study of a new phase-field model, based on an entropy equation and on a generalized version of the principle of virtual power, including microscopic forces and movements responsible for the phase transition.

Phase transition phenomena have been deeply studied in the literature of the past years and several models have been proposed, possibly accounting for dissipative effects, thermal memory, hysteresis (cf., among the others, [10, 20, 23]). Mainly, these models have been obtained generalizing the classical Stefan problem, including relaxation and dissipative effects (cf., e.g., [18]). Moreover, several different heat flux laws have been introduced, also accounting for the thermal history of the system. Analytical investigations have been developed in the direction of proving existence and uniqueness of solutions, as well as from the point of view of the long-time behaviour (cf., e.g., [1, 12, 13, 15, 19]).

The model we are concerned with has been recently introduced, and differs from the aforementioned approaches in that the equations of the resulting PDE's system are recovered by an entropy equation (cf. [5, 6]), in place of the more usual energy equation, and by a refinement of the principle of virtual power, in which microscopic forces, responsible for the phase transition, are considered (cf. [16]). This modelling approach is mainly due to M. Frémond. For the sake of completeness, before proceeding let us also mention some recent contributions on phase-field systems accounting for microscopic movements and, possibly, irreversibility of the processes (cf., among the others, [9, 14]). Furthermore, let us recall some applications of the above theory to describe special phenomena in continuum mechanics, such as damage of materials or the shape memory effect (cf. [16]). However, we point out that all the above papers are written in terms of the energy balance. The main novelty of the model here considered consists in the choice of the equation governing the temperature. Such a choice leads to some advantages both from the analytical and thermomechanical point of view, as we will discuss in the following sections.

In this paper, we aim to briefly review the derivation of the model and some related existence and uniqueness results, mainly referring to a joint paper with M. Frémond [6], and to a joint paper with P. Colli and M. Frémond [5]. In particular, the latter deals with a general situation accounting for the possibility of diffusive phase transitions and of thermal memory effects. The resulting PDE's system consists of an integro-differential equation governing the evolution of the temperature, and of a diffusive variational inclusion for the phase parameter. In particular, mainly due to the presence of a logarithmic nonlinearity in the entropy equation, positivity of the temperature directly follows from the model itself, which is solved in a suitable sense, avoiding the application of a maximum principle. Some recent results in the direction of [5] have been obtained in [2, 3, 4] for different entropy flux laws and accounting for possible anisotropic thermomechanical properties of the material.

Here, we will focus on the simplified version of the model proposed in [6], and investigate the long-time behaviour of the solutions. In particular, we aim to identify the limit points for the trajectories, i.e. to describe their ω-limit set. In [6], an existence and uniqueness result is proved for the case in which there is no thermal memory and the local interactions between the phases are neglected, i.e. the phase transition is not diffusive. Hence, we show that the cluster points of the trajectories are solutions of an associated stationary problem. The case when thermal memory and diffusive effects are accounted for (as in [5]), is more difficult, and the technique used in [6] cannot be extended to this more general situation. In particular, the problem of showing uniqueness of solutions in the general setting of thermal memory and diffusive phase transitions has not been solved up to now. However, even in this less regular framework some results concerning the long-time behavior of solutions can be obtained (cf. [7]).

Here is the outline of the paper. In Section 2 we detail the derivation of the model in the general setting of diffusive phase transitions with thermal memory. Then, we state the existence result for a related initial and boundary value problem (cf. [5]). In Section 3, we introduce the simplified version of the problem we are dealing with, and state the existence and uniqueness of the solution on the whole time interval $(0, +\infty)$ (cf. [6]), as well as some uniform bound on the absolute temperature. The related proofs are sketched in Section 4. Then, in Section 5 we find some uniform estimates on the solution on the whole time interval $(0, +\infty)$, which are used in Section 6 to characterize the ω-limit and to pass to the limit in the equations as $t \to +\infty$.

2. The model and the resulting PDE's system

In this section we present the complete model introduced and investigated in [5], and state the related existence theorem. Let us consider a two-phase system, located in a smooth bounded domain $\Omega \subset \mathbf{R}^3$, with $\Gamma := \partial\Omega$. The following notation is used

$$Q_T := \Omega \times (0, T), \quad Q_\infty := \Omega \times (0, +\infty).$$

The model is derived by continuum mechanics laws; constitutive relations are written in terms of state and dissipative variables, and they are recovered from two energy functionals: the free energy and the pseudo-potential of dissipation (cf. [21]). The first accounts for the equilibrium of the system, while the latter for its evolution.

As we develop our investigation on a macroscopic level, in order to describe the phase transition we introduce a phase parameter χ related to the microscopic state of the material. More precisely, χ stands for the concentration of one phase, as we are assuming that the two phases may coexist at each point with different proportions. Moreover, as neither voids nor overlapping can occur between the phases, for physical consistence we prescribe that (see [17] for a different approach accounting for voids)

$$\chi \in [0,1]. \tag{1}$$

Assuming that there are no macroscopic deformations, the state variables of the model are the absolute temperature θ, the phase parameter χ, and possibly the gradient $\nabla\chi$, if local interactions between the phases are considered. Finally, one may assume that the thermomechanical equilibrium of the system also depends on the past history of the gradient of the temperature (cf. [11, 20]). This leads to include

$$\widetilde{\nabla\theta}^t(s) := \int_{t-s}^{t} \nabla\theta(\tau)d\tau,$$

in the state variables, $s \in (0,+\infty)$ being the time history variable. Thus, the dissipative variables are the gradient of the temperature $\nabla\theta$ at time t, related to the heat flux, and the time derivative χ_t, related to microscopic velocities associated with the phase transition process. The equations of the model are derived from the balance laws of continuum mechanics, i.e. the energy balance and the generalized principle of virtual power written for microscopic forces and movements. In particular, our entropy equation is obtained through an approximation of the energy balance neglecting higher order dissipative contributions.

Let us now start by writing more precisely the complete model (which has been introduced in [5]), in the case when thermal memory and local microscopic interactions are allowed. Concerning thermal history, we follow the approach by Coleman and Gurtin (cf. [11]), including in the state variables $\widetilde{\nabla\theta}^t$ (cf. also [20]), and ranking the present contribution of the gradient of the temperature among the dissipative variables. Then, the free energy functional Ψ is given by a present contribution at time t (depending on θ, χ, and $\nabla\chi$), and a contribution accounting for the past history and depending on $\widetilde{\nabla\theta}^t$. More precisely, Ψ reads as follows

$$\Psi = -c_s\theta\log\theta - \frac{L}{\theta_c}(\theta-\theta_c)\chi + I_{[0,1]}(\chi) + \frac{\nu}{2}|\nabla\chi|^2 + \frac{1}{2}\int_0^{+\infty} h(s)\mathbf{L}\widetilde{\nabla\theta}^t(s)\cdot\widetilde{\nabla\theta}^t(s)ds, \tag{2}$$

where $c_s > 0$, $\nu \geq 0$, $\mathbf{L}$ is a symmetric and positive definite matrix related to the thermal properties of the material, and h is a continuous, decreasing, and positive function such that $\int_0^{+\infty} s^2h(s)ds < +\infty$. The indicator function $I_{[0,1]}(\chi)$ enforces (1), since $I_{[0,1]}(\chi) = 0$ if $\chi \in [0,1]$, and

$I_{[0,1]}(\chi) = +\infty$ otherwise. As usual, we let $\mathbf{L}$ be the identity matrix. However, one could consider more general situations (corresponding to different choices of $\mathbf{L}$), including materials which present an anisotropic behaviour w.r.t. their thermal properties (cf. [2]). Hence, the evolution of the system is governed by the so-called pseudo-potential of dissipation, which depends on the dissipative variables $(\nabla\theta, \chi_t)$, and possibly on other variables (here it is assumed that it depends also on θ). We recall that Φ is non-negative, convex w.r.t. $(\nabla\theta, \chi_t)$, and such that $\Phi = 0$ if $(\nabla\theta, \chi_t) = (\mathbf{0}, 0)$. As a consequence, the subdifferential $\partial\Phi$ (defined w.r.t. to dissipative variables) is a maximal monotone graph, and there holds

$$\partial\Phi(\nabla\theta, \chi_t) \cdot (\nabla\theta, \chi_t) \geq 0. \tag{3}$$

This property will be used to ensure the thermodynamical consistency of the model. More precisely, we consider the case

$$\Phi(\nabla\theta, \chi_t, \theta) = \frac{\mu}{2} |\chi_t|^2 + \frac{\lambda}{2\theta} |\nabla\theta|^2, \tag{4}$$

$\lambda, \mu > 0$. Let us point out that Φ accounts for non-hereditary dissipation. The presence of the factor θ^{-1} is related to the final expression of the heat flux $\mathbf{q}$, and it is somehow in agreement with the Fourier law. Possibly different choices of Φ are investigated in [3, 4]. In the sequel, the notation $(\cdot)^{nd}$ refers to the non-dissipative part and $(\cdot)^d$ to the dissipative part of a physical quantity.

Now, let us introduce the equations of the system. The equation for the absolute temperature is written in the form of the following equation for the entropy

$$s_t + \operatorname{div} \mathbf{Q} = R \text{ in } \Omega, \quad \mathbf{Q} \cdot \mathbf{n} = 0 \text{ on } \Gamma, \tag{5}$$

where

$$s = -\frac{\partial\Psi}{\partial\theta} = c_s(1 + \log\theta) + \frac{L}{\theta_c}\chi, \tag{6}$$

is the entropy (cf. (2)), $\mathbf{Q}$ the entropy flux, related to the heat flux $\mathbf{q}$ by

$$\mathbf{Q} = \frac{\mathbf{q}}{\theta}, \tag{7}$$

R is an external entropy source, and $\mathbf{n}$ the normal unit vector to the boundary. We assume that $\mathbf{Q} = \mathbf{Q}^d + \mathbf{Q}^{nd}$, where $\mathbf{Q}^d$ is the dissipative contribution (cf. (4)), given by

$$\mathbf{Q}^d = \frac{\partial\Phi}{\partial\nabla\theta} = -\frac{\lambda}{\theta}\nabla\theta = -\lambda\nabla\log\theta, \tag{8}$$

and $\mathbf{Q}^{nd}$ is the non-dissipative contribution, recovered from the free energy (2), in the case when thermal memory is considered. To write $\mathbf{Q}^{nd}$ explicitly, we introduce an auxiliary function k s.t.

$$-k' = h, \quad k, k', k'' \in L^1(0, +\infty), \quad \lim_{s \to +\infty} k(s) = 0, \tag{9}$$

h being as in (2). By the properties of h,

$$k' \leq 0, \quad k'' \geq 0. \tag{10}$$

Hence, we have (cf. [5] for a more complete derivation),

$$\mathbf{Q}^{nd} := -\int_0^{+\infty} k'(s)\widetilde{\nabla\theta}^t(s)ds = -\int_{-\infty}^t k(t-s)\nabla\theta(s)ds. \tag{11}$$

Note that (4), (7), and (8) lead to the classical Fourier law for the instantaneous (dissipative) contribution in the heat flux, i.e.

$$\mathbf{q}^d = \mathbf{Q}^d\theta = -\lambda\nabla\theta.$$

Then, by a generalization of the principle of virtual power, including microscopic movements, we recover a balance equation describing the evolution of the phase parameter from

$$B - \operatorname{div} \mathbf{H} = 0 \text{ in } \Omega, \quad \mathbf{H} \cdot \mathbf{n} = 0 \text{ on } \Gamma, \tag{12}$$

B and $\mathbf{H}$ being new microscopic interior forces. They are given by

$$B = B^{nd} + B^d = \frac{\partial\Psi}{\partial\chi} + \frac{\partial\Phi}{\partial\chi_t} = \partial I_{[0,1]}(\chi) - \frac{L}{\theta_c}(\theta - \theta_c) + \mu\chi_t, \tag{13}$$

$$\mathbf{H} = \mathbf{H}^{nd} = \frac{\partial\Psi}{\nabla\chi} = \nu\nabla\chi. \tag{14}$$

In (13) the subdifferential $\partial I_{[0,1]}(\chi)$ is defined by $\partial I_{[0,1]}(\chi) = 0$ if $\chi \in (0,1)$, $\partial I_{[0,1]}(0) = (-\infty, 0]$, and $\partial I_{[0,1]}(1) = [0, +\infty)$. Note that in (5) and (12) we have assumed that no entropy flux and no external surface forces occur on the boundary.

Before proceeding, let us discuss the relation between the entropy equation (5) and the energy balance

$$e_t + \operatorname{div} \mathbf{q} = r + B\chi_t + \mathbf{H} \cdot \nabla\chi_t, \tag{15}$$

where e is the internal energy, related to s by

$$e = \Psi + \theta s.$$

Furthermore, the right-hand side of (15) takes into account heat sources mechanically induced by microscopic forces and velocities, as well as an exterior heat source r. Substituting the constitutive relations (6), (8), (11), and (13)-(14) in (15) and applying the chain rule, (15) is rewritten as

$$\theta(s_t + \text{div } \mathbf{Q} - R) = (\mathbf{Q}^d, B^d) \cdot (\nabla\theta, \chi_t) + r^{int}, \tag{16}$$

where $R = r/\theta$, while r^{int} represents an interior rate of heat production resulting from the thermal history of the system. More precisely (cf. (8), (11) and [5]),

$$r^{int} = \frac{1}{2}\int_0^{+\infty} k''(s)\left|\widetilde{\nabla\theta}^t(s)\right|^2 ds.$$

Let us point out that, (3) (with (8) and (13)) and (10) yield that the right hand side of (16) is non-negative. Thus, by the positivity of θ the second principle of thermodynamics is verified (cf. (16)), i.e.

$$s_t + \text{div } \mathbf{Q} - R \geq 0.$$

Then, assuming that the dissipative contributions on the right hand side of (16) are negligible by small perturbations assumptions, and dividing by $\theta > 0$, the energy balance written in the form of (16) may reduce to (5).

We may now write the complete PDE's system recovered by (5) and (12) using the previous constitutive relations in terms of Ψ and Φ. Before proceeding, let us introduce the Hilbert triplet

$$V := H^1(\Omega) \hookrightarrow H := L^2(\Omega) \hookrightarrow V'.$$

As usual, H is identified with its dual space and $\langle \cdot, \cdot \rangle$ stands for the duality pairing between V' and V. By the same symbol $\|\cdot\|_X$ we denote the norm in a Banach space X and in any power X^n. We also introduce the abstract operator $A : V \to V'$, defined by

$$\langle Au, v \rangle = \int_\Omega \nabla u \cdot \nabla v, \quad u, v \in V. \tag{17}$$

Indeed, A is the realization, in the duality between V' and V, of the laplacian operator with associated homogeneous Neumann boundary conditions (cf. (5) and (12)). We may now address our problem in the abstract setting of the Hilbert triplet (V, H, V') and look for its evolution during a time interval $(0, T)$, $T > 0$. The resulting PDE's system reads, a.e. in $(0, T)$,

$$c_s(\log\theta)_t + \frac{L}{\theta_c}\chi_t + \lambda A \log\theta + k * A\theta = R \quad \text{in } V', \tag{18}$$

$$\mu\chi_t + \nu A\chi + \xi = \frac{L}{\theta_c}(\theta - \theta_c) \quad \text{in } V', \tag{19}$$

where

$$\xi \in \partial I_{[0,1]}(\chi) \quad \text{a.e. in } Q_T, \tag{20}$$

$*$ stands for the convolution operator, i.e. $(k * A\theta)(t) = \int_0^t k(t-s)A\theta(s)ds$, and R also accounts for the past history of the gradient $\int_{-\infty}^0 k(t-s)\nabla\theta(s)ds$ (cf. (11)), which is assumed to be known. The above problem (18)-(20) has been investigated in [5], where the following existence theorem has been proved.

Theorem 1: Let $T > 0$. Assume that

$$\begin{aligned}
&k \in H^1(0,T), \quad k(0) \geq 0, \quad k \equiv 0 \;\text{ if }\; k(0) = 0,\\
&\nu \geq 0,\\
&R \in L^2(0,T;H) \cap L^1(0,T;L^\infty(\Omega)),\\
&\theta_0 \in L^1(\Omega), \quad \theta_0 > 0 \text{ a.e. in } \Omega, \quad u_0 := \log\theta_0 \in H,\\
&0 \leq \chi_0 \leq 1 \text{ a.e. in } \Omega, \quad \nu\chi_0 \in V.
\end{aligned}$$

Then, there exists a quadruple $(u := \log\theta, \theta, \chi, \xi)$ with

$$\begin{aligned}
&u \in H^1(0,T;V') \cap L^2(0,T;V), \quad \theta \in L^{5/3}(Q_T),\\
&\chi \in H^1(0,T;H) \cap L^\infty(Q_T), \quad \nu\chi \in L^\infty(0,T;V),\\
&\xi \in L^{5/3}(Q_T), \quad k(0)(1 * \theta) \in L^\infty(0,T;V).
\end{aligned}$$

fulfilling

$$u(0) = u_0, \quad \theta(0) = \theta_0, \quad \chi(0) = \chi_0,$$

and (18)-(20) a.e. in $(0,T)$.

Let us point out that (18) is a nonlinear integro-differential equation, coupled with the variational inclusion (19) by a non-Lipschitz nonlinear function of u, i.e. the term $\theta = \exp u$. Theorem 1 has been proved in [5] by means of a Lipschitz approximation of the nonlinearities in (19) ($\exp u$ and the subdifferential term), combined with an a priori estimates - passage to the limit technique. This procedure relies on an intermediate existence and uniqueness result for the regularized version, which turns out to be interesting by itself. However, the question of extending the uniqueness result to the original system remains an open problem, and it seems related to the possibility of proving further regularity on θ.

Remark 2: We mention an existence and uniqueness result proved in the case when a different entropy flux, combined with different boundary conditions, is chosen. Indeed, in [3, 4] the following equation is considered in place of (18):

$$c_s u_t + \frac{L}{\theta_c}\chi_t + \lambda A\theta + k * A\theta = R \text{ in } \Omega, \quad \theta = \theta_\Gamma \text{ on } \Gamma. \tag{21}$$

Existence and uniqueness, as well as regularity and long-time behaviour of the solutions, are proved for the resulting PDE's system. However, these results are strongly related to the fact that the complete entropy flux in (21) is linear w.r.t. the temperature, and that Dirichlet boundary conditions are considered. Indeed, in this situation it is possible to prove a further regularity result for the temperature, which cannot be extended to our case.

In this paper, we are concerned with a fairly simplified version of the above problem (18)-(20), corresponding to the case when $k \equiv 0$ and $\nu = 0$ (i.e., when no thermal memory is considered and no local interactions between the phases are accounted for). This problem has been introduced and studied in [6] from the point of view of existence and uniqueness of the solutions. Here, we address the long-time behaviour of the solutions. In particular, we investigate the limit points of the solution trajectories. For the sake of simplicity, we let $c_s = \frac{L}{\theta_c} = \lambda = 1$ and mainly use the auxiliary variable $u = \log\theta$ (so that in the following notation we often use the fact that $\theta = \exp u$).

Now, we state more precisely the initial and boundary value problem we are dealing with:

$$u_t + \chi_t + Au = R \quad \text{a.e. in } Q_T, \tag{22}$$

$$\chi_t + \xi - (\exp u - \theta_c) = 0, \quad \xi \in \partial I_{[0,1]}(\chi) \quad \text{a.e. in } Q_T, \tag{23}$$

associated with the initial conditions

$$u(0) = u_0 = \log\theta(0), \quad \theta(0) = \theta_0, \quad \chi(0) = \chi_0. \tag{24}$$

In the following section, we recall the existence and uniqueness results obtained in [6] for the system (22)-(23), (24), and state our main result concerning the long-time behaviour of the solutions.

3. Main results

First, let us state precisely our assumptions on the data. We require

$$u_0 \in L^\infty(\Omega) \cap V, \tag{25}$$

$$\chi_0 \in [0,1] \quad \text{a.e. in } \Omega, \tag{26}$$

and

$$\begin{aligned} &R \in L^2(0,+\infty; H) \cap L^1(0,+\infty; L^\infty(\Omega)), \quad \text{such that there exists} \\ &\widetilde{R} \in W^{1,1}(0,+\infty; H) \text{ with } R \le \widetilde{R} \text{ a.e. in } Q_\infty. \end{aligned} \tag{27}$$

Hence, we introduce the notion of solution to Problem (22)-(23), (24) on the whole time interval $(0,+\infty)$.

Definition 3: A solution to (22)-(23) and (24) on the whole time interval $(0,+\infty)$ is a triplet (u,χ,ξ) satisfying for any $T>0$

$$u \in H^1(0,T;H) \cap L^\infty(0,T;V) \cap L^2(0,T;H^2(\Omega)), \tag{28}$$

$$\theta = \exp u \in L^\infty(Q_T), \tag{29}$$

$$\chi, \chi_t \in L^\infty(Q_T), \tag{30}$$

$$\xi \in L^\infty(Q_T), \tag{31}$$

and fulfilling (24) and (22)-(23) a.e. in $(0,T)$.

We state the following existence and uniqueness result, obtained in [6].

Theorem 4: Assume that (25)-(27) hold. Then, Problem (22)-(24) admits a unique solution (u,χ,ξ) in the sense specified by Definition 3.

The existence result stated by the above theorem is proved by means of a truncation procedure, combined with a maximum principle argument which shows that

$$\|\theta\|_{L^\infty(Q_T)} \le M, \quad M > 0. \tag{32}$$

The uniqueness result follows by a comparison principle of the solutions, proved by use of L^1-techniques (see Theorem 9 in the next section).

Let us point out that the situation considered in Theorem 4 is less general than the framework in which Theorem 1 holds. Moreover, the techniques in the proof of Theorem 4 do not apply to the case when thermal memory and diffusion in the phase transition are included. Concerning the estimate (32), in the next section we prove that M is actually independent of T, i.e. the following result holds.

Theorem 5: Under the assumptions of Theorem 4, there exists $M > 0$, independent of T, such that

$$\|\theta\|_{L^\infty(Q_\infty)} \leq M. \tag{33}$$

Let us now address the study of the long-time behaviour of the unique (as stated by Theorem 4) solution to Problem (22)-(24). More precisely, we aim to identify the cluster points, as $t \to +\infty$, of the trajectories $(u(t), \chi(t))$ in some suitable topology (cf., e.g., [13]). In the regularity setting of the solution to our problem, it turns out that the "good" topology is the topology of $H \times V'$. Thus, we introduce the ω-limit set $\omega(u_0, \chi_0)$ of the couple of solutions $(u(t), \chi(t))$ (depending on the initial data (u_0, χ_0), of course), i.e.

$$\begin{aligned} \omega(u_0, \chi_0) = \{&(u_\infty, \chi_\infty) \in H \times V' \text{ such that there exists } \{t_n\} \subset (0, +\infty) : \\ &t_n \to +\infty \text{ and } (u(t_n), \chi(t_n)) \to (u_\infty, \chi_\infty) \text{ in } H \times V'\}. \end{aligned} \tag{34}$$

Our result reads as follows.

Theorem 6: Assume (25)-(27), and let (u, χ, ξ) be the corresponding unique solution to Problem (22)-(24). Then, the ω-limit set $\omega(u_0, \chi_0)$ is a non-empty, compact, and connected subset of $H \times V'$.

Moreover, for any $(u_\infty, \chi_\infty) \in \omega(u_0, \chi_0)$ there holds, a.e. in Ω,

$$u_\infty = |\Omega|^{-1} \left(m + c_0 - \int_\Omega \chi_\infty \right), \tag{35}$$

$$\partial I_{[0,1]}(\chi_\infty) \ni \exp u_\infty - \theta_c, \tag{36}$$

where

$$c_0 := \int_\Omega (u_0 + \chi_0), \quad m := \int_0^{+\infty} \int_\Omega R. \tag{37}$$

Furthermore, for any fixed (u_0, χ_0) and R fulfilling (25)-(27), we have:

- a) if $m + c_0 > |\Omega| (\log \theta_c + 1)$, then $\omega = \{(|\Omega|^{-1} (m + c_0) - 1, 1)\}$;
- b) if $m + c_0 < |\Omega| \log \theta_c$, then $\omega = \{(|\Omega|^{-1} (m + c_0), 0)\}$;
- c) if $|\Omega| \log \theta_c \leq m + c_0 \leq |\Omega| (\log \theta_c + 1)$, then $\omega = \{(\log \theta_c, \chi_\infty)\}$, where $\chi_\infty \in [0, 1]$ and $\int_\Omega \chi_\infty = m + c_0 - |\Omega| \log \theta_c$.

Remark 7: Let us point out that, in the cases (a) and (b), ω is a singleton and the whole trajectory of the solution converges to its limit point in ω. From a physical point of view, this result ensures that, if the initial data and the entropy source R are sufficiently large (or sufficiently small) compared with the domain Ω and the phase transition temperature θ_c, the system converges as $t \to +\infty$ to a uniform situation choosing one of the two pure phases. In an intermediate situation, the absolute temperature converges to the critical phase transition temperature, while we cannot conclude that the whole trajectory of the phase parameter $\chi(t)$ converges to some limit point. Indeed, in this situation we can only state that possible limit points of $\chi(t)$ are functions $\chi_\infty \in [0,1]$, satisfying with $\int_\Omega \chi_\infty = m + c_0 - |\Omega| \log \theta_c$.

4. The existence and uniqueness result

We do not enter the details of the proofs of Theorem 4, for which we directly refer to [6]. We just sketch the main ideas of the maximum principle argument on which the proof of Theorem 5 relies. We also give some comments, which will be useful for the proof of Theorem 6 in the next section.

Remark 8: Let us point out that, as the investigation therein is restricted to a finite time interval, in [6] a less regular situation is considered as far as the assumptions on the data are concerned. In particular, a non-homogeneous boundary condition for $\log \theta$ is introduced, i.e.

$$\partial_n \log \theta = h \quad \text{on } \Gamma,$$

and source term on the right hand side of (22) is substituted by $\mathcal{R} \in V'$, given by

$$\langle \mathcal{R}, v \rangle = \int_\Omega Rv + \int_\Gamma hv_{|\Gamma}.$$

However, for the sake of simplicity, we directly deal with the assumption $h = 0$.

We will now show that the constant M in (32), occurring in the proof of Theorem 4, does not depend on T. As we have already mentioned, in the proof of Theorem 4 in [6] the authors consider an approximate version of (22)-(23), introduced by means of the following truncation operator τ:

$$\tau(y) := \max\{-M, \min\{y, M\}\}, \quad y \in \mathbf{R}, \tag{38}$$

for a suitable constant $M > 0$ to be chosen. Note that τ is contractive. The Schauder theorem yields the existence of a solution to (22), combined with (recall $\theta = \exp u$)

$$\chi_t + \partial I_{[0,1]}(\chi) \ni \tau(\theta) - \theta_c, \tag{39}$$

and (24). Hence, a maximum principle argument guarantees that (32) is verified, so that the pair (u, χ) actually solves the original problem (22)-(23) (cf. (38) and (39)).

Now, let us extend (32) to the whole time interval $(0, +\infty)$.

4.1. *Proof of Theorem 5*

Applying well-known results in the theory of parabolic equations, we infer that there exists a unique solution to

$$\widetilde{u}_t + A\widetilde{u} = \widetilde{R}, \quad \widetilde{u}(0) = 0, \tag{40}$$

with

$$\widetilde{u} \in W^{1,\infty}(0, +\infty; H) \cap H^1(0, +\infty; V) \cap L^\infty(0, +\infty; H^2(\Omega)). \tag{41}$$

Note that (41) entails

$$\widetilde{u} \in L^\infty(Q_\infty). \tag{42}$$

Hence, we choose the constant M in (38) to be

$$M = \exp(\|\widetilde{u}\|_{L^\infty(Q_\infty)} + u^*),$$

where

$$u^* = \max\{\|u_0\|_{L^\infty(\Omega)}, \log\theta_c + \|\widetilde{u}\|_{L^\infty(Q_\infty)}\}. \tag{43}$$

Let us now take the difference of (22) and (40), test it by $(\widehat{u} - u^*)^+$, where $(\cdot)^+$ stands for the positive part and $\widehat{u} := u - \widetilde{u}$. Then, we integrate over $(0, t)$. Integrating by parts in time and exploiting (43), (40), (24), and (27), we get

$$\frac{1}{2}\left\|(\widehat{u} - u^*)^+(t)\right\|_H^2 + \left\|\nabla(\widehat{u} - u^*)^+\right\|_{L^2(0,t;H)}^2 \le -\int_0^t \int_\Omega \chi_t(\widehat{u} - u^*)^+. \tag{44}$$

Now, we discuss the sign of the right hand side of (44). We first observe that it can be rewritten as

$$-\int_0^t \int_\Omega \chi_t(\widehat{u} - u^*)^+ = -\int\int_{Q_t^*} \chi_t(\widehat{u} - u^*)^+, \tag{45}$$

where $Q_t^* \subset \Omega \times (0,t)$ s.t. $\chi \in (0,1)$ and $\widehat{u} > u^*$ a.e. in Q_t^* (observe $\chi_t = 0$ where $\chi = 0$ or $\chi = 1$). Thus, after recalling that $\partial I_{[0,1]}(\chi) = 0$ if $\chi \in (0,1)$, by using (39) we can rewrite (45) as follows

$$-\int\int_{Q_t^*} \chi_t(\widehat{u} - u^*)^+ = -\int_0^t \int_\Omega (\tau(\exp u) - \theta_c)(\widehat{u} - u^*)^+. \tag{46}$$

Observe that on Q_t^*

$$\exp u > \exp(u^* - \|\widetilde{u}\|_{L^\infty(Q_\infty)}) \geq \theta_c.$$

By construction, $M \geq \theta_c$, and consequently

$$\tau(\exp u) - \theta_c \geq 0,$$

so that (cf. also (46)) $\chi_t \geq 0$ a.e. in Q_t^*. Subsequently, from (44) we deduce

$$\frac{1}{2} \left\|(\widehat{u} - u^*)^+(t)\right\|_H^2 + \|\nabla(\widehat{u} - u^*)\|_{L^2(0,t;H)}^2 \leq 0. \tag{47}$$

Thus, (47) yields $(\widehat{u} - u^*)^+ = 0$, i.e. $\widehat{u} \leq u^*$ a.e., which leads to

$$\exp u \leq \exp(u^* + \|\widetilde{u}\|_{L^\infty(Q_\infty)}) = M \quad \text{a.e. in } Q_t. \tag{48}$$

As the above argument does not actually depend on t, (32) turns out to be verified a.e. in Q_∞, and the proof is complete.

For the sake of completeness, let us recall the following proposition (for the proof refer to [6]) from which uniqueness of the solution follows.

Theorem 9: Let $T > 0$ and let (u_{0i}, χ_{0i}, R_i), $i = 1, 2$, be two families of data satisfying the assumptions of Theorem 4, and fulfilling

$$u_{01} \leq u_{02}, \quad \chi_{01} \leq \chi_{02} \quad \text{a.e. in } \Omega, \tag{49}$$

$$R_1 \leq R_2 \quad \text{a.e. in } Q_T. \tag{50}$$

Then, letting (u_i, χ_i) the corresponding solutions on $(0,T)$, for $i = 1, 2$, there holds

$$u_1 \leq u_2, \quad \chi_1 \leq \chi_2, \tag{51}$$

a.e. in Q_T.

It is worth to note that the above result can be applied to prove uniqueness of the solution on the whole interval $(0, +\infty)$. Moreover, the property of the solutions stated by Theorem 9 is interesting from a thermomechanical point of view. Indeed, if one considers two systems that are identical except for their internal energies e_1, e_2, with $e_1 \leq e_2$ at time $t = 0$, and the same inequality is satisfied for the corresponding phase parameters, then by Theorem 9 the internal energies evolve during the phase transition conserving the same monotonicity. Furthermore, from the analytical point of view this result is a sort of comparison principle between the solutions. Note that, assuming that (49)-(50) hold on the whole time interval $(0, +\infty)$, one can extend (51) for any $t \in (0, +\infty)$, and the analogous property should be verified for the limit points of the trajectories (cf. Section 6).

5. Uniform estimates on $(0, +\infty)$

In this section, we aim to prove some uniform estimates for the solution to (22)-(23) (cf. Theorem 4) on the whole time interval $(0, +\infty)$. In the sequel, we will denote by the symbol c any positive constant occurring in the following estimates, possibly depending on the data of the problem, but not on T. Before proceeding, let us point out that Theorem 4 and Theorem 5 ensure (recall that $\theta = \exp u$)

$$\|\chi\|_{L^\infty(Q_\infty)} + \|\exp u\|_{L^\infty(Q_\infty)} \leq c. \tag{52}$$

First estimate. Test (22) by $\theta = \exp u$, (23) by χ_t, sum the resulting equations and integrate over $(0, t)$. Two terms cancel out and, after integrating by parts in time and applying the Hölder inequality, we can write (cf. (52) and recall that $\chi \in [0, 1]$ a.e. in Q_t for any t)

$$\begin{aligned} &\int_\Omega \theta(t) + \int_0^t \int_\Omega \left|\nabla\theta^{1/2}\right|^2 + \int_0^t \int_\Omega |\chi_t|^2 + \theta_c \int_\Omega \chi(t) \\ &\leq \int_\Omega \exp u_0 + \theta_c \int_\Omega \chi_0 + \int_0^t \|R\|_{L^\infty(\Omega)} \|\theta\|_{L^1(\Omega)} . \end{aligned} \tag{53}$$

In particular, in (53), we have exploited the fact that, for $\xi \in \partial I_{[0,1]}(\chi)$, there holds

$$\int_0^t \int_\Omega \xi\chi_t \geq 0.$$

This can be formally proved using the definition of $\partial I_{[0,1]}(\chi)$ and applying the chain rule (cf. [8]). Hence, owing to (25)-(27), we can apply the Gronwall lemma (cf. [8]) to (53), yielding

$$\left\|\theta^{1/2}\right\|_{L^\infty(0,+\infty;H)} + \left\|\nabla\theta^{1/2}\right\|_{L^2(0,+\infty;H)} + \|\chi_t\|_{L^2(0,+\infty;H)} \leq c. \tag{54}$$

Second estimate. Let us test (22) by u_t and integrate over $(0, t)$. Using the Young inequality, we have (cf. (54) and (25), (27))

$$\begin{aligned} &\|u_t\|^2_{L^2(0,t;H)} + \frac{1}{2}\|\nabla u(t)\|^2_H \\ &\leq \frac{1}{2}\|\nabla u_0\|^2_H + \int_0^t \|R\|^2_H + \int_0^t \|\chi_t\|^2_H + \frac{1}{2}\int_0^t \|u_t\|^2_H \\ &\leq c + \frac{1}{2}\|u_t\|^2_{L^2(0,t;H)}, \end{aligned} \tag{55}$$

so that (55) leads to

$$\|u_t\|_{L^2(0,+\infty;H)} + \|\nabla u\|_{L^\infty(0,+\infty;H)} \leq c. \tag{56}$$

Third estimate. In view of (56), to estimate $\|u\|_{L^\infty(0,+\infty;V)}$ we may use the Poincaré inequality. Thus, it is sufficient to control $\left|\int_\Omega u(t)\right|$ uniformly w.r.t. t. We first test (22) by 1 and integrate over $(0,t)$. After integrating by parts in time, we get (cf. also (17))

$$\int_\Omega u(t) = \int_\Omega u_0 - \int_\Omega \chi(t) + \int_\Omega \chi_0 + \int_0^t \int_\Omega R(t), \tag{57}$$

so that exploiting (52) and (25)-(27), we have

$$\sup_{t\in(0,+\infty)} \int_\Omega u(t) \leq c. \tag{58}$$

Analogously, we can test (22) by -1 and integrate over $(0,t)$, and eventually infer that

$$\sup_{t\in(0,+\infty)} \left|\int_\Omega u(t)\right| \leq c. \tag{59}$$

After combining (59) with (56), the Poincaré inequality ensures that

$$\|u\|_{L^\infty(0,+\infty;V)} \leq c. \tag{60}$$

6. The ω-limit

As a first consequence of (52) and (60) we have that the set $\{(u(t),\chi(t)), t \geq 0\}$ is bounded in $V \times H$ and thus relatively compact in $H \times V'$ (note also that $\|\chi\|_{L^\infty(0,+\infty;H)} \leq c$). Therefore, the ω-limit set defined by (34) is a non-empty compact subset of $H \times V'$. Moreover, as $(u,\chi) \in C^0([0,+\infty); H \times V')$ (cf. Theorem 4), standard arguments in the theory of dynamical systems ensure that it is connected.

Now, let us characterize the elements of the ω-limit of $(u(t), \chi(t))$. Let $(u_\infty, \chi_\infty) \in \omega(u_0, \chi_0)$ and $\{t_n\}$ be a sequence of positive numbers such that $t_n \to +\infty$ and

$$(u(t_n), \chi(t_n)) \to (u_\infty, \chi_\infty) \quad \text{in } H \times V'. \tag{61}$$

Then, we fix $T > 0$ and define, for $t \in [0, T]$,

$$\begin{aligned} u_n(t) &:= u(t + t_n), \quad \theta_n(t) := \theta(t + t_n) = \exp u(t + t_n), \\ \chi_n(t) &:= \chi(t + t_n), \quad \xi_n(t) := \xi(t + t_n). \end{aligned} \tag{62}$$

It is straightforward to verify that, a.e. in $(0, T)$, there holds (cf. Theorem 4)

$$(u_n)_t + (\chi_n)_t + Au_n = R_n, \tag{63}$$

$$(\chi_n)_t + \xi_n = \theta_n - \theta_c, \tag{64}$$

where $\xi_n \in \partial I_{[0,1]}(\chi_n)$ and $R_n := R(t + t_n)$. Hence, (52), (54), (56), and (60) yield

$$\|u_n\|_{H^1(0,T;H)\cap L^\infty(0,T;V)} \leq c(T), \tag{65}$$

$$\|\theta_n\|_{L^\infty(Q_T)} \leq c(T), \tag{66}$$

$$\|\chi_n\|_{H^1(0,T;H)\cap L^\infty(Q_T)} \leq c(T), \tag{67}$$

for a constant $c(T)$ independent of n, but possibly depending on T. Moreover, by a comparison in (63), due to (65) and (67), we deduce

$$\|u_n\|_{L^2(0,T;H^2(\Omega))} \leq c(T), \tag{68}$$

and, analogously, a comparison in (64) leads to (cf. (66)-(67)),

$$\|\xi_n\|_{L^2(0,T;H)} \leq c(T). \tag{69}$$

Remark 10: Actually, we can deduce a further bound on ξ_n in $(0, T)$ independent of n. Indeed, proceeding as in [6] one can test (64) by $|\xi_n|^{p-2}\xi_n$ with $p > 2$. Due to monotonicity arguments one can show that $\int_0^t \int_\Omega (\chi_n)_t |\xi_n|^{p-2}\xi_n \geq 0$. Then, applying the Young inequality and exploiting (66) (which actually holds for a constant c independent of T), it follows $\|\xi_n\|_{L^p(Q_T)} \leq c(T)$ for any $p < +\infty$, from which, letting $p \to +\infty$ $\|\xi_n\|_{L^\infty(Q_T)} \leq c(T)$. Hence, by a comparison in the equation, we also deduce $\|(\chi_n)_t\|_{L^\infty(Q_T)} \leq c(T)$.

Then, weak and weak-star convergence results ensure that, at least for a suitable subsequence

$$u_n \overset{*}{\rightharpoonup} \bar{u} \quad \text{in } H^1(0,T;H) \cap L^\infty(0,T;V) \cap L^2(0,T;H^2(\Omega)), \tag{70}$$

$$\theta_n \overset{*}{\rightharpoonup} \bar{\theta} \quad \text{in } L^\infty(Q_T), \tag{71}$$

$$\chi_n \overset{*}{\rightharpoonup} \bar{\chi} \quad \text{in } H^1(0,T;H) \cap L^\infty(Q_T), \tag{72}$$

$$\xi_n \rightharpoonup \bar{\xi} \quad \text{in } L^2(0,T;H). \tag{73}$$

Moreover, by strong convergence, (70) and (72) imply

$$u_n \to \bar{u} \quad \text{in } C^0([0,T];H), \tag{74}$$

$$\chi_n \to \bar{\chi} \quad \text{in } C^0([0,T];V'). \tag{75}$$

Now, we aim to show that $\bar{u} = u_\infty$ and $\bar{\chi} = \chi_\infty$. Owing to the fact that (cf. (54) and (56))

$$\|u_t\|_{L^2(0,+\infty;H)} + \|\chi_t\|_{L^2(0,+\infty;H)} \leq c,$$

it follows

$$(u_n)_t \to 0 \quad \text{and} \quad (\chi_n)_t \to 0 \quad \text{in } L^2(0,+\infty;H), \tag{76}$$

as $n \to +\infty$. In particular, (76) and (70), (72) imply that $\bar{u}$ and $\bar{\chi}$ do not depend on time. Thus, we have (cf. (61), (62), and (74)), for any $t \in [0,T]$,

$$\bar{u}(t) = \bar{u}(0) = \lim_{n\to+\infty} u_n(0) = \lim_{n\to+\infty} u(t_n) = u_\infty, \tag{77}$$

and, analogously proceeding,

$$\bar{\chi}(t) = \chi_\infty. \tag{78}$$

We can also identify $\bar{\theta}(t) = \exp u_\infty$ and $\bar{\xi}$. Indeed, due to (74) it follows that (for a subsequence) $u_n \to u_\infty$ a.e. in Q_T. Thus, by definition of θ_n, we can infer that

$$\theta_n = \exp u_n \to \bar{\theta} = \exp u_\infty \quad \text{a.e. in } Q_T. \tag{79}$$

Moreover, by the Lebesgue theorem (cf. (66))

$$\theta_n \to \exp u_\infty \quad \text{in } L^p(Q_T), \quad \forall p < +\infty. \tag{80}$$

Now, passing to the limit in (63) as $n \to +\infty$ we get (cf. (70), (76), and (27))

$$Au_\infty = 0. \tag{81}$$

Thus, we deduce that u_∞ is a constant. Analogously, we can pass to the limit in (64) (cf. (73), (76), and (80)) and obtain

$$\bar{\xi} = \exp u_\infty - \theta_c. \tag{82}$$

Moreover, we have

$$\bar{\xi} \in \partial I_{[0,1]}(\chi_\infty). \tag{83}$$

To prove this, we consider the difference between (64) and (82), test it by $\xi_n - \bar{\xi}$ and integrate over $(0,t)$, with $t \in (0,T]$. We get, using the Young inequality,

$$\|\xi_n - \bar{\xi}\|^2_{L^2(0,t;H)} \leq c\left(\|(\chi_n)_t\|^2_{L^2(0,t;H)} + \|\exp u_n - \exp u_\infty\|^2_{L^2(0,t;H)}\right),$$

with the right hand side tending to 0 as $n \to +\infty$ due to (76) and (80). Thus, it follows

$$\xi_n \to \bar{\xi} \quad \text{in } L^2(0,T;H), \tag{84}$$

and consequently (cf. (72)), for any t,

$$\limsup_{n\to+\infty} \int_0^t \int_\Omega \xi_n \chi_n \leq \int_0^t \int_\Omega \bar{\xi}\chi_\infty. \tag{85}$$

Now, well-known monotonicity arguments and (85) ensure (cf. [8]) the desired (83). Let us set $\xi_\infty := \bar{\xi}$. Combining (82) with (83) gives (36). Let us point out that, due to (82) and the fact that u_∞ is a constant, we also deduce that ξ_∞ is a constant. To prove (35), we test (22) by 1 and integrate over $(0, t_n)$. We get

$$\int_\Omega (u(t_n) + \chi(t_n)) = \int_\Omega (u_0 + \chi_0) + \int_0^{t_n} \int_\Omega R.$$

Letting $n \to +\infty$ and recalling that u_∞ is a constant, we can infer that (cf., e.g., (61))

$$u_\infty = |\Omega|^{-1}\left(m + c_0 - \int_\Omega \chi_\infty\right). \tag{86}$$

Note that, substituting (86) in (82), we have (cf. also (83))

$$\partial I_{[0,1]}(\chi_\infty) \ni \exp\left(\left(m + c_0 - \int_\Omega \chi_\infty\right)|\Omega|^{-1}\right) - \theta_c. \tag{87}$$

Now, it remains to prove the last part of Theorem 6, namely the discussion of the cases (a)-(c). We recall that both u_∞ and $\xi_\infty \in \partial I_{[0,1]}(\chi_\infty)$ are constant in Ω. From (86) and the fact that $\chi_\infty \in [0,1]$, we can deduce

$$m + c_0 - |\Omega| \leq |\Omega|\, u_\infty \leq m + c_0. \tag{88}$$

Thus, in the case when $m + c_0 - |\Omega| > \log \theta_c |\Omega|$ (cf. (a)), it follows from (88) that

$$u_\infty > \log \theta_c, \tag{89}$$

and by (82) and the definition of $\partial I_{[0,1]}$

$$\xi_\infty > 0 \quad \Rightarrow \quad \chi_\infty = 1 \text{ a.e. in } \Omega. \tag{90}$$

Now, using (86) we get

$$u_\infty = |\Omega|^{-1} (m + c_0) - 1, \tag{91}$$

and the implication in (a) is proved. On the other hand, if we assume (89), we deduce that (90) holds (cf. (82)). Thus, in view of (91) we conclude

$$|\Omega| (\log \theta_c + 1) < (m + c_0). \tag{92}$$

The latter argument shows that (89) is indeed equivalent to (92). In particular,

$$\text{if } |\Omega| (\log \theta_c + 1) \geq (m + c_0) \quad \text{then} \quad u_\infty \leq \log \theta_c. \tag{93}$$

Now, assume that $m + c_0 < \log \theta_c |\Omega|$ (cf. (b)). Then, (88) leads to

$$u_\infty < \log \theta_c. \tag{94}$$

In this case, (82) implies

$$\xi_\infty < 0 \quad \Rightarrow \quad \chi_\infty = 0 \text{ a.e. in } \Omega, \tag{95}$$

and

$$u_\infty = |\Omega|^{-1} (m + c_0) \tag{96}$$

follows via (86), so that the implication (b) is proved. Conversely, assuming (94) yields (95), hence (96). Thus, in this case (94) actually corresponds to

$$|\Omega| \log \theta_c > (m + c_0). \tag{97}$$

In particular, we have

$$\text{if } |\Omega| \log \theta_c \leq (m + c_0) \quad \text{then} \quad u_\infty \geq \log \theta_c. \tag{98}$$

Finally, by (93) and (98) (c) follows, i.e. if

$$m + c_0 \geq |\Omega| \log \theta_c \geq m + c_0 - |\Omega|$$

necessarily we have

$$u_\infty = \log \theta_c,$$

and by (86) $\int_\Omega \chi_\infty = m + c_0 - |\Omega| \log \theta_c$. This concludes the proof of Theorem 6.

Acknowledgments

The author would like to thank the referee for his/her detailed remarks, which surely contributed to improve the final version of the paper.

References

[1] V. Barbu, P. Colli, G. Gilardi and M. Grasselli, Existence, uniqueness, and long time behavior for a nonlinear Volterra integrodifferential equation. *Differential Integral Equations* **13**, 1233–1262 (2000).

[2] E. Bonetti, A new approach to phase transitions with thermal memory via the entropy balance. To appear in: *Mathematical Models and Methods in Phase Transitions*, Ed. A. Miranville (Nova Science Publishers).

[3] E. Bonetti, P. Colli, M. Fabrizio and G. Gilardi, Global solution to a singular integrodifferential system related to the entropy balance. *Preprint I.M.A.T.I-C.N.R.* n. 5-PV, 1–34 (2005).

[4] E. Bonetti, P. Colli, M. Fabrizio and G. Gilardi, Modelling and long-time behaviour for phase transitions with entropy balance and thermal memory conductivity. *Preprint I.M.A.T.I.-C.N.R.* n. 21-PV, 1-29 (2005).

[5] E. Bonetti, P. Colli and M. Frémond, A phase field model with thermal memory governed by the entropy balance. *Math. Models Methods Appl. Sci.* **13**, 1565–1588 (2003).

[6] E. Bonetti and M. Frémond, A phase transition model with the entropy balance. *Math. Methods Appl. Sci.* **26**, 539–556 (2003).

[7] E. Bonetti and E. Rocca, Long-time beahviour for a phase-field model with thermal memory governed by the entropy balance. (in preparation).

[8] H. Brezis, *Opérateurs maximaux monotones et semi-groupes de contractions dans les espaces de Hilbert.* North-Holland, Amsterdam, 1973.

[9] G. Bonfanti, M. Frémond and F. Luterotti, Global solution to a nonlinear system for phase changes. *Adv. Math. Sci. Appl.* **10**, 1-24 (2000).

[10] M. Brokate and J. Sprekels, *Hysteresis and phase transitions.* Springer, New York, 1996.

[11] B.D. Colemann and M.E. Gurtin, Equipresence and constitutive equations for rigid heat conductors. *Z. Angew. Math. Phys.* **18**, 199-208 (1967).

[12] P. Colli, G. Gilardi and M. Grasselli, Well-posedness of the weak formulation for the phase-field model with memory. *Adv. Differential Equations* **2**, 487–508 (1997).

[13] P. Colli and Ph. Laurencot, Existence and stabilization of solutions to the phase-field model with memory. *J. Integral Equations Appl.* **10**, 169-193 (1998).

[14] P. Colli, F. Luterotti, G. Schimperna and U. Stefanelli, Global existence for a class of generalized systems of irreversible phase changes. *NoDEA Nonlinear Differentials Equations Appl.* **9**, 255-276 (2002).

[15] P. Colli and J. Sprekels, Weak solutions to some Penrose-Fife phase-field systems with temperature-dependent memory. *J. Differential Equations* **142**, 54-77 (1998).

[16] M. Frémond, *Non-smooth thermomechanics.* Springer-Verlag, Berlin, 2002.
[17] M. Frémond and E. Rocca, Global existence of solutions for a phase transition model with the possibility of voids. *Math. Models Methods Appl. Sci.* (to appear).
[18] M. Frémond and A. Visintin, Dissipation dans le changement de phase. Surfusion. Changement de phase irréversible. *C.R. Acad. Sci. Paris Sér. II* **301**, 1265–1268 (1985).
[19] C. Giorgi, M. Grasselli and V. Pata, Well-posedness and long-time behavior of the phase-field model with memory in a history space setting. *Quart. Appl. Math.* **59**, 710–736 (2001).
[20] M.E. Gurtin and A.C. Pipkin, A general theory of heat conduction with finite wave speeds. *Arch. Rational Mech. Anal.* **31**, 113–126 (1968).
[21] J.J. Moreau, Sur les lois de frottement, de viscosité et de plasticité. *C. R. Acad. Sci. Paris Sér. II* **27**, 608–611 (1970).
[22] J. Simon, Compact sets in the space $L^p(0,T;B)$. *Ann. Mat. Pura Appl. (4)* **146**, 65-96 (1987).
[23] A. Visintin, *Models of Phase transitions.* Birkhäuser, Boston, 1996.

GLOBAL SOLUTION TO A ONE DIMENSIONAL PHASE TRANSITION MODEL WITH STRONG DISSIPATION

Giovanna Bonfanti[1] and Fabio Luterotti[2]

[1] *Dipartimento di Matematica*
Università di Brescia
Via Branze 38, 25123 Brescia, Italy
e-mail: bonfanti@ing.unibs.it

[2] *Dipartimento di Matematica*
Università di Brescia
Via Branze 38, 25123 Brescia, Italy
e-mail: luterott@ing.unibs.it

Michel Frémond has recently proposed a new model for phase transition phenomena. When the process is very fast, the power of the acceleration forces cannot be neglected. This fact leads to a nonlinear parabolic-hyperbolic system. The global existence of solutions of the related Cauchy-Neumann problem is still open in the three-dimensional framework. Here we establish the well-posedness in the one-dimensional case with strong dissipation.

1. Introduction and derivation of the model

A number of papers (see, e.g., [2, 3, 20, 21, 14]) have been devoted to proving well-posedness results for strongly nonlinear parabolic systems related to a recent model for phase transitions introduced by Michel Frémond (see the book [15]). The main novelty of the model relies on assuming that the microscopic movements of the particles give rise to macroscopic effects. Such an assumption is taken into account in the energy balance equation (see (3) below). The motion equation is obtained by means of the virtual power principle, which dates back to [16]. In this concern, the power of the acceleration forces cannot be neglected when the evolution is very fast:

the resulting systems are parabolic-hyperbolic and their *only local* well-posedness was proved in [4] in the three dimensional setting. The aim of the present paper is to deal with the one dimensional system with dissipation, trying to prove some global results.

Now, we recall the basic features of Frémond's model, following the main steps of its derivation (see [15] for the complete procedure) in order to compare the present contribution with the previous ones.

Let us consider a two-phase material located in a domain Ω in $\boldsymbol{R}^n$ $(n = 1, 2, 3)$ with smooth boundary $\Gamma := \partial\Omega$. We aim to study the thermal evolution of the substance (from 0 up to a fixed time T) in Ω. We choose as state variables the fields $\theta(x, t)$ and $\chi(x, t)$, with $(x, t) \in \Omega \times [0, T]$. Indeed, θ stands for the absolute temperature, while χ denotes the order parameter (e.g., the local proportion of one of the two phases).

We are going to use the principle of virtual power to deduce the first relation of the system. For the sake of simplicity, it is assumed that the macroscopic movements have to be decoupled from the phase change. Moreover, we actually neglect the macroscopic velocities and movements. Since it is supposed that the phase transition results from microscopic movements, we take into account their power. The power of the interior forces in a generic subdomain $\mathcal{D}$ contained in Ω is

$$\mathcal{P}_i(\mathcal{D}, \frac{\partial\chi}{\partial t}) = -\int_{\mathcal{D}} \left\{ B\frac{\partial\chi}{\partial t} + \nabla\frac{\partial\chi}{\partial t} \cdot \boldsymbol{H} \right\} d\Omega,$$

where B and $\boldsymbol{H}$ are the interior forces (see [15]). The quantity B is a volume density of energy. The vector $\boldsymbol{H}$ is an energy flux vector.

The power of the exterior forces in a generic subdomain $\mathcal{D}$ contained in Ω is

$$\mathcal{P}_e(\mathcal{D}, \frac{\partial\chi}{\partial t}) = \int_{\mathcal{D}} A\frac{\partial\chi}{\partial t} d\Omega + \int_{\partial\mathcal{D}} a\frac{\partial\chi}{\partial t} d\Gamma,$$

where A (resp. a) is a volume (resp. surface) density of energy supplied to the material by microscopic actions without macroscopic motions.

The actual power $\mathcal{P}_a$ of the acceleration forces in $\mathcal{D}$ is

$$\mathcal{P}_a(\mathcal{D}, \frac{\partial\chi}{\partial t}) = \int_{\mathcal{D}} \rho_0\gamma_0\frac{\partial\chi}{\partial t}\, d\Omega,$$

where the quantity $\rho_0\gamma_0 = \rho_0\left(\partial^2\chi/\partial t^2\right)$ stands for the inertia forces of the microscopic motions, and ρ_0 is a positive constant, proportional to the mass of the microscopic links between the grains or elements of the material which evolve during the phase change.

The equation of motion for the microscopic movements results from the principle of virtual power

$$\forall \text{ virtual velocity } c \quad \forall \mathcal{D}, \quad \mathcal{P}_a(\mathcal{D}, c) = \mathcal{P}_i(\mathcal{D}, c) + \mathcal{P}_e(\mathcal{D}, c),$$

which yields

$$\rho_0 \gamma_0 = \rho_0 \frac{\partial^2 \chi}{\partial t^2} = \mathrm{div} \boldsymbol{H} - B + A, \quad \text{in } \Omega \times]0, T[, \tag{1}$$

$$\boldsymbol{H} \cdot \boldsymbol{n} = a, \quad \text{in } \Gamma \times]0, T[, \tag{2}$$

where $\boldsymbol{n}$ denotes the unit normal vector pointing outward Γ.

The effects of the acceleration forces $\rho_0 \gamma_0$ in the equation (1) have been investigated in the damage phenomena, see [8]. These effects are very important in fast evolutions.

Since the analytical treatment is not affected by non-zero data, for the sake of simplicity we assume $A = a = 0$.

Taking into account the power of the microscopic movements, we address the following energy balance equation

$$\frac{de}{dt} + \mathrm{div} \boldsymbol{q} = B \frac{\partial \chi}{\partial t} + \boldsymbol{H} \cdot \nabla \frac{\partial \chi}{\partial t}, \quad \text{in } \Omega \times]0, T[, \tag{3}$$

where e denotes the internal energy and $\boldsymbol{q} = -k \nabla \theta$ stands for the heat flux vector (i.e., the Fourier law is assumed with constant conductivity k). If we suppose the system insulated from the exterior, we also have the no-flux condition

$$-\boldsymbol{q} \cdot \boldsymbol{n} = 0, \quad \text{in } \Gamma \times]0, T[. \tag{4}$$

Now, we are going to introduce some constitutive relations to define the quantities e, B, $\boldsymbol{H}$. To this end, we choose the free energy Ψ as

$$\Psi(\chi, \nabla \chi, \theta) = -c_s \theta \log \theta - \frac{\lambda(\chi)}{\theta_c}(\theta - \theta_c) + \frac{1}{4}(\chi^2 - 1)^2 + \frac{\nu}{2} |\nabla \chi|^2, \tag{5}$$

where λ is a smooth function related to the latent heat at the critical transition temperature $\theta_c > 0$, $c_s > 0$ represents the specific heat, the parameter $\nu > 0$ is the factor of the interfacial energy term. Let us point out that (5) is one of the standard choices for the free energy in phase field models, (see, e.g., [10]). Next, we set for the pseudo-potential of dissipation

$$\Phi\left(\frac{\partial \chi}{\partial t}, \nabla \frac{\partial \chi}{\partial t}\right) = \frac{\mu}{2}\left(\frac{\partial \chi}{\partial t}\right)^2 + \frac{h}{2}\left|\nabla \frac{\partial \chi}{\partial t}\right|^2 + I_{[0,+\infty[}\left(\frac{\partial \chi}{\partial t}\right), \tag{6}$$

where μ and h are positive coefficients, while $I_{[0,+\infty[}$ denotes the indicator function of the interval $[0, +\infty[$. The latter contribution entails that we are actually dealing with an *irreversible* evolution of the phase variable χ: it means that, e.g., the substance does not re-melt after solidification. Of course, if do not impose such a constraint on $\frac{\partial \chi}{\partial t}$, the usual reversible evolution is addressed.

Our choices for the constitutive laws for B and $\boldsymbol{H}$ are provided by

$$B = \frac{\partial \Psi}{\partial \chi} + \frac{\partial \Phi}{\partial \left(\frac{\partial \chi}{\partial t}\right)}, \qquad \boldsymbol{H} = \frac{\partial \Psi}{\partial (\nabla \chi)} + \frac{\partial \Phi}{\partial \left(\nabla \frac{\partial \chi}{\partial t}\right)}, \tag{7}$$

while the internal energy e is related to the free energy Ψ and to the entropy $s = -\partial \Psi / \partial \theta$ by the rather classical relationship

$$e = \Psi + s\,\theta = \Psi - \theta \frac{\partial \Psi}{\partial \theta}. \tag{8}$$

Actually, it is not difficult to show [12] that in this setting the Second Principle of Thermodynamics is satisfied in the form of the Clausius-Duhem inequality. Setting for the sake of brevity $\frac{\partial}{\partial t} = \partial_t$, the relations (1), (3), (5), (6), (7), and (8) yield the following system

$$c_s\,\partial_t \theta + \lambda'(\chi)\,\partial_t \chi - k\,\Delta\theta$$
$$= -\frac{\lambda'(\chi)}{\theta_c}\,(\theta - \theta_c)\,\partial_t\chi + \mu(\partial_t\chi)^2 + h|\nabla\partial_t\chi|^2 + \eta\partial_t\chi \;\text{ in } \Omega\times]0,T[, \tag{9}$$

$$\rho_0\partial_{tt}\chi + \mu\,\partial_t\chi + \eta - h\Delta\partial_t\chi - \nu\Delta\chi + \chi^3 - \chi \ni \frac{\lambda'(\chi)}{\theta_c}\,(\theta - \theta_c) \;\text{ in } \Omega\times]0,T[, \tag{10}$$

$$\eta \in \partial I_{[0,+\infty[}(\partial_t\chi) \quad \text{in } \Omega\times]0,T[\,, \tag{11}$$

where $\partial I_{[0,+\infty[}$ denotes the subdifferential of $I_{[0,+\infty[}$ and it is actually a maximal monotone graph in $\boldsymbol{R}\times\boldsymbol{R}$ (see, e.g., [9]). Due to the properties of $\partial I_{[0,+\infty[}$, $\eta\partial_t\chi$ vanishes. Next, (2) and (4) give the no-flux boundary conditions

$$\boldsymbol{H}\cdot\boldsymbol{n} = \nu\nabla\chi\cdot\boldsymbol{n} + h\nabla\partial_t\chi\cdot\boldsymbol{n} = 0 \qquad -\boldsymbol{q}\cdot\boldsymbol{n} = 0 \quad \text{in } \Gamma\times]0,T[\,. \tag{12}$$

Finally, we complement our problem by the initial conditions

$$\theta(\cdot,0) = \theta_0 \qquad \chi(\cdot,0) = \chi_0 \qquad \partial_t\chi(\cdot,0) = \chi_1 \quad \text{in } \Omega\,. \tag{13}$$

Now, we consider the main differences between (9)-(11) and the model considered in [2]. First, the power of the microscopic accelerations was neglected in [2] and, obviously, the term $\rho_0\partial_{tt}\chi$ didn't occur. On the other hand, a constraint on the order parameter χ (actually a phase proportion) could be taken into account there by introducing the indicator function of an interval in the free energy expression (in place of the double well potential). Following the same steps in the derivation of the model, a term like $\partial I_{[0,1]}(\chi)$ replaces $\chi^3 - \chi$ in (10). Here we cannot deal with the term $\partial I_{[0,1]}(\chi)$, because it is rather well-known that the simultaneous presence of $\partial_{tt}\chi$ and of a constraint on χ in (10) would lead to a unilateral problem, which may be faced only in (other) particular cases.

We have to remark that, whenever the right-hand side of (9) is negligible if compared with the left-hand side quantities (i.e., within the small perturbations assumption: θ close to θ_c, $\mu(\partial_t\chi)^2 \approx 0$, and $h|\nabla\partial_t\chi|^2 \approx 0$), (9) actually reduces to

$$c_s\,\partial_t\theta + \lambda'(\chi)\,\partial_t\chi - k\,\Delta\theta = 0, \tag{14}$$

which is close to the standard expression of the energy balance equation in phase transition phenomena with the Fourier heat flux law (see, e.g., [10]).

The analysis of the system (9)-(11) under the above initial and boundary conditions was performed in [4] in the case λ is a linear function of χ, i.e. $\lambda(\chi) = L\chi$ with L positive constant and for two different choices of the pseudo-potential of dissipation. Namely, $\Phi_1(\partial_t\chi, \nabla\partial_t\chi) = \frac{\mu}{2}(\partial_t\chi)^2 + \frac{h}{2}|\nabla\partial_t\chi|^2$ was taken to describe a reversible evolution with strong dissipation, whereas $\Phi_2(\partial_t\chi) = \frac{\mu}{2}(\partial_t\chi)^2 + I_{[0,+\infty[}(\partial_t\chi)$ was chosen to model an irreversible evolution without dissipation. In both situations, local (in time) existence and uniqueness results were proved by using fixed point arguments and regularization tools. For the sake of completeness, we stress that –in these local results– the final time depends on the coefficient ρ_0, and the life-time of the solution vanishes as $\rho_0 \to 0^+$. In the papers [5, 6] the asymptotic analysis as $\rho_0 \to 0^+$ was addressed in the case of both choices of the pseudo-potential of dissipation. As outlined above, this cannot be achieved in the functional framework of [4]. Indeed, the papers [5] and [6] deal with a stronger setting, obtaining (of course, local) proper existence and uniqueness results for $\rho_0 > 0$, with a final time independent of $\rho_0 > 0$, so that the asymptotic analysis can be done.

On the other hand, reviewing the contributions on the problem without the acceleration, one can see that most of them deal with more or less reduced versions (no diffusion in [14], some kind of small perturbations in [2, 3, 20], an *a priori* speed limit of phase change in [21]). Whenever the previous restrictions are not imposed, the existence result is only local (see [23]), except for the one dimensional setting (see [18, 19]). The latter suggests to address the one dimensional case for the problem with the term $\rho_0\partial_{tt}\chi$. One step was successfully made in the paper [7], in the case without dissipation. Here we aim to tackle a slightly different functional framework, considering both the dissipation and the irreversible evolution of the order parameter. Moreover, we are able to deal with a quite general maximal monotone graph α acting on $\partial_t\chi$ in place of $\mu Id + \partial I_{[0,+\infty[}$ (where Id denotes, of course, the identity operator), so that our results still hold when $\alpha = \partial I_{[0,+\infty[}$, or $\alpha = \mu Id$, or even in a more complicated case.

Normalizing most of the physical constants in (9)-(11) and keeping to the one dimensional framework, the PDE system we are concerned with turns out to be

$$\partial_t\theta - \partial_{xx}\theta + \lambda'(\chi)\theta\,\partial_t\chi = (\partial_t\chi)^2 + (\partial_x\partial_t\chi)^2 \qquad \text{a.e. in } \Omega\times]0,T[, \quad (15)$$

$$\partial_{tt}\chi - \partial_{xx}\partial_t\chi - \partial_{xx}\chi + \xi + \chi^3 - \chi = \lambda'(\chi)(\theta - \theta_c) \qquad \text{a.e. in } \Omega\times]0,T[, \quad (16)$$

$$\xi \in \alpha(\partial_t\chi) \qquad \text{a.e. in } \Omega\times]0,T[\,. \quad (17)$$

Our paper is organized as follows. The next Section is devoted to the notation, the assumptions, and the statements of the main results. Section 3 contains the proof of the existence result given by Theorem 1: it is performed by means of a fixed point procedure, which provides the existence of solutions locally in time, and of an *a priori* estimates argument allowing to extend such solutions in the whole time interval. Section 4 is concerned with the uniqueness proof, which is achieved by proper contracting estimates.

2. Notation and main results

We set

$$\Omega :=]0,\ell[, \quad Q_t :=]0,\ell[\times]0,t[\quad \forall t \in]0,T], \quad Q := Q_T.$$

Next, we let

$$H := L^2(\Omega), \quad V := H^1(\Omega), \quad W := \{H^2(\Omega) \text{ such that } \partial_x u(0) = \partial_x u(\ell) = 0\}, \quad (18)$$

and identify H with its dual space H', so that

$$V \subset H \subset V',$$

with dense, compact, and continuous embeddings. Besides, let the symbol $\|\cdot\|$ denote the standard norm of H, while $\|\cdot\|_E$ stands for the norm of the general normed space E. Moreover, we denote by $< \cdot, \cdot >$ the duality pairing between V' and V, by $(\cdot,\cdot)$ the scalar product in H, and by $J : V \to V'$ the Riesz isomorphism of V onto V'.

Note that, thanks to the one dimensional framework of our problems, we have the continuous injections

$$L^1(\Omega) \subset V', \qquad V \subset L^\infty(\Omega). \tag{19}$$

Hence, there exist two positive constants c_1 and c_2 such that the following relations hold

$$\begin{aligned} \|u\|_{V'} &\leq c_1 \|u\|_{L^1(\Omega)}\,, \quad \forall u \in L^1(\Omega), \\ \|u\|_{L^\infty(\Omega)} &\leq c_2 \|u\|_V\,, \quad \forall u \in V. \end{aligned} \tag{20}$$

Finally, we also remark that there exists a positive constant c_3, depending only on T, such that the following holds for any $u \in H^1(0,T;H)$

$$\|u\|^2_{L^2(0,t;H)} \leq c_3 \left(\|u(0)\|^2 + \int_0^t \|\partial_t u\|^2_{L^2(0,s;H)} ds \right) \quad \forall t \in (0,T]. \tag{21}$$

We give here the precise statement of our problem, introducing the following assumptions on the data.

$$\theta_c > 0 \text{ is a prescribed constant,} \tag{22}$$

$$\varphi : \boldsymbol{R} \to [0,+\infty] \quad \text{is proper, convex, lower semicontinuous, and} \quad \varphi(0) = 0,$$

$$\alpha := \partial\varphi\,, \tag{23}$$

$$\lambda \in C^2(\boldsymbol{R}), \quad \lambda'' \in L^\infty(\boldsymbol{R}), \tag{24}$$

$$\theta_0 \in V \text{ and } \theta_0 > 0 \quad \text{in } \overline{\Omega}, \tag{25}$$

$$\chi_0 \in W\,, \tag{26}$$

$$\chi_1 \in V \text{ and } \varphi(\chi_1) \in L^1(\Omega)\,. \tag{27}$$

Now, we are ready to state the main results of the paper.

Theorem 1: Under the assumptions (22)-(27), there exists a triplet (θ, χ, ξ) such that

$$\theta \in H^1(0,T;H) \cap C^0([0,T];V) \cap L^2(0,T;W), \tag{28}$$

$$\chi \in H^2(0,T;H) \cap W^{1,\infty}(0,T;V) \cap H^1(0,T;W), \tag{29}$$

$$\xi \in L^2(0,T;H), \tag{30}$$

and the following relations hold

$$\partial_t\theta - \partial_{xx}\theta + \lambda'(\chi)\theta\,\partial_t\chi = (\partial_t\chi)^2 + (\partial_x\partial_t\chi)^2 \quad \text{a.e. in } Q, \tag{31}$$

$$\partial_{tt}\chi - \partial_{xx}\partial_t\chi - \partial_{xx}\chi + \xi + \chi^3 - \chi = \lambda'(\chi)(\theta - \theta_c) \quad \text{a.e. in } Q, \tag{32}$$

$$\xi \in \alpha(\partial_t\chi) \quad \text{a.e. in } Q, \tag{33}$$

$$\theta(\cdot,0) = \theta_0 \quad \chi(\cdot,0) = \chi_0 \quad \partial_t\chi(\cdot,0) = \chi_1 \quad \text{a.e. in } \Omega. \tag{34}$$

Moreover, there exists a positive constant θ^* such that

$$\theta \geq \theta^* > 0 \quad \text{a.e. in } Q. \tag{35}$$

Furthermore, the following uniqueness result holds.

Theorem 2: Let (θ, χ, ξ) be a triplet given by Theorem 1 satisfying (28)-(34). Then, such a triplet is unique.

The proof of these results will be carried out throughout the remainder of the paper: the existence of a global solution is derived by means of a fixed point technique combined with prolongation tools; the uniqueness result is established by proper contracting estimates, while the positivity of the temperature is proved via a maximum principle argument.

3. Proof of Theorem 1

Now, we detail the existence result in Theorem 1. To this aim, we apply the Schauder theorem to a suitable operator $\mathcal{T}$, which we construct later on.

For $R > 0$, let us consider the set

$$Y(\tau, R) = \{v \in H^1(0,\tau; W^{1,4}(\Omega)) \text{ such that } \|v\|_{H^1(0,\tau;W^{1,4}(\Omega))} \leq R\}\,, \tag{36}$$

(i.e., the closed ball of $H^1(0,\tau;W^{1,4}(\Omega))$ with center 0 and radius R), where $\tau \in]0,T]$ will be determined later, in such a way that the below-defined operator $\mathcal{T} : Y(\tau,R) \to Y(\tau,R)$ is compact and continuous.

We consider the following auxiliary problems, whose well-posedness is guaranteed by standard arguments (hence, for the sake of brevity, we omit any details).

Let $\widehat{\chi} \in Y(\tau,R)$ be fixed and let $\overline{\theta} := \mathcal{T}_1(\widehat{\chi})$ be the unique solution to the following problem.

Problem 1. Given $\widehat{\chi} \in Y(\tau,R)$, find $\overline{\theta}$ such that

$$\overline{\theta} \in [W^{1,1}(0,\tau;H) + H^1(0,\tau;V')] \cap C^0([0,\tau];H) \cap L^2(0,\tau;V)\,, \tag{37}$$

$$< \partial_t\overline{\theta}, v > + (\partial_x\overline{\theta}, \partial_x v) + (\lambda'(\widehat{\chi})\overline{\theta}\partial_t\widehat{\chi}, v) = ((\partial_t\widehat{\chi})^2 + (\partial_x\partial_t\widehat{\chi})^2, v) \quad \forall v \in V \ \text{ a.e. in }]0,\tau[\,, \tag{38}$$

$$\overline{\theta}(\cdot,0) = \theta_0 \quad \text{a.e. in } \Omega\,. \tag{39}$$

Now, given such $\overline{\theta}$, let $(\overline{\chi}, \overline{\xi})$, with $\overline{\chi} := \mathcal{T}_2(\overline{\theta})$, be the unique solution of the following problem.

Problem 2. Given $\overline{\theta} \in L^2(0,\tau;H)$, find $(\overline{\chi}, \overline{\xi})$ such that

$$\overline{\chi} \in H^2(0,\tau;H) \cap W^{1,\infty}(0,\tau;V) \cap H^1(0,\tau;W)\,, \tag{40}$$

$$\overline{\xi} \in L^2(0,\tau;H)\,, \tag{41}$$

$$\partial_{tt}\overline{\chi} - \partial_{xx}\partial_t\overline{\chi} - \partial_{xx}\overline{\chi} + \overline{\xi} + \overline{\chi}^3 - \overline{\chi} = \lambda'(\overline{\chi})(\overline{\theta} - \theta_c) \quad \text{a.e. in } Q_\tau\,, \tag{42}$$

$$\overline{\xi} \in \alpha(\partial_t\overline{\chi}) \quad \text{a.e. in } Q_\tau\,, \tag{43}$$

$$\overline{\chi}(\cdot,0) = \chi_0 \quad \text{a.e. in } \Omega\,, \tag{44}$$

$$\partial_t\overline{\chi}(\cdot,0) = \chi_1 \quad \text{a.e. in } \Omega\,. \tag{45}$$

Finally, we define the operator $\mathcal{T}$ as the composition $\mathcal{T}_2 \circ \mathcal{T}_1$. Our aim is to show that, for a suitable $\tau > 0$, the Schauder theorem applies to the map $\mathcal{T}$ from $Y(\tau, R)$ into itself: indeed, by construction any fixed point of $\mathcal{T}$ yields a solution to Problem (28)-(34). Thus, we will prove that there exists $\tau > 0$ such that $\mathcal{T}$ satisfies the following properties

$\mathcal{T}$ maps $Y(\tau, R)$ into itself;
$\mathcal{T}$ is compact;
$\mathcal{T}$ is continuous.

We start by deriving some *a priori* bounds on $\overline{\theta}$ and $\overline{\chi}$. We warn that in the proofs we will employ the same symbol c for different constants, even in the same formula, for the sake of simplicity. Moreover, we recall here some properties which will be useful in the sequel. Thanks to (24), there exists a positive constant c_4 such that

$$|\lambda'(u_1) - \lambda'(u_2)| \le c_4|u_1 - u_2| \quad \forall u_1,\ u_2 \in \boldsymbol{R}\,. \tag{46}$$

Moreover, we will use the elementary inequality

$$ab \le (\delta/2)a^2 + (2\delta)^{-1}b^2 \quad \forall a, b \in \boldsymbol{R}, \quad \delta > 0\,. \tag{47}$$

Now, in order to obtain *a priori* bounds on $\overline{\theta}$, we choose $v = \overline{\theta}$ in (38) and we integrate from 0 to t, with $0 < t < \tau$. Owing to (46), the Hölder inequality and the continuous embedding $V \hookrightarrow L^\infty(\Omega)$ (cf. (20)), and recalling the definition of $Y(\tau, R)$, we have

$$
\begin{aligned}
&\frac{1}{2}\|\overline{\theta}(t)\|^2 + \|\partial_x\overline{\theta}\|^2_{L^2(0,t;H)} \leq \frac{1}{2}\|\theta_0\|^2 \\
&+ c\int_0^t \Big(\|\widehat{\chi}(s)\|_{L^4(\Omega)} + 1\Big)\|\overline{\theta}(s)\|_{L^\infty(\Omega)}\|\widehat{\partial}_t\chi(s)\|_{L^4(\Omega)}\|\overline{\theta}(s)\|\,ds \\
&+ \int_0^t \Big(\|\partial_t\widehat{\chi}(s)\|^2_{L^4(\Omega)} + \|\partial_x\partial_t\widehat{\chi}(s)\|^2_{L^4(\Omega)}\Big)\|\overline{\theta}(s)\|\,ds \\
&\leq \frac{1}{2}\|\theta_0\|^2 + c\int_0^t \Big((R+1)\|\partial_t\widehat{\chi}(s)\|_{L^4(\Omega)}\|\overline{\theta}(s)\|_V \\
&+ \|\partial_t\widehat{\chi}(s)\|^2_{L^4(\Omega)} + \|\partial_x\partial_t\widehat{\chi}(s)\|^2_{L^4(\Omega)}\Big)\|\overline{\theta}(s)\|\,ds\,. \qquad (48)
\end{aligned}
$$

Next, in order to recover the full V-norm of $\overline{\theta}$ on the left-hand side, we add $\|\overline{\theta}\|^2_{L^2(0,t;H)}$ to (48). Then, we use (47) and we get

$$
\begin{aligned}
&\frac{1}{2}\|\overline{\theta}(t)\|^2 + \|\overline{\theta}\|^2_{L^2(0,t;V)} \leq \frac{1}{2}\|\theta_0\|^2 + \frac{1}{2}\|\overline{\theta}\|^2_{L^2(0,t;V)} \\
&+ c\int_0^t (1 + \|\partial_t\widehat{\chi}(s)\|^2_{L^4(\Omega)})\|\overline{\theta}(s)\|^2\,ds \\
&+ c\int_0^t \Big(\|\partial_t\widehat{\chi}(s)\|^2_{L^4(\Omega)} + \|\partial_x\partial_t\widehat{\chi}(s)\|^2_{L^4(\Omega)}\Big)\|\overline{\theta}(s)\|\,ds\,. \qquad (49)
\end{aligned}
$$

Recalling that, by the definition of $Y(\tau, R)$, the term $\Big(\|\partial_t\widehat{\chi}\|^2_{L^4(\Omega)} + \|\partial_x\partial_t\widehat{\chi}\|^2_{L^4(\Omega)}\Big)$ is bounded in $L^1(0,\tau)$, we apply to (49) a generalized version of the Gronwall lemma introduced in [1], and we deduce that there exists a positive constant c_5 depending on c_4, T, Ω, and R such that

$$
\|\overline{\theta}\|_{L^\infty(0,\tau;H)\cap L^2(0,\tau;V)} \leq c_5. \qquad (50)
$$

Next, in order to obtain *a priori* bounds on $\overline{\chi}$, we multiply (42) by $\partial_t\overline{\chi}$ and we integrate over Q_t. Taking the monotonicity of α into account, applying the Hölder inequality, using some standard Sobolev embeddings, and (47), we have

$$\frac{1}{2}\|\partial_t\overline{\chi}(t)\|^2 + \|\partial_x\partial_t\overline{\chi}\|^2_{L^2(0,t;H)} + \frac{1}{2}\|\partial_x\overline{\chi}(t)\|^2 + \frac{1}{4}\int_\Omega (\overline{\chi}^2(t)-1)^2\,dx \leq$$
$$\frac{1}{2}\|\chi_1\|^2 + \frac{1}{2}\|\partial_x\chi_0\|^2 + \frac{1}{4}\int_\Omega (\chi_0^2-1)^2\,dx$$
$$+ c\int_0^t \Big(\|\overline{\chi}(s)\|_{L^4(\Omega)}+1\Big)\|\overline{\theta}(s)-\theta_c\|_{L^4(\Omega)}\|\partial_t\overline{\chi}(s)\|\,ds$$
$$\leq \frac{1}{2}\|\chi_1\|^2 + \frac{1}{2}\|\partial_x\chi_0\|^2 + \frac{1}{4}\int_\Omega (\chi_0^2-1)^2\,dx + c + c\|\overline{\chi}\|^2_{L^2(0,t;V)}$$
$$+ c\int_0^t \|\overline{\theta}(s)-\theta_c\|_V^2\,\|\partial_t\overline{\chi}(s)\|^2\,ds\,. \tag{51}$$

Thanks to (50), $\|\overline{\theta}-\theta_c\|_V^2$ is bounded in $L^1(0,\tau)$. Thus, we apply to (51) the Gronwall lemma and we deduce that there exists a positive constant c_6 such that

$$\|\overline{\chi}\|_{W^{1,\infty}(0,\tau;H)\cap H^1(0,\tau;V)} \leq c_6. \tag{52}$$

On account of (19), from (52) it follows

$$\|\overline{\chi}\|_{L^\infty(Q_\tau)} \leq c_7\,, \tag{53}$$

for some positive constant c_7.

Next, we multiply (42) by $-\partial_{xx}\partial_t\overline{\chi}$, we integrate over Q_t, and we obtain

$$\frac{1}{2}\|\partial_x\partial_t\overline{\chi}(t)\|^2 + \|\partial_{xx}\partial_t\overline{\chi}\|^2_{L^2(0,t;H)} + \frac{1}{2}\|\partial_{xx}\overline{\chi}(t)\|^2 - \iint_{Q_t}\overline{\xi}\,\partial_{xx}\partial_t\overline{\chi}$$
$$\leq \frac{1}{2}\|\partial_x\chi_1\|^2 + \frac{1}{2}\|\partial_{xx}\chi_0\|^2 + \sum_{i=1}^{2}|I_i(t)|, \tag{54}$$

where

$$I_1(t) := \iint_{Q_t}(\overline{\chi}^3-\overline{\chi})\,\partial_{xx}\partial_t\overline{\chi}, \tag{55}$$

$$I_2(t) := \iint_{Q_t}\lambda'(\overline{\chi})\,(\overline{\theta}-\theta_c)\,\partial_{xx}\partial_t\overline{\chi}. \tag{56}$$

Now, the term $-\iint_{Q_t}\overline{\xi}\,\partial_{xx}\partial_t\overline{\chi}$ in the left-hand side of (54) is non-negative, thanks to the monotonicity of α (see [21, Lemma 4.1] for a rigorous justification). Moreover, we estimate the previous integrals owing to (53), (46), (47), and (50). We obtain

$$|I_1(t)| + |I_2(t)| \le c + c\|\overline{\theta} - \theta_c\|^2_{L^2(0,t;H)} + \frac{1}{2}\|\partial_{xx}\partial_t\overline{\chi}\|^2_{L^2(0,t;H)}$$
$$\le c + \frac{1}{2}\|\partial_{xx}\partial_t\overline{\chi}\|^2_{L^2(0,t;H)} \,. \tag{57}$$

Thus, on account of (57), from (54) we deduce the further bound

$$\|\overline{\chi}\|_{W^{1,\infty}(0,\tau;V)\cap H^1(0,\tau;W)} \le c_8 \,, \tag{58}$$

for some positive constant c_8. Finally, we multiply (42) by $\partial_{tt}\overline{\chi}$ and we integrate over Q_t. Proceeding formally (see [11], Lemma 4.1 for a rigorous justification), we have

$$\|\partial_{tt}\overline{\chi}\|^2_{L^2(0,t;H)} + \frac{1}{2}\|\partial_x\partial_t\overline{\chi}(t)\|^2 + \int_\Omega \varphi(\partial_t\overline{\chi}(t))\,dx \le \frac{1}{2}\|\partial_x\chi_1\|^2$$
$$+ \int_\Omega \varphi(\chi_1)\,dx + \|\partial_{xx}\overline{\chi} + \overline{\chi} - \overline{\chi}^3 + \lambda'(\overline{\chi})\,(\overline{\theta} - \theta_c)\|_{L^2(0,t;H)}\,\|\partial_{tt}\overline{\chi}\|_{L^2(0,t;H)} \,. \tag{59}$$

Using (47), on account of (27), (50) and (58), we readily deduce

$$\|\partial_{tt}\overline{\chi}\|_{L^2(0,\tau;H)} \le c_9 \tag{60}$$

and, by comparison in (42),

$$\|\overline{\xi}\|_{L^2(0,\tau;H)} \le c_{10} \,, \tag{61}$$

for some positive constants c_9 and c_{10}.
Now, our aim is to find $\tau > 0$ such that the operator $\mathcal{T} : Y(\tau, R) \to Y(\tau, R)$ turns out to be well-defined. Exploiting the previous estimates (cf. (58)) and by using standard interpolation tools (see, e.g., [22]), we have

$$\|\overline{\chi}\|_{W^{1,4}(0,\tau;W^{1,4}(\Omega))} \le c_{11}. \tag{62}$$

Thus, by the Hölder inequality, we find

$$\|\overline{\chi}\|_{H^1(0,\tau;W^{1,4}(\Omega))} \le c(\tau)^{1/4}\|\overline{\chi}\|_{W^{1,4}(0,\tau;W^{1,4}(\Omega))} \le c_{12}(\tau)^{1/4}. \tag{63}$$

Hence, we can choose τ so small that

$$c_{12}(\tau)^{1/4} \le R \tag{64}$$

and ensure that $\overline{\chi}$ belongs to $Y(\tau, R)$.

Next, we observe that the above arguments (cf. (58) and (60)) lead to the estimate

$$\|\overline{\chi}\|_{H^2(0,\tau;H)\cap W^{1,\infty}(0,\tau;V)\cap H^1(0,\tau;W)} \le c, \tag{65}$$

for some positive constant c independent of the choice of $\widehat{\chi}$ in $Y(\tau, R)$, which ensures that $\mathcal{T}$ is a compact operator.

It remains to show that the operator $\mathcal{T}$ is continuous with respect the natural topology induced in $Y(\tau, R)$ by $H^1(0, \tau; W^{1,4}(\Omega))$. To this aim, we consider a sequence $\widehat{\chi}_n$ in $Y(\tau, R)$ such that

$$\widehat{\chi}_n \to \widehat{\chi} \text{ strongly in } Y(\tau, R)\,, \tag{66}$$

as $n \to +\infty$. Now, we denote by $\overline{\theta}_n$ the sequence of the solutions to Problem 1 once $\widehat{\chi}$ is substituted by $\widehat{\chi}_n$, i.e.,

$$\overline{\theta}_n := \mathcal{T}_1(\widehat{\chi}_n). \tag{67}$$

Arguing as in the derivation of (50), we can find a positive constant c not depending on n such that

$$\|\overline{\theta}_n\|_{L^\infty(0,\tau;H)\cap L^2(0,\tau;V)} \leq c\,. \tag{68}$$

By well-known compactness results, there exists a subsequence of $\{\overline{\theta}_n\}$, which we do not relabel, such that

$$\overline{\theta}_n \rightharpoonup^* \overline{\theta} \quad \text{in } L^\infty(0, \tau; H) \cap\ L^2(0, \tau; V)\,, \tag{69}$$

as $n \to +\infty$. We first show that $\overline{\theta}$ in (69) is the solution to Problem 1, i.e., $\overline{\theta} = \mathcal{T}_1(\widehat{\chi})$. To this end, we take the difference between the corresponding equations (38), we test it by $\overline{\theta}_n - \mathcal{T}_1(\widehat{\chi})$ and we integrate it from 0 to t, with $0 < t < \tau$. We have

$$\frac{1}{2}\|(\overline{\theta}_n - \mathcal{T}_1(\widehat{\chi}))(t)\|^2 + \|\partial_x(\overline{\theta}_n - \mathcal{T}_1(\widehat{\chi}))\|^2_{L^2(0,t;H)} \leq \sum_{i=3}^{7} |I_i(t)|, \tag{70}$$

where

$$I_3(t) := \iint_{Q_t} \left((\partial_t\widehat{\chi}_n)^2 - (\partial_t\widehat{\chi})^2\right)(\overline{\theta}_n - \mathcal{T}_1(\widehat{\chi})) \tag{71}$$

$$I_4(t) := \iint_{Q_t} \left((\partial_x\partial_t\widehat{\chi}_n)^2 - (\partial_x\partial_t\widehat{\chi})^2\right)(\overline{\theta}_n - \mathcal{T}_1(\widehat{\chi})) \tag{72}$$

$$I_5(t) := \iint_{Q_t} \lambda'(\widehat{\chi})\,\partial_t\widehat{\chi}_n\,(\overline{\theta}_n - \mathcal{T}_1(\widehat{\chi}))^2 \tag{73}$$

$$I_6(t) := \iint_{Q_t} \lambda'(\widehat{\chi})\,\mathcal{T}_1(\widehat{\chi})(\partial_t\widehat{\chi}_n - \partial_t\widehat{\chi})\,(\overline{\theta}_n - \mathcal{T}_1(\widehat{\chi})) \tag{74}$$

$$I_7(t) := \iint_{Q_t} \overline{\theta}_n\partial_t\widehat{\chi}_n(\lambda'(\widehat{\chi}_n) - \lambda'(\widehat{\chi}))(\overline{\theta}_n - \mathcal{T}_1(\widehat{\chi}))\,. \tag{75}$$

We control the previous integrals taking (66) into account and exploiting the regularity specified by (37). Owing to the Hölder inequality, we get

$$|I_3(t)| + |I_4(t)| \le c \int_0^t \Big(\|(\partial_t \widehat{\chi}_n)^2(s) - (\partial_t \widehat{\chi})^2(s)\| + \|(\partial_x \partial_t \widehat{\chi}_n)^2(s) - (\partial_x \partial_t \widehat{\chi})^2(s)\| \Big) \|(\overline{\theta}_n - \mathcal{T}_1(\widehat{\chi}))(s)\| \, ds \,. \tag{76}$$

Moreover, using also (46) and some standard Sobolev embeddings, we have

$$\begin{aligned} |I_5(t)| &\le c \int_0^t \|\partial_t \widehat{\chi}(s)\|_{L^\infty(\Omega)} \|(\overline{\theta}_n - \mathcal{T}_1(\widehat{\chi}))(s)\|^2 \, ds \\ &\le c \int_0^t \|\partial_t \widehat{\chi}(s)\|_{W^{1,4}(\Omega)} \|(\overline{\theta}_n - \mathcal{T}_1(\widehat{\chi}))(s)\|^2 \, ds \,. \end{aligned} \tag{77}$$

Analogously,

$$\begin{aligned} |I_6(t)| &\le c \int_0^t \|\mathcal{T}_1(\widehat{\chi})(s)\|_{L^\infty(\Omega)} \|\partial_t \widehat{\chi}_n(s) - \partial_t \widehat{\chi}(s)\| \, \|(\overline{\theta}_n - \mathcal{T}_1(\widehat{\chi}))(s)\| \, ds \\ &\le c \|\partial_t \widehat{\chi}_n - \partial_t \widehat{\chi}\|^2_{L^2(0,t;H)} + c \int_0^t \|\mathcal{T}_1(\widehat{\chi})(s)\|_V^2 \, \|(\overline{\theta}_n - \mathcal{T}_1(\widehat{\chi}))(s)\|^2 \, ds \,. \end{aligned} \tag{78}$$

Next, recalling also (68), we obtain

$$\begin{aligned} &|I_7(t)| \\ &\le c \int_0^t \|\widehat{\chi}_n(s) - \widehat{\chi}(s)\|_{L^\infty(\Omega)} \, \|\overline{\theta}_n(s)\| \, \|\partial_t \widehat{\chi}_n(s)\|_{L^\infty(\Omega)} \, \|(\overline{\theta}_n - \mathcal{T}_1(\widehat{\chi}))(s)\| \, ds \\ &\le c \|\widehat{\chi}_n - \widehat{\chi}\|^2_{L^2(0,t;W^{1,4}(\Omega))} + c \int_0^t \|\partial_t \widehat{\chi}_n(s)\|^2_{W^{1,4}(\Omega)} \, \|(\overline{\theta}_n - \mathcal{T}_1(\widehat{\chi}))(s)\|^2 \, ds \,. \end{aligned} \tag{79}$$

Finally, we apply the Gronwall lemma to (70) on account of (76)-(79): the convergence specified by (66) allows us to deduce

$$\overline{\theta}_n \to \mathcal{T}_1(\widehat{\chi}) \ \text{ in } L^\infty(0,\tau;H) \cap \ L^2(0,\tau;V) \,, \tag{80}$$

as $n \to +\infty$. Thus, $\overline{\theta}$ in (69) can be identified with $\mathcal{T}_1(\widehat{\chi})$. Thanks to the uniqueness result holding for Problem 1, we actually deduce that the whole sequence $\overline{\theta}_n$ converges to $\overline{\theta}$, as $n \to +\infty$.

Secondly, we consider the sequence $(\overline{\chi}_n, \overline{\xi}_n)$ of the solutions to Problem 2 once $\overline{\theta}$ is substituted by $\overline{\theta}_n$, i.e., in particular,

$$\overline{\chi}_n := \mathcal{T}_2(\overline{\theta}_n) = \mathcal{T}_2 \circ \mathcal{T}_1(\widehat{\chi}_n) = \mathcal{T}(\widehat{\chi}_n). \tag{81}$$

Proceeding as in the previous estimates (cf. (58), (60), and (61)), we can find a positive constant c, not depending on n, such that

$$\|\overline{\chi}_n\|_{H^2(0,\tau;H)\cap W^{1,\infty}(0,\tau;V)\cap H^1(0,\tau;W)} \leq c\,, \tag{82}$$

$$\|\overline{\xi}_n\|_{L^2(0,\tau;H)} \leq c. \tag{83}$$

Hence, $\{\overline{\chi}_n\}$ and $\{\overline{\xi}_n\}$ admit subsequences, which we do not relabel, such that

$$\overline{\chi}_n \rightharpoonup^* \overline{\chi} \quad \text{in} \quad H^2(0,\tau;H)\cap W^{1,\infty}(0,\tau;V)\cap H^1(0,\tau;W)\,, \tag{84}$$

$$\overline{\xi}_n \rightharpoonup \overline{\xi} \quad \text{in} \quad L^2(0,\tau;H)\,, \tag{85}$$

as $n \to +\infty$. In particular, by compactness (see [17], [24]), we have

$$\overline{\chi}_n \to \overline{\chi} \quad \text{in} \quad H^1(0,\tau;W^{1,4}(\Omega))\,. \tag{86}$$

The above convergences, as well as (80), allow us to pass to the limit in the relations (42)-(43). Thus, thanks to the uniqueness result holding for Problem 2, we can identify $\overline{\chi}$ with $\mathcal{T}_2(\overline{\theta}) = \mathcal{T}_2\circ\mathcal{T}_1(\widehat{\chi}) = \mathcal{T}(\widehat{\chi})$. Furthermore, we have that the whole sequence $\overline{\chi}_n$ converges to $\mathcal{T}(\widehat{\chi})$, so that (86) entails

$$\mathcal{T}(\widehat{\chi}_n) \to \mathcal{T}(\widehat{\chi}) \quad \text{in} \quad H^1(0,\tau;W^{1,4}(\Omega)), \tag{87}$$

which concludes the proof of the continuity of the operator $\mathcal{T}$.

Thus, we have proved that $\mathcal{T}$ has a fixed point in $Y(\tau,R)$, i.e., there exists a local in time solution of the system (31)-(34), defined on the interval $]0,\tau[$. Note that, at the moment, (31) is satisfied only in a weak sense (cf. (38)). Actually, performing some standard parabolic estimates and exploiting (25) and (40) (cf. also (62)), we can prove the further regularity for θ specified in (28), so that (31) is satisfied a.e. in Q_τ.

Now, we have to discuss the extension of the latter solution to the whole interval $]0,T[$ as well. To this end, we derive some additional a priori estimates, which yield suitable *global* bounds on the solution.

Positivity of the temperature. We aim to establish the lower bound (35) for θ. We suppose that the regularity (28)-(29) is fulfilled by the pair (θ,χ) satisfying (31). We have

$$\partial_t\theta - \partial_{xx}\theta = -\lambda'(\chi)\theta\partial_t\chi + (\partial_t\chi)^2 + (\partial_x\partial_t\chi)^2 := a\theta + b. \tag{88}$$

Now, owing to the regularity (29), we easily deduce that $a \in L^\infty(Q)$. Moreover $b \geq 0$ a.e. in Q. Let us set $\theta_0^* := \min\theta_0$ (recall (25)) and $C_* := \|a\|_{L^\infty(Q)}$. It suffices to multiply (31) by the function

$$\vartheta := (\theta - \theta_0^* e^{-C_* t})^- \in L^2(Q),$$

and take the integral over Q_t, for $t \in (0,T)$. We obtain

$$\int_0^t \int_\Omega \left(\partial_t(\theta_0^* e^{-C_* s} - \vartheta)\vartheta - \vartheta_x^2 - a(\theta_0^* e^{-C_* s} - \vartheta)\vartheta\right) dx\, ds \geq 0,$$

hence

$$\frac{1}{2}\|\vartheta(t)\|^2 + \int_0^t \int_\Omega \left(|\partial_x \vartheta(x,s)|^2 + \theta_0^*(C_* + a(x,s))e^{-C_* s}\vartheta\right) dx\, ds \leq \frac{1}{2}\|\vartheta(0)\|^2 + C_* \int_0^t \|\vartheta(s)\|^2 ds.$$

Since $\vartheta(0) = 0$, an application of the Gronwall lemma implies that $\vartheta = 0$ a.e. on Q. Hence, relation (35) is proved with $\theta^* := \theta_0^* e^{-C_* T}$.

First estimate. We multiply (32) by $\partial_t \chi$; we add the resulting equation to (31) and we integrate over Q_t. Performing some cancellations and taking the monotonicity of α into account, we have

$$\begin{aligned}
&\int_\Omega \theta(t)\, dx + \frac{1}{2}\|\partial_t \chi(t)\|^2 + \frac{1}{2}\|\partial_x \chi(t)\|^2 + \frac{1}{4}\int_\Omega (\chi^2(t) - 1)^2\, dx \\
&\leq \int_\Omega \theta_0\, dx + \frac{1}{2}\|\chi_1\|^2 + \frac{1}{2}\|\partial_x \chi_0\|^2 \\
&+ \frac{1}{4}\int_\Omega (\chi_0^2 - 1)^2\, dx + \|\partial_t \chi\|^2_{L^2(0,t;H)} - \iint_{Q_t} \theta_c \lambda'(\chi)\partial_t \chi \\
&\leq \int_\Omega \theta_0\, dx + \frac{1}{2}\|\chi_1\|^2 + \frac{1}{2}\|\partial_x \chi_0\|^2 + \frac{1}{4}\int_\Omega (\chi_0^2 - 1)^2\, dx \\
&+ c\|\partial_t \chi\|^2_{L^2(0,t;H)} + c\,,
\end{aligned} \tag{89}$$

where we also used (46) and (21). Then, owing to (35), applying to (89) the Gronwall lemma, we deduce the following upper bounds

$$\|\theta\|_{L^\infty(0,T;L^1(\Omega))} \leq c\,, \tag{90}$$

$$\|\chi\|_{W^{1,\infty}(0,T;H) \cap L^\infty(0,T;V)} \leq c\,, \tag{91}$$

and hence, thanks to (20), it follows

$$\|\chi\|_{L^\infty(Q)} \leq c\,. \tag{92}$$

Second estimate. We add θ to both sides of (31), we multiply by $J^{-1}\theta$ (where J denotes the Riesz isomorphism of V onto V', see sect. 2 above) in the duality pairing between V' and V, and we integrate from 0 to t. We obtain

$$\frac{1}{2}\|\theta(t)\|_{V'}^2 + \|\theta\|_{L^2(0,t;H)}^2 \leq \frac{1}{2}\|\theta_0\|_{V'}^2 + \|\theta\|_{L^2(0,t;V')}^2 + \sum_{i=8}^{10} |I_i(t)|, \tag{93}$$

where

$$I_8(t) := -\int_0^t < \lambda'(\chi)\theta\partial_t\chi, J^{-1}\theta > ds \tag{94}$$

$$I_9(t) := \int_0^t < (\partial_t\chi)^2, J^{-1}\theta > ds \tag{95}$$

$$I_{10}(t) := \int_0^t < (\partial_x\partial_t\chi)^2, J^{-1}\theta > ds\,. \tag{96}$$

Now, we estimate the integrals in the right-hand side of (93). Thanks to the properties of J, (20), (24), the Hölder inequality, we deduce

$$\begin{aligned} |I_8(t)| &\leq \int_0^t \|\lambda'(\chi)\theta\partial_t\chi(s)\|_{V'}\|\theta(s)\|_{V'}\,ds \\ &\leq c_1^2 \int_0^t \|\lambda'(\chi)\theta\partial_t\chi(s)\|_{L^1(\Omega)}\|\theta(s)\|_{L^1(\Omega)}\,ds \\ &\leq c\int_0^t (\|\chi(s)\|_{L^\infty(\Omega)}+1)\|\theta(s)\|\,\|\partial_t\chi(s)\|\,\|\theta(s)\|_{L^1(\Omega)}\,ds\,. \end{aligned} \tag{97}$$

Using (47), (90), (91), and (92), we have

$$|I_8(t)| \leq c + \frac{1}{4}\|\theta\|_{L^2(0,t;H)}^2\,. \tag{98}$$

Again, (20), (90), and (91) give

$$\begin{aligned} |I_9(t)| &\leq \int_0^t \|(\partial_t\chi(s))^2\|_{V'}\|\theta(s)\|_{V'}\,ds \\ &\leq c_1^2 \int_0^t \|(\partial_t\chi(s))^2\|_{L^1(\Omega)}\|\theta(s)\|_{L^1(\Omega)}\,ds \leq c\,. \end{aligned} \tag{99}$$

Analogously,

$$|I_{10}(t)| \leq c_1^2 \int_0^t \|(\partial_x\partial_t\chi(s))^2\|_{L^1(\Omega)}\|\theta(s)\|_{L^1(\Omega)}\,ds \leq c\int_0^t \|\partial_x\partial_t\chi(s)\|^2\,ds\,. \tag{100}$$

Taking into account (98), (99), and (100), (93) yields

$$\frac{1}{2}\|\theta(t)\|_{V'}^2+\frac{3}{4}\|\theta\|_{L^2(0,t;H)}^2 \le \frac{1}{2}\|\theta_0\|_{V'}^2+\|\theta\|_{L^2(0,t;V')}^2+c\int_0^t \|\partial_x\partial_t\chi(s)\|^2\,ds+c\,. \tag{101}$$

Next, we multiply (32) by $-\partial_{xx}\partial_t\chi$, we integrate over Q_t, and we obtain

$$\frac{1}{2}\|\partial_x\partial_t\chi(t)\|^2 + \|\partial_{xx}\partial_t\chi\|_{L^2(0,t;H)}^2 + \frac{1}{2}\|\partial_{xx}\chi(t)\|^2 - \iint_{Q_t} \xi\ \partial_{xx}\partial_t\chi$$
$$\le \frac{1}{2}\|\partial_x\chi_1\|^2 + \frac{1}{2}\|\partial_{xx}\chi_0\|^2 + \sum_{i=11}^{12} |I_i(t)|, \tag{102}$$

where

$$I_{11}(t) := \iint_{Q_t} (\chi^3 - \chi)\ \partial_{xx}\partial_t\chi, \tag{103}$$

$$I_{12}(t) := \iint_{Q_t} \lambda'(\chi)\ (\theta-\theta_c)\ \partial_{xx}\partial_t\chi. \tag{104}$$

Again, we remark that the term $-\iint_{Q_t}\xi\ \partial_{xx}\partial_t\chi$ in the left-hand side of (102) is non-negative, thanks to the monotonicity of α. Now, we estimate the integrals on the right-hand side of (102). Using (92), we have

$$|I_{11}(t)| \le c + \delta\|\partial_{xx}\partial_t\chi\|_{L^2(0,t;H)}^2\,, \tag{105}$$

and

$$|I_{12}(t)| \le c\int_0^t (\|\chi(s)\|_{L^\infty(\Omega)}+1)(\|\theta(s)\|+1)\ \|\partial_{xx}\partial_t\chi(s)\|\,ds$$
$$\le \delta\|\partial_{xx}\partial_t\chi\|_{L^2(0,t;H)}^2 + c(\delta)\|\theta\|_{L^2(0,t;H)}^2 + c\,, \tag{106}$$

where $0<\delta<\frac{1}{4}$. Thus, we infer from (102)

$$\frac{1}{2}\|\partial_x\partial_t\chi(t)\|^2 + (1-2\delta)\|\partial_{xx}\partial_t\chi\|_{L^2(0,t;H)}^2 + \frac{1}{2}\|\partial_{xx}\chi(t)\|^2$$
$$\le \frac{1}{2}\|\partial_x\chi_1\|^2 + \frac{1}{2}\|\partial_{xx}\chi_0\|^2 + c + c(\delta)\|\theta\|_{L^2(0,t;H)}^2\,. \tag{107}$$

We multiply both sides of (107) by $\frac{1}{4c(\delta)}$ and we add the resulting inequality to (101). We apply the Gronwall lemma. Thanks to (25), (26), (27), we deduce the following upper bounds

$$\|\theta\|_{L^\infty(0,T;V')\cap L^2(0,T;H)} \le c\,, \tag{108}$$

$$\|\chi\|_{W^{1,\infty}(0,T;V)\cap H^1(0,T;W)} \le c\,. \tag{109}$$

Further estimates. Exploiting (108)-(109) and arguing as in the fixed point procedure (see, in particular, the derivation of (50) and (60)), we get the regularity prescribed in (28)-(29).

Thanks to the previous *global* estimates, the local solution to the system (31)-(34) provided by the fixed point procedure can be extended on the whole interval $]0, T[$ by standard prolongation arguments. Thus, the proof of Theorem 1 is complete.

4. Proof of Theorem 2

Now, we prove that the triplet (θ, χ, ξ) given by Theorem 2.1 is unique. We proceed by contradiction. Let $(\theta_1, \chi_1, \xi_1)$ and $(\theta_2, \chi_2, \xi_2)$ be two solutions to problem (28)-(34). We set $\tilde{\theta} = \theta_1 - \theta_2$ and $\tilde{\chi} = \chi_1 - \chi_2$.

We start by considering the difference between the corresponding equations (31); we multiply it by $\tilde{\theta}$ and we integrate over Q_t with $0 < t < T$. We have

$$\frac{1}{2}\|\tilde{\theta}(t)\|^2 + \|\partial_x \tilde{\theta}\|^2_{L^2(0,t;H)} \leq \sum_{i=13}^{17} |I_i(t)|, \tag{110}$$

where

$$I_{13}(t) := \iint_{Q_t} \partial_t \tilde{\chi}(\partial_t \chi_1 + \partial_t \chi_2)\, \tilde{\theta} \tag{111}$$

$$I_{14}(t) := \iint_{Q_t} \partial_x \partial_t \tilde{\chi}(\partial_x \partial_t \chi_1 + \partial_x \partial_t \chi_2)\, \tilde{\theta} \tag{112}$$

$$I_{15}(t) := \iint_{Q_t} \lambda'(\chi_2)\, \partial_t \chi_2\, \tilde{\theta}^2 \tag{113}$$

$$I_{16}(t) := \iint_{Q_t} \lambda'(\chi_1)\, \theta_1\, \partial_t \tilde{\chi}\, \tilde{\theta} \tag{114}$$

$$I_{17}(t) := \iint_{Q_t} \theta_1 \partial_t \chi_2 (\lambda'(\chi_1) - \lambda'(\chi_2))\tilde{\theta}. \tag{115}$$

We control the previous integrals owing to the regularity of the pairs (θ_1, χ_1) and (θ_2, χ_2) specified by (28)-(29). We get

$$\begin{aligned} |I_{13}(t)| &\leq \Big(\|\partial_t \chi_1\|_{L^\infty(Q)} + \|\partial_t \chi_2\|_{L^\infty(Q)}\Big) \int_0^t \|\partial_t \tilde{\chi}(s)\|\, \|\tilde{\theta}(s)\|\, ds \\ &\leq c\|\partial_t \tilde{\chi}\|^2_{L^2(0,t;H)} + c\|\tilde{\theta}\|^2_{L^2(0,t;H)} \end{aligned} \tag{116}$$

and

$$\begin{aligned}|I_{14}(t)| &\leq \int_0^t \|\partial_x\partial_t\tilde{\chi}(s)\| \, \|\partial_x\partial_t\chi_1(s) + \partial_x\partial_t\chi_2(s)\|_{L^\infty(\Omega)} \, \|\tilde{\theta}(s)\| \, ds \\ &\leq \frac{1}{2}\|\partial_x\partial_t\tilde{\chi}\|^2_{L^2(0,t;H)} + c\int_0^t \|\partial_t\chi_1(s) + \partial_t\chi_2(s)\|^2_W \, \|\tilde{\theta}(s)\|^2 \, ds \,. \end{aligned} \tag{117}$$

Analogously, using also (46), we have

$$|I_{15}(t)| \leq c\|\partial_t\chi_2\|_{L^\infty(Q)}\|\tilde{\theta}\|^2_{L^2(0,t;H)} \leq c\|\tilde{\theta}\|^2_{L^2(0,t;H)} \tag{118}$$

and

$$\begin{aligned}|I_{16}(t)| &\leq c\|\theta_1\|_{L^\infty(Q)} \int_0^t \|\partial_t\tilde{\chi}(s)\| \, \|\tilde{\theta}(s)\| \, ds \\ &\leq c\|\partial_t\tilde{\chi}\|^2_{L^2(0,t;H)} + c\|\tilde{\theta}\|^2_{L^2(0,t;H)} \,. \end{aligned} \tag{119}$$

Using again (46) and (21), we obtain

$$\begin{aligned}|I_{17}(t)| &\leq c\|\theta_1\|_{L^\infty(Q)} \, \|\partial_t\chi_2\|_{L^\infty(Q)} \int_0^t \|\tilde{\chi}(s)\| \, \|\tilde{\theta}(s)\| \, ds \\ &\leq c\|\tilde{\chi}\|^2_{L^2(0,t;H)} + c\|\tilde{\theta}\|^2_{L^2(0,t;H)} \\ &\leq c\|\partial_t\tilde{\chi}\|^2_{L^2(0,t;H)} + c\|\tilde{\theta}\|^2_{L^2(0,t;H)} \,. \end{aligned} \tag{120}$$

Next, we consider the difference between the corresponding equations (32), we multiply it by $\partial_t\tilde{\chi}$ and we integrate over Q_t. We get

$$\begin{aligned}\frac{1}{2}\|\partial_t\tilde{\chi}(t)\|^2 + \|\partial_x\partial_t\tilde{\chi}\|^2_{L^2(0,t;H)} + \frac{1}{2}\|\partial_x\tilde{\chi}(t)\|^2 + \iint_{Q_t}(\xi_1 - \xi_2)\partial_t\tilde{\chi} \\ \leq \sum_{i=18}^{20} |I_i(t)| \,, \end{aligned} \tag{121}$$

where

$$I_{18}(t) := \iint_{Q_t} \tilde{\chi}\big(1 - \chi_1^2 - \chi_1\chi_2 - \chi_2^2\big)\,\partial_t\tilde{\chi} \tag{122}$$

$$I_{19}(t) := \iint_{Q_t} (\lambda'(\chi_1) - \lambda'(\chi_2))(\theta_1 - \theta_c)\partial_t\tilde{\chi} \tag{123}$$

$$I_{20}(t) := \iint_{Q_t} \lambda'(\chi_2)\,\tilde{\theta}\,\partial_t\tilde{\chi} \,. \tag{124}$$

Now, the term $\iint_{Q_t}(\xi_1 - \xi_2)\partial_t\tilde{\chi}$ in the left-hand side of (121) is non-negative, thanks to the monotonicity of α. Arguing as before (owing to the regularity of (θ_1, χ_1) and (θ_2, χ_2) specified by (28)-(29)), we estimate the latter integrals. We find

$$|I_{18}(t)| \le c \int_0^t \|\tilde{\chi}(s)\| \, \|\partial_t \tilde{\chi}(s)\| \, ds \le c\|\tilde{\chi}\|^2_{L^2(0,t;H)} + c\|\partial_t \tilde{\chi}\|^2_{L^2(0,t;H)} \le c\|\partial_t \tilde{\chi}\|^2_{L^2(0,t;H)} \tag{125}$$

$$|I_{19}(t)| \le c\|\theta_1 - \theta_c\|_{L^\infty(Q)} \int_0^t \|\tilde{\chi}(s)\| \, \|\partial_t \tilde{\chi}(s)\| \, ds \le c\|\partial_t \tilde{\chi}\|^2_{L^2(0,t;H)} \tag{126}$$

$$|I_{20}(t)| \le c \int_0^t \|\tilde{\theta}(s)\| \, \|\partial_t \tilde{\chi}(s)\| \, ds \le c\|\tilde{\theta}\|^2_{L^2(0,t;H)} + c\|\partial_t \tilde{\chi}\|^2_{L^2(0,t;H)} . \tag{127}$$

Then, we add (110) and (121), on account of (116)-(120) and (125)-(127); we apply the Gronwall lemma and we conclude $\tilde{\theta} = \tilde{\chi} = 0$ a.e. in Q. Finally, by comparison in (32), it also follows that $\xi_1 = \xi_2$ a.e. in Q.

References

[1] C. Baiocchi, Sulle equazioni differenziali astratte lineari del primo e del secondo ordine negli spazi di Hilbert. *Ann. Mat. Pura Appl.* **76**, 233–304 (1967).

[2] G. Bonfanti, M. Frémond, and F. Luterotti, Global solution to a nonlinear system for irreversible phase changes. *Adv. Math. Sci. Appl.* **10**, 1–24 (2000).

[3] G. Bonfanti, M. Frémond, and F. Luterotti, Local solutions to the full model of phase transitions with dissipation. *Adv. Math. Sci. Appl.* **11**, 791-810 (2001).

[4] G. Bonfanti, M. Frémond, and F. Luterotti, Existence and uniqueness results to a phase transition model based on microscopic accelerations and movements. *Nonlinear Anal. Real World Appl.* **5**, 123-140 (2004).

[5] G. Bonfanti and F. Luterotti, Convergence results for a phase transition model with vanishing microscopic acceleration. *Math. Models Methods Appl. Sci.* **14**, 375-392 (2004).

[6] G. Bonfanti and F. Luterotti, Well-posedness results and asymptotic behaviour for a phase transition model accounting for microscopic accelerations. *Commun. Pure Appl. Anal.* (to appear).

[7] G. Bonfanti and F. Luterotti, Global solution to a phase transition model with microscopic movements and accelerations in one space dimension. *Quad. Sem. Mat. Brescia* n. 20/2004.

[8] R. Boutenkak, *Endommagement et gradient d'endommagement. Modélisation en calculs dynamiques.* Stage de DESS de mathématiques appliquées, Université Pierre et Marie Curie, Paris, 1997.

[9] H. Brézis, *Opérateurs Maximaux Monotones et Semi-groupes de Contractions dans les Espaces de Hilbert.* North-Holland, Amsterdam, 1973.

[10] G. Caginalp, An analysis of a phase field model of a free boundary. *Arch. Rational Mech. Anal.* **92**, 205–245 (1986).
[11] P. Colli, On some doubly nonlinear evolution equations in Banach spaces. *Japan J. Indust. Appl. Math.* **9**, 181–203 (1992).
[12] P. Colli, M. Frémond and O. Klein, Global existence of a solution to phase field model for supercooling. *Nonlinear Anal. Real World Appl.* **2**, 523–539 (2001).
[13] P. Colli and Ph. Laurençot, Existence and stabilization of solutions to the phase-field model with memory. *J. Integral Equations Appl.* **10**, 169–194 (1998).
[14] P. Colli, F. Luterotti, G. Schimperna and U. Stefanelli, Global existence for a class of generalized systems for irreversible phase changes. *NoDEA Nonlinear Differential Equations Appl.* **9**, 255–276 (2002).
[15] M. Frémond, *Non-smooth Thermomechanics.* Springer-Verlag, Berlin, 2002.
[16] P. Germain, *Mécanique des Milieux Continus.* Masson, Paris, 1973.
[17] J.L. Lions, *Quelques méthodes de résolution des problèmes aux limites non linéaires.* Dunod Gauthier–Villars, Paris, 1969.
[18] Ph. Laurençot, G. Schimperna and U. Stefanelli, Global existence of a strong solution to the one-dimensional full model for phase transitions. *J. Math. Anal. Appl.* **271**, 426-442 (2002).
[19] F. Luterotti and U. Stefanelli, Existence result for the one-dimensional full model of phase transitions. *Z. Anal. Anwendungen* **21**, 335-350 (2002).
[20] F. Luterotti, G. Schimperna and U. Stefanelli, Existence result for a nonlinear model related to irreversible phase changes. *Math. Models Methods Appl. Sci.* **11**, 808–825 (2001).
[21] F. Luterotti, G. Schimperna, and U. Stefanelli, Global solution to a phase field model with irreversible and constrained phase evolution. *Quart. Appl. Math.* **60**, 301–316 (2002).
[22] L. Nirenberg, On elliptic partial differential equations. *Ann. Scuola Norm. Sup. Pisa (3)* **13**, 115–162 (1959).
[23] G. Schimperna, F. Luterotti and U. Stefanelli, Local solution to Frémond's full model for irreversible phase transitions. In: *Proceedings of the Workshop "Modelli Matematici e Problemi Analitici per Materiali Speciali", INdAM meeting in Cortona, June 2001*, Ed. M. Fabrizio, B. Lazzari and A. Morro (World Sci. Publishing, River Edge NJ, 2002) p. 323.
[24] J. Simon, Compact sets in the space $L^p(0,T;B)$. *Ann. Mat. Pura Appl. (4)* **146**, 65–96 (1987).

A GLOBAL IN TIME RESULT FOR AN INTEGRO-DIFFERENTIAL PARABOLIC INVERSE PROBLEM IN THE SPACE OF BOUNDED FUNCTIONS

Fabrizio Colombo[1], Davide Guidetti[2] and Vincenzo Vespri[3]

[1]*Dipartimento di Matematica*
Politecnico di Milano
Via Bonardi 9
20133 Milano, Italy
E-mail: fabcol@mate.polimi.it

[2]*Dipartimento di Matematica*
Università di Bologna
Piazza di Porta S. Donato 5
40126 Bologna, Italy
E-mail: guidetti@dm.unibo.it

[3]*Dipartimento di Matematica*
Università degli Studi di Firenze
Viale Morgagni 67/a
50134 Firenze, Italy
E-mail: vespri@math.unifi.it

The problem we deal with arises in the theory of heat propagation in materials with memory. We consider the identification of both the relaxation kernel and the time dependence of the heat source for an integro–differential equation of parabolic type. We prove an existence and uniqueness theorem, *global in time*, for the abstract version of the problem and we give an application to the concrete case. The novelties of this work are: the choice of the functional setting (spaces of bounded functions with values in an interpolation space) and the *existence and uniqueness results, global in time,* which are not easy to obtain for inverse problems.

1. Introduction

Integro–differential parabolic inverse problems constitute an important class, since one of the classical models is represented by the heat equation for materials with memory. In several problems in physics and chemistry, the heat equation with memory is coupled with one or more parabolic or hyperbolic equations, ruling further state variables; in particular, in the theory of phase transition for materials with memory, we find several models that have been recently studied from several view points: for instance, as direct problems in Hilbert spaces and as dynamical systems (long time behavior). We list here, without claim at completeness, some papers and books [1, 2, 3, 4, 5, 6, 7, 8, 18, 19, 20, 22, 26, 27, 28] in which one can find some models and results involving the heat equation. The list of references related to inverse integro-differential parabolic problems is long, and we recall just some of the papers that share our strategies and methods, based on analytic semigroup theory, fixed point arguments and optimal regularity theorems: such techniques where first used by Lorenzi and Sinestrari [24].

In the papers [12, 13, 16, 17] the authors study inverse problems for the heat equation with memory in non smooth domains and with memory kernels depending on time and on one space variable, in Hölder and Sobolev (of fractional order) spaces. In the paper[15], a unified approach for fully non linear problems has been proposed, containing as particular cases the theory of combustion of materials with memory and some models of spread of disease, again in the setting of Hölder and Sobolev spaces. In the papers [9, 10, 11, 21], the authors treat inverse problems for a model in population dynamics and for phase-field models in the Hölder setting.

In [14], the authors have used for the first time the same spaces used here, but with no weights, so that they obtain local in time results for a population dynamic model. In the recent and interesting paper [23], the authors consider an inverse problem for a phase field model and prove an existence and uniqueness (under conditions) result global in time in a Hilbert space.

The novelties of this work are: the choice of the functional setting (spaces of bounded functions with values in an interpolation space) and *existence and uniqueness results, global in time,* which are not easy to obtain for inverse problems. We point out that the functional setting we have assumed leads to solutions of the problem which are less regular in time and more regular in space, compared with the existent literature on this topic. Our global in time result is proved for an abstract version of the problem. Hence, we give an application to the concrete case.

For the sake of simplicity, we restrict our analysis to the heat equation with memory but the estimates we have obtained in Section 3 can be applied to get global in time results for more general linear models with convolution kernels.

We now formulate the problem we are going to investigate. Let Ω be an open bounded set in $\mathbb{R}^3$ with a suitably regular boundary (to be specified in the sequel), and $T > 0$. We can easily deduce the evolution equation for the temperature u from the continuity equation (for $(t,x) \in [0,T] \times \Omega$)

$$D_t u(t,x) + \operatorname{div} J(t,x) - f_0(t,x) = 0, \tag{1}$$

where the vector J denotes the density of heat flow per unit surface area per unit time, and f_0 is the heat source per unit volume per unit time in Ω. We recall that the well-known Fourier's law for materials with memory is given by

$$J(t,x) = -k\nabla u(t,x) - k' \int_0^t h(t-s)\nabla u(s,x)\,ds, \tag{2}$$

where we suppose that the diffusion coefficient k is a positive real number, and for the sake of simplicity the coefficient k' is assumed to be equal to 1. The convolution kernel h, which accounts for the thermal memory, is supposed to depend on time only. To obtain the equation ruling the evolution of the temperature, we replace (2) into the continuity equation (1), and we get

$$D_t u(t,x) = \operatorname{div}\left[k\nabla u(t,x) + \int_0^t h(t-s)\nabla u(s,x)\,ds\right] + f_0(t,x). \tag{3}$$

We now have to focus on the fact that the memory kernel h *is not* a physical observable. This is the fact that motivates the inverse problem: h has to be considered unknown. An additional difficulty arises when the heat source is placed in a fixed position of the material, but its time dependence is not known. We suppose that

$$f_0(t,x) = f(t)g(x) \tag{4}$$

where $g(x)$ is a given datum, but $f(t)$ has to be considered a further unknown of the problem.

In order to determine, simultaneously, both the unknown functions h, f, and the temperature u, we need, for example, additional measurements on the temperature on suitable parts of the material, that can be represented in integral form (see (6)). We can now state our problem.

Problem 1: (The Inverse Problem (IP)) Determine the temperature $u : [0,T]\times\Omega \longrightarrow \mathbb{R}$, the diffusion coefficient k and the functions $h : [0,T] \longrightarrow \mathbb{R}$, $f : [0,T] \longrightarrow \mathbb{R}$ satisfying the system

$$\begin{cases} D_t u(t,x) = k\Delta u(t,x) + \int_0^t h(t-s)\Delta u(s,x)\,ds + f(t)g(x), \\ u(0,x) = u_0(x), \quad x \in \Omega, \\ \dfrac{\partial u}{\partial \nu}(t,x) = 0, \quad (t,x) \in [0,T] \times \partial\Omega, \end{cases} \tag{5}$$

with the additional conditions

$$\int_\Omega u(t,x)\mu_j(dx) = G_j(t), \quad \forall t \in [0,T], \quad j = 1,2, \tag{6}$$

where g, u_0, G_1, G_2 are given data, and μ_1 and μ_2 are finite Borel measures in $C(\overline{\Omega})$.

The plan of the paper is the following.

- Section 2 is devoted to the functional setting of our problem and the statements of the main abstract and concrete results.
- In Section 3, we prove some fundamental estimates in weighted spaces which are of crucial importance to get global in time results.
- Section 4 contains a suitably equivalent reformulation of the Inverse Abstract Problem in terms of Volterra integral equations of the second kind.
- In Section 5 we prove, via a fixed point argument, the main result of this paper, i.e. Theorem 6, which is a consequence of the preliminary results obtained in the previous sections.

2. Definitions and main results

The results that we are going to recall in this section hold in the case X is a Banach space with norm $\|\cdot\|$. For $T > 0$, we denote by $C([0,T];X)$ the usual space of continuous functions with values in X, while we denote by $\mathcal{B}([0,T];X)$ the space of bounded functions with values in X. $\mathcal{B}([0,T];X)$ will be endowed with the sup–norm

$$\|u\|_{\mathcal{B}([0,T];X)} := \sup_{0\le t\le T} \|u(t)\| \tag{7}$$

and $C([0,T];X)$ will be considered a closed subspace of $\mathcal{B}([0,T];X)$.

We shall use the notation $C([0,T];\mathbb{R}) = C([0,T])$ and $\mathcal{B}([0,T];\mathbb{R}) = \mathcal{B}([0,T])$.

By $\mathcal{L}(X)$ we denote the space of all bounded linear operators from X into itself, equipped with the sup–norm, while $\mathcal{L}(X;\mathbb{R}) = X'$ is the space of all bounded linear functionals on X considered with the natural norm. We set $\mathbb{N}_0 := \mathbb{N} \cup \{0\}$.

Definition 2: Let $A : \mathcal{D}(A) \subseteq X \to X$, be a linear operator, possibly with $\overline{\mathcal{D}(A)} \neq X$. The operator A is said to be *sectorial* if it satisfies the following assumptions:

- there exist $\theta \in (\pi/2, \pi)$ and $\omega \in \mathbb{R}$, such that any $\lambda \in \mathbb{C} \setminus \{\omega\}$ with $|\arg(\lambda - \omega)| \leq \theta$ belongs to the resolvent set of A.
- there exists $M > 0$ such that $\|(\lambda - \omega)(\lambda I - A)^{-1}\|_{\mathcal{L}(X)} \leq M$ for any $\lambda \in \mathbb{C} \setminus \{\omega\}$ with $|\arg(\lambda - \omega)| \leq \theta$.

The fact that the resolvent set of A is not empty implies that A is closed, so that $\mathcal{D}(A)$ endowed with the graph norm becomes a Banach space.

According to the definition of sectorial operator, it is possible to introduce the semigroup $\{e^{tA}\}_{t\geq 0}$ of bounded linear operators in $\mathcal{L}(X)$, so that $t \to e^{tA}$ is an analytic function from $(0, \infty)$ to $\mathcal{L}(X)$ satisfying for $k \in \mathbb{N}$ the relations

$$\frac{d^k}{dt^k} e^{tA} = A^k e^{tA}, \quad t > 0, \tag{8}$$

and

$$Ae^{tA}x = e^{tA}Ax, \quad x \in \mathcal{D}(A), \quad t > 0. \tag{9}$$

Moreover, for $k \in \mathbb{N}_0$, $\varepsilon > 0$, there exist positive constants $M_{k,\epsilon}$, such that

$$\|t^k A^k e^{tA}\|_{\mathcal{L}(X)} \leq M_{k,\epsilon} e^{(\omega+\varepsilon)t}, \qquad t > 0. \tag{10}$$

For more details see for example Lunardi, Pazy and Sinestrari [25, 29, 30]. Let us define the family of interpolation spaces (see the books [25, 32]) $\mathcal{D}_A(\theta, \infty)$, $\theta \in (0, 1)$, between $\mathcal{D}(A)$ and X by

$$\mathcal{D}_A(\theta, \infty) := \Big\{ x \in X : \ |x|_{\mathcal{D}_A(\theta,\infty)} := \sup_{0<t<1} t^{1-\theta} \|Ae^{tA}x\| < \infty \Big\}, \tag{11}$$

with the norm

$$\|x\|_{\mathcal{D}_A(\theta,\infty)} := \|x\| + |x|_{\mathcal{D}_A(\theta,\infty)}. \tag{12}$$

It is straightforward to verify that, if $T > 0$, $k \in \mathbb{N}$, $f \in \mathcal{D}_A(\theta, \infty)$, $t \in (0, T]$, then

$$\|A^k e^{tA} f\| \leq C(T, k) \|f\|_{\mathcal{D}_A(\theta,\infty)} t^{\theta - k}. \tag{13}$$

We also set

$$\mathcal{D}_A(1+\theta,\infty) := \{x \in \mathcal{D}(A) : Ax \in \mathcal{D}_A(\theta,\infty)\}, \tag{14}$$

$\mathcal{D}_A(1+\theta,\infty)$ turns out to be a Banach space when equipped with the norm

$$\|x\|_{\mathcal{D}_A(1+\theta,\infty)} := \|x\| + \|Ax\|_{\mathcal{D}_A(\theta,\infty)}. \tag{15}$$

If $\theta \leq \xi \leq 1+\theta$ and $x \in \mathcal{D}_A(1+\theta,\infty)$,

$$\|x\|_{\mathcal{D}_A(\xi,\infty)} \leq C(\theta,\xi)\|x\|^{1+\theta-\xi}_{\mathcal{D}_A(\theta,\infty)}\|x\|^{\xi-\theta}_{\mathcal{D}_A(\theta+1,\infty)}, \tag{16}$$

with $C(\theta,\xi)$ independent of x. Optimal regularity results and the analytic semigroup theory are fundamental tools in the study of direct and inverse parabolic problems. Our strategy is to formulate the abstract version of the inverse problem in terms of a system of equivalent fixed point equations. Indeed, we can dispose of several optimal regularity results for such equivalent formulation. For instance, consider the following optimal regularity result for the Cauchy Problem (CP):

$$\begin{cases} u'(t) = Au(t) + f(t), & t \in [0,T], \\ u(0) = u_0. \end{cases} \tag{17}$$

Let $A : \mathcal{D}(A) \to X$ be a sectorial operator and $\theta \in (0,1)$. In [25, 30] we can find the proofs of the following

Theorem 3: (Strict solution in spaces $\mathcal{B}([0,T];\mathcal{D}_A(\theta,\infty))$) For any $f \in C([0,T];X) \cap \mathcal{B}([0,T];\mathcal{D}_A(\theta,\infty))$, $u_0 \in \mathcal{D}_A(\theta+1,\infty)$ the Cauchy problem (CP) admits a unique solution $u \in C^1([0,T];X) \cap C([0,T];\mathcal{D}(A)) \cap \mathcal{B}([0,T];\mathcal{D}_A(\theta+1,\infty))$.

Remark 4: We point out that the key point in the proof of our global in time results is the introduction of the weighted spaces $\mathcal{B}_\lambda([0,T];\mathcal{D}_A(\theta,\infty))$. In the next section we give all the necessary estimates to prove them.

2.1. *The main abstract result*

Problem 5: (Inverse Abstract Problem (IAP)) Let A be a sectorial operator in X. Determine a real number k and three functions u, h, f, such that

$$(\alpha) \begin{cases} u \in C^2([0,T];X) \cap C^1([0,T];\mathcal{D}(A)), \\ D_t u \in \mathcal{B}([0,T];\mathcal{D}_A(1+\theta,\infty)), \quad D_t^2 u \in \mathcal{B}([0,T];\mathcal{D}_A(\theta,\infty)), \end{cases}$$

(β) $h \in C([0,T])$,

(γ) $f \in C^1([0,T])$,

and they satisfy system

$$\begin{cases} u'(t) = kAu(t) + \int_0^t h(t-s)Au(s)ds + f(t)g, & t \in [0,T], \\ u(0) = u_0, \end{cases} \tag{18}$$

with the additional conditions:

$$\langle u(t), \phi_j \rangle = G_j(t), \quad t \in [0,T], \quad j = 1,2, \tag{19}$$

where ϕ_j $(j = 1,2)$ are given bounded linear functionals on X, and G_j, u_0, g are given data.

We study the (IAP) under the following assumptions:

(H1) $\theta \in (0,1)$, X is a Banach space and A is a sectorial operator in X.
(H2) $u_0 \in \mathcal{D}_A(1+\theta+\varepsilon, \infty)$, for some $\varepsilon \in (0, 1-\theta)$.
(H3) $g \in \mathcal{D}_A(\theta+\varepsilon, \infty)$, for some $\varepsilon \in (0, 1-\theta)$.
(H4) $\phi_j \in X'$, for $j = 1,2$.
(H5) $G_j \in C^2([0,T])$, for $j = 1,2$.
(H6) We set

$$M := \begin{pmatrix} < Au_0, \phi_1 > & < g, \phi_1 > \\ < Au_0, \phi_2 > & < g, \phi_2 > \end{pmatrix}, \tag{20}$$

and we suppose that the matrix M is invertible. We define

$$M^{-1} := \begin{pmatrix} a_{11} & a_{12} \\ a_{21} & a_{22} \end{pmatrix}. \tag{21}$$

Observe that, owing to (H6), the system

$$\begin{cases} k_0 < Au_0, \phi_1 > + f_0 < g, \phi_1 > = G_1'(0), \\ k_0 < Au_0, \phi_2 > + f_0 < g, \phi_2 > = G_2'(0), \end{cases} \tag{22}$$

has a unique solution (k_0, f_0).
(H7) We require that

$$k_0 > 0.$$

(H8) $v_0 := k_0 Au_0 + f_0 g \in \mathcal{D}_A(1+\theta, \infty)$.
(H9) $< u_0, \phi_j > = G_j(0), \quad < v_0, \phi_j > = G_j'(0), \quad j = 1,2.$

The main abstract result is the following:

Theorem 6: Assume that conditions (H1)-(H9) are fulfilled. Then Problem 5 has a unique (global in time) solution (k, u, h, f), with $k \in \mathbb{R}^+$, and u, h, f satisfying conditions (α), (β) and (γ).

Proof: The proof of Theorem 6 is developed in Section 5. □

2.2. An application

We choose as reference space

$$X := C(\overline{\Omega}), \tag{23}$$

where Ω is an open bounded set in $\mathbb{R}^3$ with boundary of class $C^{2(1+\theta+\varepsilon)}$, for some $\theta \in (0,1/2)$, $\varepsilon \in (0,(1/2)-\theta)$. We define

$$\begin{cases} \mathcal{D}(A) := \{u \in \bigcap_{1\leq p<+\infty} W^{2,p}(\Omega) \ : \Delta u \in C(\overline{\Omega}), \ \ D_\nu u|_{\partial\Omega} = 0\}, \\ Au := \Delta u, \ \ \forall u \in \mathcal{D}(A). \end{cases} \tag{24}$$

It was proved by Stewart[31] that A is a sectorial operator in X. Then, we recall the following characterizations of the interpolation spaces related to A (see [25]):

$$\mathcal{D}_A(\xi,\infty) = \begin{cases} C^{2\xi}(\overline{\Omega}), \ \text{if} \ \ 0<\xi<1/2, \\ \{u \in C^{2\xi}(\overline{\Omega}) : \ D_\nu u|_{\partial\Omega} = 0\}, \ \ \text{if} \ \ 1/2<\xi<1. \end{cases} \tag{25}$$

It is known that, if $0 < \xi \leq \theta+\varepsilon$, we have

$$\mathcal{D}_A(1+\xi,\infty) = \{u \in C^{2(1+\xi)}(\overline{\Omega}) : D_\nu u|_{\partial\Omega} = 0\}. \tag{26}$$

Thus, we consider the Inverse Problem 1 under the following assumptions:

(K1) Ω is an open bounded set in $\mathbb{R}^3$ with boundary of class $C^{2(1+\theta+\varepsilon)}$, for some $\theta \in (0,1/2)$, $\varepsilon \in (0,1/2-\theta)$.
(K2) $u_0 \in C^{2(1+\theta+\varepsilon)}(\overline{\Omega})$, $\quad D_\nu u_0|_{\partial\Omega} = 0$.
(K3) $g \in C^{2(\theta+\varepsilon)}(\overline{\Omega})$.
(K4) For $j = 1,2$, μ_j is a bounded Borel measure in $\overline{\Omega}$. We set, for $\psi \in X$,

$$< \psi, \phi_j > := \int_{\overline{\Omega}} \psi(x)\mu_j(dx). \tag{27}$$

(K5) Suppose that (H5) holds.
(K6) Suppose that (H6) holds with ϕ_j ($j = 1,2$) defined in (27) and A defined in (24).
(K7) Suppose that (H7) holds.
(K8) $v_0 := k_0\Delta u_0 + f_0 g \in C^{2(1+\theta)}(\overline{\Omega})$, $\quad D_\nu v_0|_{\partial\Omega} = 0$.
(K9) Suppose that (H9) holds.

Applying the main result of the paper (Theorem 6), we deduce the following:

Theorem 7: Assume that conditions (K1)-(K9) are satisfied. Then, the Inverse Problem 1 has a unique (global in time) solution (k,u,h,f), such that

$$\begin{aligned}&u \in C^2([0,T]; C(\overline{\Omega})) \cap C^1([0,T]; \mathcal{D}(A)),\\&D_t u \in \mathcal{B}([0,T]; C^{2(1+\theta)}(\overline{\Omega})), \quad D_t^2 u \in \mathcal{B}([0,T]; C^{2\theta}(\overline{\Omega})),\\&h \in C([0,T]),\\&f \in C^1([0,T]),\\&k \in \mathbb{R}^+,\end{aligned}$$

with A defined in (24).

3. The weighted spaces

In this section, we introduce some crucial estimates that will be essential to obtain existence and uniqueness of a global in time solution for our inverse problem. Let $\lambda > 0$, $T > 0$, $\theta \in (0,1)$. If $f \in \mathcal{B}([0,T]; X)$, we set

$$\|u\|_{\mathcal{B}_\lambda([0,T];X)} := \sup_{0 \le t \le T} e^{-\lambda t}\|u(t)\|. \tag{28}$$

We denote by $C_\lambda([0,T]; X)$ the space $C([0,T]; X)$, equipped with the norm $\|\cdot\|_{\mathcal{B}_\lambda([0,T];X)}$. We will use the notation $C_\lambda([0,T]; \mathbb{R}) = C_\lambda([0,T])$ and $\mathcal{B}_\lambda([0,T]; \mathbb{R}) = \mathcal{B}_\lambda([0,T])$. We now prove some useful estimates in these weighted spaces for the solution of the Cauchy problem given by Theorem 3.

Theorem 8: Let $A : \mathcal{D}(A) \to X$ be a sectorial operator, $\theta \in (0,1)$. Let us suppose that $f \in C([0,T]; X) \cap \mathcal{B}([0,T]; \mathcal{D}_A(\theta,\infty))$. Then, the following estimates hold:

$$\|e^{tA} * f\|_{C_\lambda([0,T];X)} \le \frac{C_0}{1+\lambda}\|f\|_{C_\lambda([0,T];X)}; \tag{29}$$

if $\theta \le \xi \le 1 + \theta$,

$$\|e^{tA} * f\|_{\mathcal{B}_\lambda([0,T];\mathcal{D}_A(\xi,\infty))} \le \frac{C(\theta,\xi)}{(1+\lambda)^{1+\theta-\xi}}\|f\|_{\mathcal{B}_\lambda([0,T];\mathcal{D}_A(\theta,\infty))}, \tag{30}$$

with C_0 and $C(\theta, \xi)$ independent of f and λ.

Proof: In the following section the symbol C, sometimes with an index, will denote different positive constants. We show (29). Owing to (10), we have

$$\begin{aligned}e^{-\lambda t}\|e^{tA} * f(t)\| &= e^{-\lambda t}\| \textstyle\int_0^t e^{(t-s)A} f(s)\, ds\|\\&\le M_0\|f\|_{C_\lambda([0,T];X)} \textstyle\int_0^t e^{(\omega+1-\lambda)s}\, ds\\&\le \tfrac{C_0}{1+\lambda}\|f\|_{C_\lambda([0,T];X)}.\end{aligned} \tag{31}$$

To get (30), owing to (16), it suffices to consider the cases $\xi = \theta$ and $\xi = 1 + \theta$. If $\xi = \theta$, we have:

$$e^{-\lambda t}|e^{tA} * f(t)\,|_{\mathcal{D}_A(\theta,\infty)} \\ = e^{-\lambda t}\sup_{0<\tau<1}\tau^{1-\theta}\|\int_0^t Ae^{(t+\tau-s)A}f(s)ds\|.$$

If $t > s > 0$ and $\tau \in (0,1)$, we have

$$\begin{aligned} e^{-\lambda t}\tau^{1-\theta}\|Ae^{(t+\tau-s)A}f(s)\| &\leq e^{-\lambda t}|e^{(t-s)A}f(s)|_{\mathcal{D}_A(\theta,\infty)} \\ &\leq C_1 e^{-\lambda t}|f(s)|_{\mathcal{D}_A(\theta,\infty)} \\ &\leq C_1 e^{-\lambda(t-s)}\|f\|_{\mathcal{B}_\lambda([0,T];\mathcal{D}_A(\theta,\infty))}, \end{aligned}$$

so we get

$$\begin{aligned} e^{-\lambda t}|e^{tA} * f(t)|_{\mathcal{D}_A(\theta,\infty)} &\leq C_1\|f\|_{\mathcal{B}_\lambda([0,T];\mathcal{D}_A(\theta,\infty))}\int_0^t e^{-\lambda(t-s)}ds \\ &\leq \frac{C_2}{\lambda}\|f\|_{\mathcal{B}_\lambda([0,T];\mathcal{D}_A(\theta,\infty))}. \end{aligned}$$

Next, we consider the case $\xi = 1 + \theta$. We start by estimating $A(e^{tA} * f)$ in $C_\lambda([0,T];X)$:

$$\begin{aligned} e^{-\lambda t}\|A(e^{tA} * f)(t)\| &\leq e^{-\lambda t}\int_0^t \|Ae^{(t-s)A}f(s)\|\,ds \\ &\leq C_1\|f\|_{\mathcal{B}_\lambda([0,T];\mathcal{D}_A(\theta,\infty))}\int_0^t e^{-\lambda(t-s)}(t-s)^{-1+\theta}\,ds. \end{aligned} \tag{32}$$

If $\lambda > 0$, we observe that

$$\begin{aligned} \int_0^t e^{-\lambda(t-s)}(t-s)^{-1+\theta}ds &= \int_0^t e^{-\lambda s}s^{-1+\theta}ds \\ &\leq \int_0^{+\infty} e^{-\lambda s}s^{\theta-1}ds \\ &= \lambda^{-\theta}\,\Gamma(\theta). \end{aligned}$$

So, we have obtained

$$\|A(e^{tA} * f)\|_{\mathcal{B}_\lambda([0,T];X)} \leq \frac{C}{(1+\lambda)^\theta}\|f\|_{\mathcal{B}_\lambda([0,T];\mathcal{D}_A(\theta,\infty))}. \tag{33}$$

It remains to estimate $A(e^{tA} * f)$ in $\mathcal{B}_\lambda([0,T];\mathcal{D}_A(\theta,\infty))$; if $\lambda > 0$, $\tau \in (0,1)$, $0 < s < t \leq T$, we have

$$\begin{aligned} &e^{-\lambda t}\tau^{1-\theta}\,\|A^2 e^{(\tau+t-s)A}f(s)\| \\ &\leq e^{-\lambda t}\tau^{1-\theta}\|A^2 e^{(\tau+t-s)A}\|_{\mathcal{L}(\mathcal{D}_A(\theta,\infty);X)}\|f(s)\|_{\mathcal{D}_A(\theta,\infty)} \\ &\leq e^{-\lambda(t-s)}\tau^{1-\theta}\|A^2 e^{(\tau+t-s)A}\|_{\mathcal{L}(\mathcal{D}_A(\theta,\infty);X)}\|f\|_{\mathcal{B}_\lambda([0,T];\mathcal{D}_A(\theta,\infty))} \\ &\leq Ce^{-\lambda(t-s)}\tau^{1-\theta}(\tau+t-s)^{-2+\theta}\|f\|_{\mathcal{B}_\lambda([0,T];\mathcal{D}_A(\theta,\infty))}. \end{aligned}$$

So we get the estimates

$$\begin{aligned} &e^{-\lambda t}|\,A(e^{tA} * f)(t)|_{\mathcal{D}_A(\theta,\infty)} \\ &= e^{-\lambda t}\sup_{0<\tau<1}\tau^{1-\theta}\|Ae^{\tau A}A\int_0^t e^{(t-s)A}f(s)ds\| \\ &\leq C\|f\|_{\mathcal{B}_\lambda([0,T];\mathcal{D}_A(\theta,\infty))}\sup_{0<\tau<1}\tau^{1-\theta}\int_0^t e^{-\lambda s}(\tau+s)^{-2+\theta}ds \end{aligned} \tag{34}$$

and

$$\begin{aligned}\tau^{1-\theta}\int_0^t e^{-\lambda s}(\tau+s)^{-2+\theta}ds &= \int_0^{t/\tau} e^{-\lambda\tau s}(1+s)^{-2+\theta}ds \\ &\leq \int_0^{+\infty}(1+s)^{-2+\theta}ds.\end{aligned}$$

This concludes the proof. □

Theorem 9: Let $\lambda \geq 0$, $h \in C([0,T])$ and $u \in C([0,T];X) \cap \mathcal{B}([0,T];\mathcal{D}_A(\theta,\infty))$. Then, there exists $C > 0$, independent of T, λ, h, u, such that, setting

$$h * u(t) := \int_0^t h(t-s)u(s)ds, \tag{35}$$

$h*u \in C([0,T];X)\cap\mathcal{B}([0,T];\mathcal{D}_A(\theta,\infty))$ and satisfies the following estimate:

$$\begin{aligned}&\|h * u\|_{\mathcal{B}_\lambda([0,T];\mathcal{D}_A(\theta,\infty))} \\ &\leq C\min\Big\{T\|h\|_{\mathcal{B}_\lambda([0,T])}\|u\|_{\mathcal{B}_\lambda([0,T];\mathcal{D}_A(\theta,\infty))}, \\ &(1+\lambda)^{-1}\|h\|_{\mathcal{B}_\lambda([0,T])}\|u\|_{\mathcal{B}([0,T];\mathcal{D}_A(\theta,\infty))}, \\ &(1+\lambda)^{-1}\|h\|_{\mathcal{B}([0,T])}\|u\|_{\mathcal{B}_\lambda([0,T];\mathcal{D}_A(\theta,\infty))}\Big\}.\end{aligned} \tag{36}$$

Proof: If $0 \leq t \leq T$, we have

$$\begin{aligned}\|e^{-\lambda t}(h*u)(t)\| &\leq \int_0^t e^{-\lambda(t-s)}|h(t-s)|e^{-\lambda s}\|u(s)\|ds \\ &\leq \min\Big\{T\|h\|_{\mathcal{B}_\lambda([0,T])}\|u\|_{\mathcal{B}_\lambda([0,T];X)}, \\ &\int_0^t e^{-\lambda s}ds\|h\|_{\mathcal{B}_\lambda([0,T])}\|u\|_{\mathcal{B}([0,T];X)}, \\ &\int_0^t e^{-\lambda s}ds\|h\|_{\mathcal{B}([0,T])}\|u\|_{\mathcal{B}_\lambda([0,T];X)}\Big\} \\ &\leq C\min\Big\{T\|h\|_{\mathcal{B}_\lambda([0,T])}\|u\|_{\mathcal{B}_\lambda([0,T];X)}, \\ &(1+\lambda)^{-1}\|h\|_{\mathcal{B}_\lambda([0,T])}\|u\|_{\mathcal{B}([0,T];X)}, \\ &(1+\lambda)^{-1}\|h\|_{\mathcal{B}([0,T])}\|u\|_{\mathcal{B}_\lambda([0,T];X)}\Big\}.\end{aligned}$$

Moreover, if $\tau \in (0,1)$, we have

$$\begin{aligned}e^{-\lambda t}\|\tau^{1-\theta}Ae^{\tau A}(h*u)(t)\| &\leq \int_0^t e^{-\lambda(t-s)}|h(t-s)|e^{-\lambda s}\|\tau^{1-\theta}Ae^{\tau A}u(s)\|ds \\ &\leq \min\{T\|h\|_{\mathcal{B}_\lambda([0,T])}\|u\|_{\mathcal{B}_\lambda([0,T];\mathcal{D}_A(\theta,\infty))}, \\ &\int_0^t e^{-\lambda s}ds\|h\|_{\mathcal{B}_\lambda([0,T])}\|u\|_{\mathcal{B}([0,T];\mathcal{D}_A(\theta,\infty))}, \\ &\int_0^t e^{-\lambda s}ds\|h\|_{\mathcal{B}([0,T])}\|u\|_{\mathcal{B}_\lambda([0,T];\mathcal{D}_A(\theta,\infty))}\} \\ &\leq C\min\{T\|h\|_{\mathcal{B}_\lambda([0,T])}\|u\|_{\mathcal{B}_\lambda([0,T];\mathcal{D}_A(\theta,\infty)}, \\ &(1+\lambda)^{-1}\|h\|_{\mathcal{B}_\lambda([0,T])}\|u\|_{\mathcal{B}([0,T];\mathcal{D}_A(\theta,\infty))}, \\ &(1+\lambda)^{-1}\|h\|_{\mathcal{B}([0,T])}\|u\|_{\mathcal{B}_\lambda([0,T];\mathcal{D}_A(\theta,\infty))}\}.\end{aligned}$$

These estimates imply the conclusions immediately. □

Theorem 10: Let $f \in \mathcal{B}([0,T];X)$. Assume that $f(0) = 0$ and suppose that f is continuous in 0. Then

$$\lim_{\lambda \to +\infty} \|f\|_{\mathcal{B}_\lambda([0,T];X)} = 0.$$

In particular, if h and u satisfy the assumptions of Theorem 9, then

$$\lim_{\lambda \to +\infty} \|h * u\|_{\mathcal{B}_\lambda([0,T];\mathcal{D}_A(\theta,\infty))} = 0.$$

Proof: Let $\varepsilon > 0$ and $\delta > 0$ be such that, if $0 \leq t \leq \delta$, then $\|f(t)\| \leq \varepsilon$. So, if $0 \leq t \leq \delta$, we have $e^{-\lambda t}\|f(t)\| \leq \varepsilon$. If $\delta \leq t \leq T$,

$$e^{-\lambda t}\|f(t)\| \leq e^{-\delta\lambda}\|f\|_{\mathcal{B}([0,T];X)} \leq \varepsilon$$

if λ is sufficiently large. If h and u satisfy the conditions of Theorem 9, we have, from Theorem 9, for $0 < \delta \leq T$,

$$\|h * u\|_{\mathcal{B}([0,\delta];\mathcal{D}_\theta(A))} \leq C\delta\|h\|_{\mathcal{B}([0,T])}\|u\|_{\mathcal{B}([0,T];\mathcal{D}_A(\theta,\infty))},$$

so that the first part of the statement is applicable. □

4. An equivalent fixed point system

In this section, we reformulate Problem 5 in terms of the equivalent nonlinear fixed point system (46). In proving Theorem 12, we find out a set of regularity and compatibility conditions on the data that makes the inverse problem well-posed. Moreover, starting from the equivalent fixed point system, we obtain existence and uniqueness results for system (IAP) via the Contraction Principle.

To this aim, we start by introducing some notation. We set

$$A_0 := k_0 A. \tag{37}$$

As $k_0 > 0$, see (H7), A_0 is a sectorial operator in X. Next, for $t \in [0,T]$ we set

$$\overline{h}_1(t) := a_{11}G_1''(t) + a_{12}G_2''(t), \tag{38}$$

$$\overline{w}_1(t) := a_{21}G_1''(t) + a_{22}G_2''(t), \tag{39}$$

$$\overline{v}(t) := e^{tA_0}v_0. \tag{40}$$

We immediately observe that $\overline{h}_1$ and $\overline{w}_1$ belong to $C([0,T])$ and, owing to Theorem 3, $\overline{v} \in C([0,T];\mathcal{D}(A)) \cap \mathcal{B}([0,T];\mathcal{D}_A(1+\theta,\infty))$. Next, we define, again for $t \in [0,T]$,

$$\overline{h}(t) := \overline{h}_1(t) - k_0 a_{11} < A\overline{v}(t), \phi_1 > -k_0 a_{12} < A\overline{v}(t), \phi_2 >, \tag{41}$$

$$\overline{w}(t) := \overline{w}_1(t) - k_0 a_{21} < A\overline{v}(t), \phi_1 > -k_0 a_{22} < A\overline{v}(t), \phi_2 > . \tag{42}$$

Of course, $\overline{h}$ and $\overline{w}$ belong to $C([0,T])$. Finally, we introduce the following (nonlinear) operators, defined for every $(v,h,w) \in [C([0,T];\mathcal{D}(A)) \cap \mathcal{B}([0,T];\mathcal{D}_A(1+\theta,\infty))] \times C([0,T]) \times C([0,T])$:

$$\mathcal{R}_1(v,h,w)(t) := e^{tA_0} * (hAu_0 + wg + h * Av)(t), \tag{43}$$

$$\begin{aligned} \mathcal{R}_2(v,h,w)(t) := & -k_0[a_{11} < A\mathcal{R}_1(v,h,w)(t), \phi_1 > \\ & + a_{12} < A\mathcal{R}_1(v,h,w)(t), \phi_2 >] \\ & - a_{11} < h * A\overline{v}(t), \phi_1 > \\ & - a_{12} < h * A\overline{v}(t), \phi_2 > \\ & - a_{11} < h * A\mathcal{R}_1(v,h,w)(t), \phi_1 > \\ & - a_{12} < h * A\mathcal{R}_1(v,h,w)(t), \phi_2 >, \end{aligned} \tag{44}$$

$$\begin{aligned} \mathcal{R}_3(v,h,w)(t) := & -k_0[a_{21} < A\mathcal{R}_1(v,h,w)(t), \phi_1 > \\ & + a_{22} < A\mathcal{R}_1(v,h,w)(t), \phi_2 >] \\ & - a_{21} < h * A\overline{v}(t), \phi_1 > \\ & - a_{22} < h * A\overline{v}(t), \phi_2 > \\ & - a_{21} < h * A\mathcal{R}_1(v,h,w)(t), \phi_1 > \\ & - a_{22} < h * A\mathcal{R}_1(v,h,w)(t), \phi_2 > . \end{aligned} \tag{45}$$

We observe that $\mathcal{R}_2$ and $\mathcal{R}_3$ are well-defined, because

$$\mathcal{R}_1(v,h,w) \in C([0,T];\mathcal{D}(A)) \cap \mathcal{B}([0,T];\mathcal{D}_A(1+\theta,\infty)).$$

Moreover, $\mathcal{R}_2(v,h,w)$ and $\mathcal{R}_3(v,h,w)$ both belong to $C([0,T])$.

Now, we can introduce the following problem:

Problem 11: Determine three functions v, h, w, such that

(α') $v \in C([0,T];\mathcal{D}(A)) \cap \mathcal{B}([0,T];\mathcal{D}(1+\theta,\infty))$,
(β') $h \in C([0,T])$,
(γ') $w \in C([0,T])$,

satisfying the system

$$\begin{cases} v = \overline{v} + \mathcal{R}_1(v,h,w), \\ h = \overline{h} + \mathcal{R}_2(v,h,w), \\ w = \overline{w} + \mathcal{R}_3(v,h,w). \end{cases} \tag{46}$$

Theorem 12: (Equivalence) Let A be a sectorial operator in the Banach space X and $\theta \in (0,1)$. Let us assume that the data g, u_0, ϕ_j and G_j $(j = 1,2)$ satisfy the conditions (H1)-(H9) . Suppose that k, u, f, h solve Problem 5, with $k \in \mathbb{R}$ and u, f, h fulfilling the regularity conditions (α), (β), (γ). Then,

(I) $k = k_0$;

(II) the triplet (v, h, w), where $v = u'$ and $w = f'$, satisfies the conditions (α'), (β'), (γ') and solves Problem 11;

(III) conversely, if (v, h, w), with the above regularity, is a solution of the Problem 11, then the triplet (u, h, f), where $u = u_0 + 1 * v$ and $f = f_0 + 1 * w$, satisfies the regularity conditions (α), (β), (γ) and solves Problem 5 with $k = k_0$.

Proof: Let (u, h, f) satisfy the regularity conditions (α), (β), (γ) and solve Problem 5, with $k \in \mathbb{R}$. Applying ϕ_j $(j = 1,2)$ to the first equation in (18) and taking $t = 0$, we get

$$k < Au_0, \phi_j > + f(0) < g, \phi_j >= G_j'(0).$$

Employing (H6), we immediately obtain

$$k = k_0, \qquad f(0) = f_0. \tag{47}$$

Thus, (I) is proved. Now, we set

$$v := u', \qquad w := f'.$$

Then $v \in C^1([0,T]; X) \cap C([0,T]; \mathcal{D}(A)) \cap \mathcal{B}([0,T]; \mathcal{D}_A(1+\theta, \infty))$, $v' \in \mathcal{B}([0,T]; \mathcal{D}_A(1+\theta, \infty))$, $w \in C([0,T])$. In particular, (v, h, w) complies with (α'), (β'), (γ'). Differentiating the first equation in (18) and, using (47) and (H8), we obtain

$$\begin{cases} v'(t) = k_0 Av(t) + h(t)Au_0 + h * Av(t) + w(t)g, & t \in [0,T], \\ v(0) = v_0. \end{cases} \tag{48}$$

Applying ϕ_j $(j = 1,2)$ to the first equation in (48), we get, for $t \in [0,T]$,

$$\begin{aligned} h(t) < Au_0, \phi_j > + w(t) < g, \phi_j > = G_j''(t) - k_0 < Av(t), \phi_j > \\ - < h * Av(t), \phi_j > . \end{aligned} \tag{49}$$

It follows that

$$\begin{aligned} h(t) = \overline{h}_1(t) - k_0[a_{11} < Av(t), \phi_1 > + a_{12} < Av(t), \phi_2 >] \\ - a_{11} < h * Av(t), \phi_1 > - a_{12} < h * Av(t), \phi_2 >, \end{aligned} \tag{50}$$

$$\begin{aligned} w(t) = \overline{w}_1(t) - k_0[a_{21} < Av(t), \phi_1 > + a_{22} < Av(t), \phi_2 >] \\ -a_{21} < h * Av(t), \phi_1 > -a_{22} < h * Av(t), \phi_2 > . \end{aligned} \tag{51}$$

Now we apply the variation of parameter formula to (48), to get

$$\begin{aligned} v(t) &= \overline{v}(t) + e^{tA_0} * (hAu_0 + h * Av + wg)(t) \\ &= \overline{v}(t) + \mathcal{R}_1(v, h, w)(t), \end{aligned} \tag{52}$$

hence, from (50)-(51), we obtain

$$h = \overline{h} + \mathcal{R}_2(v, h, w),$$

$$w = \overline{w} + \mathcal{R}_3(v, h, w).$$

Therefore, (v, h, w) satisfies Problem 11.

On the other hand, assume that v, h, w satisfy conditions (α'), (β'), (γ') and system (46). As $v_0 \in \mathcal{D}_A(1+\theta, \infty)$ and $hAu_0 + h * Av + wg \in C([0,T]; X) \cap \mathcal{B}([0,T]; \mathcal{D}_A(\theta, \infty))$, we have that $v \in C^1([0,T]; X) \cap C([0,T]; \mathcal{D}(A)) \cap \mathcal{B}([0,T]; \mathcal{D}_A(1+\theta, \infty))$ and solves system (48). From (46), we also obtain that (50), (51) and (49) are satisfied. Applying ϕ_j $(j = 1, 2)$ to the first equation in (48), we get

$$< v'(t), \phi_j > = G_j''(t), \qquad t \in [0, T]. \tag{53}$$

Now we set $u := u_0 + 1 * v$ and $f := f_0 + 1 * w$. Then, owing to (H2), u satisfies (α), while f satisfies (γ). Observe also that the first equation in (48) can be written in the form

$$u'' = D_t(k_0 Au + h * Au + fg). \tag{54}$$

As $u'(0) = v_0$ and $f(0) = f_0$, from (H8) we obtain that system (18) is satisfied, with $k = k_0$. Finally, applying (H9) and (53), we can conclude that (19), too, is satisfied. □

5. Proof of Theorem 6

To prove Theorem 6, we shall employ the following:

Lemma 13: *Assume that conditions (H1)-(H9) are satisfied.*

Let $\mathcal{R}_1$, $\mathcal{R}_2$, $\mathcal{R}_3$ be the operators defined by (43), (44), (45), respectively. Then, there exists $C > 0$ such that, $\forall \lambda \geq 0$ $\forall v, v_1, v_2 \in C([0,T]; \mathcal{D}(A)) \cap \mathcal{B}([0,T]; \mathcal{D}_A(1+\theta, \infty))$ and $\forall$ $h, h_1, h_2 \in C([0,T])$, $\forall w, w_1, w_2 \in C([0,T])$ we have the estimates:

$$\begin{aligned}\|\mathcal{R}_1(v, h\ , w)\|_{\mathcal{B}_\lambda([0,T];\mathcal{D}_A(1+\theta,\infty))} &\leq C[\|\overline{h} * A\overline{v}\|_{\mathcal{B}_\lambda([0,T];\mathcal{D}_A(\theta,\infty))} \\ &+ (1+\lambda)^{-\varepsilon}(\|h\|_{\mathcal{B}_\lambda([0,T])} + \|w\|_{\mathcal{B}_\lambda([0,T])}) \\ &+ (1+\lambda)^{-1}(\|h-\overline{h}\|_{\mathcal{B}_\lambda([0,T])} + \|v-\overline{v}\|_{\mathcal{B}_\lambda([0,T];\mathcal{D}_A(1+\theta,\infty))}) \\ &+ \|h-\overline{h}\|_{\mathcal{B}_\lambda([0,T])}\|v-\overline{v}\|_{\mathcal{B}_\lambda([0,T];\mathcal{D}_A(1+\theta,\infty)}]; \end{aligned} \tag{55}$$

$$\begin{aligned}\|\mathcal{R}_1(v_1, h_1, w_1) - \mathcal{R}_1(v_2, h_2, w_2)\|_{\mathcal{B}_\lambda([0,T];\mathcal{D}_A(1+\theta,\infty))} \\ \leq C[\|h_1 - h_2\|_{\mathcal{B}_\lambda([0,T])}\|v_1-\overline{v}\|_{\mathcal{B}_\lambda([0,T];\mathcal{D}_A(1+\theta,\infty))} \\ +\|h_2-\overline{h}\|_{\mathcal{B}_\lambda([0,T])}\|v_1-v_2\|_{\mathcal{B}_\lambda([0,T];\mathcal{D}_A(1+\theta,\infty))} \\ +(1+\lambda)^{-\varepsilon}(\|h_1-h_2\|_{\mathcal{B}_\lambda([0,T])} + \|w_1-w_2\|_{\mathcal{B}_\lambda([0,T])}) \\ +(1+\lambda)^{-1}\|v_1-v_2\|_{\mathcal{B}_\lambda([0,T];\mathcal{D}_A(1+\theta,\infty))}]; \end{aligned} \tag{56}$$

and, if $i \in \{2,3\}$,

$$\begin{aligned}\|\mathcal{R}_i(v, h\ , w)\|_{\mathcal{B}_\lambda([0,T])} &\leq C(\|\mathcal{R}_1(v,h,w)\|_{\mathcal{B}_\lambda([0,T];\mathcal{D}_A(1+\theta,\infty))} \\ &+(1+\lambda)^{-1}\|h\|_{\mathcal{B}_\lambda([0,T])} \\ &+\|h-\overline{h}\|_{\mathcal{B}_\lambda([0,T])}\|\mathcal{R}_1(v,h,w)\|_{\mathcal{B}_\lambda([0,T];\mathcal{D}_A(1+\theta,\infty))}); \end{aligned} \tag{57}$$

$$\begin{aligned}&\|\mathcal{R}_i(v_1,h_1,w_1) - \mathcal{R}_i(v_2,h_2,w_2)\|_{\mathcal{B}_\lambda([0,T])} \\ &\leq C(\|\mathcal{R}_1(v_1,h_1,w_1) - \mathcal{R}_1(v_2,h_2,w_2)\|_{\mathcal{B}_\lambda([0,T];\mathcal{D}_A(1+\theta,\infty))} \\ &+(1+\lambda)^{-1}\|h_1-h_2\|_{\mathcal{B}_\lambda([0,T])} \\ &+\|h_1-h_2\|_{\mathcal{B}_\lambda([0,T])}\|\mathcal{R}_1(v_1,h_1,w_1)\|_{\mathcal{B}_\lambda([0,T];\mathcal{D}_A(1+\theta,\infty))} \\ &+\|h_2-\overline{h}\|_{\mathcal{B}_\lambda([0,T])}\|\mathcal{R}_1(v_1,h_1,w_1) - \mathcal{R}_1(v_2,h_2,w_2)\|_{\mathcal{B}_\lambda([0,T];\mathcal{D}_A(1+\theta,\infty))}). \end{aligned} \tag{58}$$

Proof: Owing to (H2)-(H3) and Theorem 8 (applied replacing θ with $\theta+\varepsilon$ and ξ with $1+\theta$), we have, for some $C_1 > 0$, independent of λ, h, w,

$$\begin{aligned}\|e^{tA_0} * (hAu_0 + wg)\|_{\mathcal{B}_\lambda([0,T];\mathcal{D}_A(1+\theta,\infty))} \\ \leq C_1(1+\lambda)^{-\varepsilon}(\|h\|_{\mathcal{B}_\lambda([0,T])} + \|w\|_{\mathcal{B}_\lambda([0,T])}). \end{aligned} \tag{59}$$

Moreover, employing again Theorem 8 and Theorem 9, we have

$$\begin{aligned}&\|e^{tA_0} * (h * Av)\|_{\mathcal{B}_\lambda([0,T];\mathcal{D}_A(1+\theta,\infty))} \\ &\leq C_2\|h * Av\|_{\mathcal{B}_\lambda([0,T];\mathcal{D}_A(\theta,\infty))} \\ &\leq C_2(\|\overline{h} * A\overline{v}\|_{\mathcal{B}_\lambda([0,T];\mathcal{D}_A(\theta,\infty))} + \|(h-\overline{h}) * A\overline{v}\|_{\mathcal{B}_\lambda([0,T];\mathcal{D}_A(\theta,\infty))} \\ &+\|\overline{h} * A(v-\overline{v})\|_{\mathcal{B}_\lambda([0,T];\mathcal{D}_A(1+\theta,\infty))} \\ &+\|(h-\overline{h}) * A(v-\overline{v})\|_{\mathcal{B}_\lambda([0,T];\mathcal{D}_A(\theta,\infty))}) \\ &\leq C_3[\|\overline{h} * A\overline{v}\|_{\mathcal{B}_\lambda([0,T];\mathcal{D}_A(\theta,\infty))} \\ &+(1+\lambda)^{-1}(\|A\overline{v}\|_{\mathcal{B}([0,T];\mathcal{D}_A(\theta,\infty))}\|h-\overline{h}\|_{\mathcal{B}_\lambda([0,T])} \\ &+\|\overline{h}\|_{\mathcal{B}([0,T])}\|v-\overline{v}\|_{\mathcal{B}_\lambda([0,T];\mathcal{D}_A(1+\theta,\infty))}) \\ &+\|h-\overline{h}\|_{\mathcal{B}_\lambda([0,T])}\|v-\overline{v}\|_{\mathcal{B}_\lambda([0,T];\mathcal{D}_A(1+\theta,\infty))}], \end{aligned} \tag{60}$$

with C_2 and C_3 independent of λ, v, h. Thus, (55) follows from (59) and (60). Concerning (56), we have

$$\begin{aligned}&\|e^{tA_0} *((h_1-h_2)Au_0+(w_1-w_2)g)\|_{\mathcal{B}_\lambda([0,T];\mathcal{D}_A(1+\theta,\infty))}\\ &\quad\le C_1(1+\lambda)^{-\varepsilon}(\|h_1-h_2\|_{\mathcal{B}_\lambda([0,T])}+\|w_1-w_2\|_{\mathcal{B}_\lambda([0,T])}).\end{aligned}\tag{61}$$

Next, always using Theorems 8 and 9, we have

$$\begin{aligned}\|e^{tA_0} * (h_1 * Av_1 - {}&h_2 * Av_2)\|_{\mathcal{B}_\lambda([0,T];\mathcal{D}_A(1+\theta,\infty))}\\ &\le C_2\|h_1 * Av_1 - h_2 * Av_2\|_{\mathcal{B}_\lambda([0,T];\mathcal{D}_A(\theta,\infty))}\\ &\le C_2(\|(h_1-h_2) * A\overline{v}\|_{\mathcal{B}_\lambda([0,T];\mathcal{D}_A(\theta,\infty))}\\ &+\|\overline{h} * A(v_1-v_2)\|_{\mathcal{B}_\lambda([0,T];\mathcal{D}_A(\theta,\infty))}\\ &+\|(h_1-h_2) * A(v_1-\overline{v})\|_{\mathcal{B}_\lambda([0,T];\mathcal{D}_A(\theta,\infty))}\\ &+\|(h_2-\overline{h}) * A(v_1-v_2)\|_{\mathcal{B}_\lambda([0,T];\mathcal{D}_A(\theta,\infty))})\\ &\le C_4[(1+\lambda)^{-1}(\|A\overline{v}\|_{\mathcal{B}([0,T];\mathcal{D}_A(\theta,\infty))}\\ &\|h_1-h_2\|_{\mathcal{B}_\lambda([0,T])}\\ &+\|\overline{h}\|_{\mathcal{B}([0,T])}\|v_1-v_2\|_{\mathcal{B}_\lambda([0,T];\mathcal{D}_A(1+\theta,\infty))})\\ &+\|h_1-h_2\|_{\mathcal{B}_\lambda([0,T])}\|v_1-\overline{v}\|_{\mathcal{B}_\lambda([0,T];\mathcal{D}_A(1+\theta,\infty))})\\ &+\|h_2-\overline{h}\|_{\mathcal{B}_\lambda([0,T])}\|v_1-v_2\|_{\mathcal{B}_\lambda([0,T];\mathcal{D}_A(1+\theta,\infty))})],\end{aligned}\tag{62}$$

with C_4 independent of λ, v_1,v_2, h_1, h_2. Hence, (56) follows from (61) and (62). The formulae (57) and (58) can be shown analogously. Concerning (57), it is convenient to observe that

$$h * A\mathcal{R}_1(v,h,w) = (h-\overline{h}) * A\mathcal{R}_1(v,h,w) + \overline{h} * A\mathcal{R}_1(v,h,w),$$

while (58) follows from the identity

$$\begin{aligned}h_1 * A\mathcal{R}_1(v_1,h_1,w_1) - {}&h_2 * A\mathcal{R}_1(v_2,h_2,w_2)\\ &= (h_1-h_2) * A\mathcal{R}_1(v_1,h_1,w_1)\\ &+(h_2-\overline{h}) * A[\mathcal{R}_1(v_1,h_1,w_1)-\mathcal{R}_1(v_2,h_2,w_2)]\\ &+\overline{h} * A[\mathcal{R}_1(v_1,h_1,w_1)-\mathcal{R}_1(v_2,h_2,w_2)].\end{aligned}\tag{63}$$

□

Proof of Theorem 6. Let $\lambda \ge 0$. We set

$$Y(\lambda) := (C_\lambda([0,T];\mathcal{D}(A)) \cap \mathcal{B}_\lambda([0,T];\mathcal{D}_A(1+\theta,\infty)) \times C_\lambda([0,T])^2, \tag{64}$$

endowed with the norm

$$\|(v,h,w)\|_{Y(\lambda)} := \max\{\|v\|_{\mathcal{B}_\lambda([0,T];\mathcal{D}_A(1+\theta,\infty))}, \|h\|_{\mathcal{B}_\lambda([0,T])}, \|w\|_{\mathcal{B}_\lambda([0,T])}\}, \tag{65}$$

with $v \in C_\lambda([0,T];\mathcal{D}(A)) \cap \mathcal{B}_\lambda([0,T];\mathcal{D}_A(1+\theta,\infty))$, $h \in C_\lambda([0,T])$, $w \in \mathcal{B}_\lambda([0,T])$. Note that $Y(\lambda)$, endowed with the norm (65), is a Banach space. Let $\lambda \geq 0$, $\rho > 0$ and set

$$Y(\lambda,\rho) := \{(v,h,w) \in Y(\lambda) : \|(v,h,w) - (\overline{v},\overline{h},\overline{w})\|_{Y(\lambda)} \leq \rho\}. \tag{66}$$

Then, for every $\rho > 0$ $Y(\lambda,\rho)$ is a closed subset of $Y(\lambda)$.

Now we introduce the following operator N: for $(v,h,w) \in Y(\lambda)$ we set

$$N(v,h,w) := (\overline{v} + \mathcal{R}_1(v,h,w), \overline{h} + \mathcal{R}_2(v,h,w), \overline{w} + \mathcal{R}_3(v,h,w)). \tag{67}$$

Clearly, N is a nonlinear operator in $Y(\lambda)$.

Next, we show that Problem 5 has a solution (k,u,h,f), with the regularity properties (α), (β), (γ). Applying Theorem 12, we are reduced to look for a solution (v,h,w) of system (46), satisfying the conditions (α'), (β'), (γ'). This is equivalent to look for a fixed point of N in $Y(\lambda)$, for some $\lambda \geq 0$.

First step : We show that there exists $\overline{\rho}_1 > 0$ *such that, for any* $0 < \rho \leq \overline{\rho}_1$, *there exists* $\lambda_1(\rho) \geq 0$, *so that, for* $\lambda \geq \lambda_1(\rho)$, $N(Y(\lambda,\rho)) \subseteq Y(\lambda,\rho)$.

In fact, if $(v,h,w) \in Y(\lambda,\rho)$, we have, applying (55) and (57),

$$\begin{aligned}\|\mathcal{R}_1(v,h,w)\|_{\mathcal{B}_\lambda([0,T];\mathcal{D}_A(1+\theta,\infty))} &\leq C[\|\overline{h} * A\overline{v}\|_{\mathcal{B}_\lambda([0,T];\mathcal{D}_A(\theta,\infty))} \\ &\quad +(1+\lambda)^{-\varepsilon}(2\rho + \|\overline{h}\|_{C([0,T])} + \|\overline{w}\|_{C([0,T])}) \\ &\quad +2(1+\lambda)^{-1}\rho + \rho^2] \\ &\leq C_1[\|\overline{h} * A\overline{v}\|_{\mathcal{B}_\lambda([0,T];\mathcal{D}_A(\theta,\infty))} \\ &\quad +(\rho+1)(1+\lambda)^{-\varepsilon} + \rho^2],\end{aligned}$$

for $i \in \{2,3\}$,

$$\begin{aligned}\|\mathcal{R}_i(v,h,w)\|_{\mathcal{B}_\lambda([0,T])} &\leq C_2\{(1+\rho)[\|\overline{h} * A\overline{v}\|_{\mathcal{B}_\lambda([0,T];\mathcal{D}_A(\theta,\infty))} \\ &\quad +(\rho+1)(1+\lambda)^{-\varepsilon} + \rho^2] \\ &\quad +(1+\lambda)^{-1}(\rho + \|\overline{h}\|_{C([0,T])})\} \\ &\leq C_3(1+\rho)[\|\overline{h} * A\overline{v}\|_{\mathcal{B}_\lambda([0,T];\mathcal{D}_A(\theta,\infty))} \\ &\quad +(\rho+1)(1+\lambda)^{-\varepsilon} + \rho^2],\end{aligned}$$

with C_1, C_2, C_3 independent of λ and ρ. Now, we choose $\overline{\rho}_1 > 0$, such that, if $0 < \rho \leq \overline{\rho}_1$,

$$\max\{\ C_1\rho^2,\ C_3(1+\rho)\rho^2\ \} < \rho. \tag{68}$$

Then, owing to Theorem 10, for any $\rho \in (0,\overline{\rho}_1]$ there exists $\lambda(\rho) \geq 0$ such that, if $\lambda \geq \lambda(\rho)$ and $(v,h,w) \in Y(\lambda,\rho)$

$$\begin{aligned}\|N(v,h,w) - (\overline{v},\overline{h},\overline{w})\|_{Y(\lambda)} = \max\{\ &\|\mathcal{R}_1(v,h,w)\|_{\mathcal{B}_\lambda([0,T];\mathcal{D}_A(1+\theta,\infty))}, \\ &\|\mathcal{R}_2(v,h,w)\|_{\mathcal{B}_\lambda([0,T])}, \\ &\|\mathcal{R}_3(v,h,w)\|_{\mathcal{B}_\lambda([0,T])}\ \} \\ &\leq \rho.\end{aligned}$$

Second step : We show that there exists $\overline{\rho}_2 > 0$*, such that, if* $0 < \rho \leq \overline{\rho}_2$*, there exists* $\lambda_2(\rho) \geq 0$*, so that, if* $\lambda \geq \lambda_2(\rho)$*, and* (v_1, h_1, w_1) *and* (v_2, h_2, w_2) *are elements of* $Y(\lambda, \rho)$*, we have the estimate*

$$\|N(v_1, h_1, w_1) - N(v_2, h_2, w_2)\|_{Y(\lambda)} \leq \frac{1}{2}\|(v_1, h_1, w_1) - (v_2, h_2, w_2)\|_{Y(\lambda)}. \tag{69}$$

Indeed, let $\rho \in (0, \overline{\rho}_1]$ and $\lambda \geq \lambda_1(\rho)$. Then, by (56) and (58), we have

$$\begin{aligned}
\|\mathcal{R}_1(v_1, h_1, w_1) - & \mathcal{R}_1(v_2, h_2, w_2)\|_{\mathcal{B}_\lambda([0,T];\mathcal{D}_A(1+\theta,\infty))} \\
& \leq C[\rho(\|h_1 - h_2\|_{\mathcal{B}_\lambda([0,T])} + \|v_1 - v_2\|_{\mathcal{B}_\lambda([0,T];\mathcal{D}_A(1+\theta,\infty))}) \\
& +(1+\lambda)^{-\varepsilon}(\|h_1 - h_2\|_{\mathcal{B}_\lambda([0,T])} + \|w_1 - w_2\|_{\mathcal{B}_\lambda([0,T])}) \\
& +(1+\lambda)^{-1}\|v_1 - v_2\|_{\mathcal{B}_\lambda([0,T];\mathcal{D}_A(1+\theta,\infty))}] \\
& \leq 2C[\rho + (1+\lambda)^{-\varepsilon} + (1+\lambda)^{-1}] \\
& \times\|(v_1 - v_2, h_1 - h_2, w_1 - w_2)\|_{Y(\lambda)}.
\end{aligned}$$

and, for $i \in \{2, 3\}$,

$$\begin{aligned}
\|\mathcal{R}_i(v_1, h_1, w_1) - & \mathcal{R}_i(v_2, h_2, w_2)\|_{\mathcal{B}_\lambda([0,T])} \\
& \leq C_4(1+\rho)[\rho + (1+\lambda)^{-\varepsilon} + (1+\lambda)^{-1}] \\
& \times\|(v_1 - v_2, h_1 - h_2, w_1 - w_2)\|_{Y(\lambda)},
\end{aligned}$$

with C_4 independent of λ and ρ. Now, we choose $\overline{\rho}_2 \in (0, \overline{\rho}_1]$ in such a way that

$$\max\{2C\rho,\ C_4\rho(1+\rho)\} < 1/2. \tag{70}$$

Then, for any $\rho \in (0, \overline{\rho}_2]$ there exists $\lambda_2(\rho) \geq 0$, such that, if $\lambda \geq \lambda_2(\rho)$ and (v_1, h_1, w_1) and (v_2, h_2, w_2) in $Y(\lambda, \rho)$, we have (69).
Third step: existence of a solution. We apply the results of the two previous steps: we fix $\rho \in (0, \overline{\rho}_2\})$ and $\lambda \geq \max\{\lambda_1(\rho), \lambda_2(\rho)\}$. Then N maps $Y(\lambda, \rho)$ into itself and, restricted to this subset, is a contraction. Then Banach's fixed point theorem allows to conclude that there exists a unique $(v, h, w) \in Y(\lambda, \rho)$, which is a fixed point for N. From the equivalence Theorem 12, we conclude that a solution exists.
*Fourth step: uniqueness of the solution.*Let (k_j, u_j, h_j, f_j) $(j = 1, 2)$ be solutions of Problem 5. We want to show that they coincide. We already know that $k_1 = k_2 = k_0$. We set

$$v_j := u_j', \qquad w_j := f_j', \qquad j = 1, 2. \tag{71}$$

Then, owing to Theorem 12, v_1 and v_2 belong to $C([0,T];\mathcal{D}(A)) \cap \mathcal{B}([0,T];\mathcal{D}_A(1+\theta,\infty))$, w_1 and w_2 are in $C([0,T])$, and (v_1, h_1, w_1) and

(v_2, h_2, w_2) are both fixed points of N. For $j = 1, 2$, we have, owing to (55),

$$\begin{aligned}\|v_j - \overline{v}\,\|_{\mathcal{B}_\lambda([0,T];\mathcal{D}_A(1+\theta,\infty))} &= \|\mathcal{R}_1(v_j, h_j, w_j)\|_{\mathcal{B}_\lambda([0,T];\mathcal{D}_A(1+\theta,\infty))}\\ &\le C[\|\overline{h} * A\overline{v}\|_{\mathcal{B}_\lambda([0,T];\mathcal{D}_A(\theta,\infty))}\\ &+(1+\lambda)^{-\varepsilon}(\|h_j\|_{\mathcal{B}_\lambda([0,T])} + \|w_j\|_{\mathcal{B}_\lambda([0,T])})\\ &+(1+\lambda)^{-1}(\|h_j - \overline{h}\|_{\mathcal{B}_\lambda([0,T])} + \|v_j - \overline{v}\|_{\mathcal{B}_\lambda([0,T];\mathcal{D}_A(1+\theta,\infty))})\\ &+\|h_j - \overline{h}\|_{\mathcal{B}_\lambda([0,T])}\|v_j - \overline{v}\|_{\mathcal{B}_\lambda([0,T];\mathcal{D}_A(1+\theta,\infty))}].\end{aligned}$$

Applying Theorem 10 and observing that $\mathcal{R}_2(v_j, h_j, w_j)(0) = 0$, we have

$$\lim_{\lambda\to+\infty}\|h_j - \overline{h}\|_{\mathcal{B}_\lambda([0,T])} = 0,$$

so we conclude that

$$\lim_{\lambda\to+\infty}\|v_j - \overline{v}\|_{\mathcal{B}_\lambda([0,T];\mathcal{D}_A(1+\theta,\infty))}$$

$$= \lim_{\lambda\to+\infty}\|\mathcal{R}_1(v_j, h_j, w_j)\|_{\mathcal{B}_\lambda([0,T];\mathcal{D}_A(1+\theta,\infty))} = 0.$$

From (57) we obtain, for $i \in \{2, 3\}$,

$$\begin{aligned}\|\mathcal{R}_i\,(v_j, h_j, w_j)\|_{\mathcal{B}_\lambda([0,T])} &\le C(\|\mathcal{R}_1(v_j, h_j, w_j)\|_{\mathcal{B}_\lambda([0,T];\mathcal{D}_A(1+\theta,\infty))}\\ &+(1+\lambda)^{-1}\|h_j\|_{\mathcal{B}_\lambda([0,T])}\\ &+\|h_j - \overline{h}\|_{\mathcal{B}_\lambda([0,T])}\|\mathcal{R}_1(v_j, h_j, w_j)\|_{\mathcal{B}_\lambda([0,T];\mathcal{D}_A(1+\theta,\infty))}),\end{aligned}$$

so that we conclude that, for $j = 1, 2$,

$$\lim_{\lambda\to+\infty}\|(v_j, h_j, w_j) - (\overline{v}, \overline{h}, \overline{w})\|_{Y(\lambda)} = 0. \tag{72}$$

Now, we choose λ sufficiently large, so that

$$\max\{\|(v_j, h_j, w_j) - (\overline{v}, \overline{h}, \overline{w})\|_{Y(\lambda)} : j = 1, 2\} \le \rho \le \overline{\rho}_2, \tag{73}$$

and, if necessary, we increase λ in such a way that $\lambda \ge \max\{\lambda_1(\rho), \lambda_2(\rho)\}$ and (73) continues to hold. Then, for such ρ (v_1, h_1, w_1) and (v_2, h_2, w_2) are both fixed points of N in $Y(\lambda, \rho)$. We conclude that they coincide.

References

[1] M. Brokate and J. Sprekels, *Hysteresis and phase transitions.* Springer, New York, 1996.

[2] D. Brandon and W.J. Hrusa, Construction of a class of integral models for heat flow in materials with memory. *J. Integral Equations Appl.* **1**, 175–201 (1988).

[3] G. Caginalp, An analysis of a phase field model of a free boundary. *Arch. Rational Mech. Anal.* **92**, 205–245 (1986).

[4] G. Caginalp, Stefan and Hele-Shaw type models as asymptotic limits of the phase-field equations. *Phys. Rev. A* **39**, 5887–5896 (1989).

[5] G. Caginalp, The dynamics of a conserved phase field system: Stefan-like, Hele-Shaw, and Cahn-Hilliard models as asymptotic limits. *IMA J. Appl. Math.* **44**, 77–94 (1990).

[6] G. Caginalp and X. Chen, Convergence of the phase field model to its sharp interface limits. *European J. Appl. Math.* **9**, 417–445 (1998).

[7] G. Caginalp and X. Chen, Phase field equations in the singular limit of sharp interface problems. In: *On the Evolution of Phase Boundaries*, Ed. M.E. Gurtin and G.B. McFadden, (Springer, New York, 1992), p. 1.

[8] B.D. Coleman and M.E. Gurtin, Equipresence and constitutive equations for rigid heat conductors. *Z. Angew. Math. Phys.* **18**, 199–208 (1967).

[9] F. Colombo, Direct and inverse problems for a phase-field model with memory. *J. Math. Anal. Appl.* **260**, 517–545 (2001).

[10] F. Colombo, An inverse problem for a generalized Kermack-McKendrick model. *J. Inverse Ill-Posed Probl.* **10**, 221–241 (2002).

[11] F. Colombo, D. Guidetti and V. Vespri, Identification of two memory kernels and the time dependence of the heat source for a parabolic conserved phase-field model. *Math. Methods Appl. Sci.* **28**, 2085–2115 (2005).

[12] F. Colombo and A. Lorenzi, Identification of time and space dependent relaxation kernels for materials with memory related to cylindrical domains. I. *J. Math. Anal. Appl.* **213**, 32–62 (1997).

[13] F. Colombo and A. Lorenzi, Identification of time and space dependent relaxation kernels for materials with memory related to cylindrical domains. II. *J. Math. Anal. Appl.* **213**, 63–90 (1997).

[14] F. Colombo and V. Vespri, A semilinear integrodifferential inverse problem. In: *Evolution equations*, Ed. J. Goldstein, R. Nagel and S. Romanelli (Dekker, New York, 2003), p. 91.

[15] F. Colombo and D. Guidetti, A unified approach to nonlinear integro-differential inverse problems of parabolic type. *Z. Anal. Anwendungen* **21**, 431–464 (2002).

[16] F. Colombo, D. Guidetti and A. Lorenzi, Integro-differential identification problems for the heat equation in cylindrical domains. *Adv. Math. Sci. Appl.* **13**, 639–662 (2003).

[17] F. Colombo, D. Guidetti and A. Lorenzi, Integrodifferential identification problems for thermal materials with memory in non-smooth plane domains. *Dynam. Systems Appl.* **12**, 533–559 (2003).

[18] P. Colli, G. Gilardi and M. Grasselli, Global smooth solution to the standard phase-field model with memory. *Adv. Differential Equations* **2**, 453–486 (1997).

[19] P. Colli, G. Gilardi and M. Grasselli, Well-posedness of the weak formulation for the phase-field model with memory. *Adv. Differential Equations* **2**, 487–508 (1997).

[20] C. Giorgi, M. Grasselli and V. Pata, Uniform attractors for a phase-field model with memory and quadratic nonlinearity. *Indiana Univ. Math. J.* **48**, 1395–1445 (1999).

[21] M. Grasselli, An inverse problem in population dynamics. *Num. Funct. Anal. Optim.* **18**, 311–323 (1997).

[22] M. Grasselli, V. Pata and F.M. Vegni, Longterm dynamics of a conserved phase-field system with memory. *Asymptot. Analysis* **33**, 261–320 (2003).

[23] A. Lorenzi, E. Rocca and G. Schimperna, Direct and inverse problems for a parabolic integro-differential system of Caginalp type. *Adv. Math. Sci Appl.* **15**, 227–263 (2005).

[24] A. Lorenzi and E. Sinestrari, An inverse problem in the theory of materials with memory. *Nonlinear Anal.* **12**, 1317–1335 (1988).

[25] A. Lunardi, *Analytic semigroups and optimal regularity in parabolic problems.* Birkhäuser, Basel, 1995.

[26] A. Novick–Cohen, The Cahn-Hilliard equation: mathematical and modeling perspectives. *Adv. Math. Sci. Appl.* **8**, 965–985 (1998).

[27] A. Novick–Cohen, Conserved phase-field equations with memory. In: *Curvature Flows and related Topics*, Ed. A. Damlamian, J. Spruck and A. Visintin. (Gakkōtosho, Tokyo, 1995), p. 179.

[28] C.V. Pao, *Nonlinear parabolic and elliptic equations.* Plenum Press, New York, 1992.

[29] A. Pazy, *Semigroups of linear operators and applications to partial differential equations.* Springer-Verlag, New York, 1983.

[30] E. Sinestrari, On the abstract Cauchy problem of parabolic type in spaces of continuous functions. *J. Math. Anal. Appl.* **107**, 16–66 (1985).

[31] H.B. Stewart, Generation of analytic semigroups by strongly elliptic operators. *Trans. Amer. Math. Soc.* **199**, 141–162 (1974).

[32] H. Triebel, *Interpolation theory, function spaces, differential operators.* North-Holland Publishing Co., Amsterdam-New York, 1978.

WEAK SOLUTIONS FOR STEFAN PROBLEMS WITH CONVECTIONS

Takesi Fukao

General Education, Gifu National College of Technology
2236-2 Kamimakuwa, Motosu-shi, Gifu 501-0495 Japan
E-mail: fukao@gifu-nct.ac.jp

In this paper we introduce some nonlinear partial differential equations and discuss the solvability of their initial and boundary value problems. These are called the enthalpy formulations of the Stefan problem, which describes the solid-liquid phase transitions. Firstly, we tackle a problem with convection governed by Navier-Stokes equations in the unknown liquid region. The weak solution, which is a limit of an approximate procedure of the solid-liquid interface in some sense, is constructed. Secondly, we introduce a coupled system of Stefan and Navier-Stokes equations. This is a transmission problem between the solid-liquid and gas regions.

1. Introduction

The Stefan problem is a mathematical model which describes the dynamics of solid-liquid phase transition dynamics. The research of this problem started from the middle of the 19th century. In 1889, J. Stefan, an Austrian physicist, proposed a simplified model for the solid-liquid phase transition, which is called the one phase Stefan problem. Subsequently, several kinds of Stefan problems, classical solutions, enthalpy formulations and so on, have been proposed from various points of view. For instance, we refer to the book of Visintin [16] for the literature on the various Stefan problems.

More specifically, in the paper [5], DiBenedetto and O'Leary they treated a heat conduction problem with convective effects due to the flow of a material governed by the Stokes equation in the liquid region. For other research papers on Stefan problems with convection, we also refer, for instance, to Cannon, DiBenedetto and Knightly [2], Rodrigues [12],

Starovoitov [14], Casella and Giangi [3]. In these models, the enthalpy formulation of the Stefan problem is employed, and the movement of the material configuration is neglected, namely the problems are formulated in a fixed material domain.

Here we introduce an interesting subject, which is related to our problem of Stefan type. It is called "the Czochralski pulling method". Such a method is widely used for the production of a column of single crystal from various melt substances. For the literature on the Czochralski method, we refer to Pawłow [11]. In this phenomenon, the phase transition occurs at the interface between the melt and the crystal, and the dynamics of the phase transition interface is one of our interesting research subjects.

Our idea is to treat the Czochralski pulling method as a problem of Stefan type. In the Czochralski method, the crystallization of the melt yields a smooth change of the material configuration in time and a smooth flow in the material domain. When these effects are taken into account, the modeling should be made in a time-dependent domain, namely in a non-cylindrical domain. The determination of the convection is also one of the important questions in the mathematical modeling of the Czochralski method. It is reasonable to postulate that the convection is equal to the pulling velocity in the crystal (solid) and is a solution of the incompressible Navier-Stokes, or simply of the Stokes equation, in the melt (liquid). But, in general, the liquid region of the material should be determined by the classical solution of the Stefan problem. In our setting, we make it up by the weak solution via the usual regular approximation. In this case, the mathematical difficulty stems not only from the nonlinear system, but also from the time-dependence of the domain.

Let $0 < T < +\infty$ and $B(t)$ be the open ball in $\mathbb{R}^3$ with smooth boundary $\partial B(t)$ for all $t \in [0, T]$. Now, we assume that the shape of the ball $B(0)$ changes to $B(t)$ smoothly in time with e velocity $\mathbf{v} := \mathbf{v}(t,x) := (v_1(t,x), v_2(t,x), v_3(t,x))$. In the sequel, the function $\theta := \theta(t,x)$ stands for the temperature, the function $e := e(t,x)$ stands for the "internal energy" density and the energy functional $\mathcal{E}^t$, defined by

$$\mathcal{E}^t(e) := \int_{B(t)} \rho e(t,x)dx,$$

where $\rho > 0$, which is not necessarily constant, stands for the density. $\mathcal{E}^t$ is called the total "internal energy". Furthermore, consider a vector field, the so-called heat flux, $\mathbf{q} := \mathbf{q}(t,x) := (q_1(t,x), q_2(t,x), q_3(t,x))$ at point $t \in (0,T)$ and $x \in B(t)$, where each component $q_i(t,x)$ is the heat conduction in the direction i, for $i = 1, 2, 3$. Let us recall here three important physical requirements. The first one is that the quantity $\mathcal{E}^t$ is controlled by the source f and $\mathbf{q}$ in the following way:

$$\frac{d}{dt}\mathcal{E}^t(e) = \int_{B(t)} f(t,x)dx - \int_{\partial B(t)} \mathbf{q}(t,x) \cdot \boldsymbol{\nu}(t,x) d\partial B(t), \tag{1}$$

where $\boldsymbol{\nu} = \boldsymbol{\nu}(t,x) := (\nu_1(t,x), \nu_2(t,x), \nu_3(t,x))$ is the 3-dimensional unit outward normal vector on $\partial B(t)$, so that $\mathbf{q}(t) \cdot \boldsymbol{\nu}(t)$ is the amount of heat going to the exterior of $B(t)$ through the boundary $\partial B(t)$. In general, we call equation (1) the energy balance law. Secondly, we have to take into account the Fourier law $\mathbf{q} = -k\nabla\theta$, where, and hereafter, $\nabla\theta := (\partial\theta/\partial x_1, \partial\theta/\partial x_2, \partial\theta/\partial x_3)$ and $k > 0$, not necessarily constant, stands for the heat conductivity of the material. Finally, let us recall the following formula: $e = c\theta$ where $c > 0$, also not necessarily constant, stands for the specific heat.

Note that the left-hand side of (1) equals the sum of the "rate of the change of energy in $B(t)$" and of the "convective influx of $\mathcal{E}^t$ into $B(t)$ through its boundary $\partial B(t)$", that is,

$$\frac{d}{dt}\mathcal{E}^t(e) = \int_{B(t)} \rho\frac{\partial e}{\partial t}(t,x)dx + \int_{\partial B(t)} \rho e(t,x) v_b(t,x) d\partial B(t),$$

where $v_b(t,\cdot)$ is the normal speed of the boundary $\partial B(t)$, defined by $v_b := \mathbf{v} \cdot \boldsymbol{\nu}$. Mass conservation is expressed through the condition $\text{div}\mathbf{v}(t) = 0$ for all $t \in [0,T]$ So, using Gauss' divergence theorem we obtain

$$\int_{B(t)} c\rho\frac{\partial\theta}{\partial t}dx + \int_{B(t)} c\rho\mathbf{v}\cdot\nabla\theta dx = \int_{B(t)} f dx - \int_{B(t)} \text{div}\Big(-k\nabla\theta\Big)dx. \tag{2}$$

Using that $B(t)$ is arbitrary, we can give meaning to the following heat equation with convection:

$$c\rho\left(\frac{\partial\theta}{\partial t} + \mathbf{v}\cdot\nabla\theta\right) - k\Delta\theta = f, \tag{3}$$

where and hereafter $\Delta\theta := \sum_{i=1}^{3}(\partial^2\theta/\partial x_i^2)$. The first two terms on the left-hand side yield the material derivative of θ.

In this paper, we shall give a survey on the existence results of the Stefan problem with convection governed by the Navier-Stokes equation in section 2, and on the transmission-Stefan problem in the solid-liquid and the gas regions in section 3. Finally, we shall propose some mathematical models of Stefan problems with convection of the transmission type.

2. Stefan problem in non-cylindrical domain with convection governed by Navier-Stokes equations

In this section, we introduce the Stefan problem with convection governed by the Navier-Stokes equation in the unknown liquid region. The problem is to find the unknown interface accounting for the convection.

Let $\Omega_m(t) \subset \mathbb{R}^3$ be a time dependent bounded domain with smooth boundary $\Gamma_m(t) := \partial\Omega_m(t)$, such that $\Omega_m(t)$ is occupied by a material having two phases: the liquid state $\Omega_\ell(t)$ and the solid state $\Omega_s(t)$. These regions are separated by an unknown interface $S(t)$, namely $\Omega_m(t) := \Omega_\ell(t) \cup S(t) \cup \Omega_s(t)$ for each $t \in [0,T]$. Moreover, we define non-cylindrical domains and their lateral boundaries by

$$Q_i := \bigcup_{t\in(0,T)} \{t\} \times \Omega_i(t) \quad \text{for all } i = m, \ell, s, \qquad \Sigma_m := \bigcup_{t\in(0,T)} \{t\} \times \Gamma_m(t).$$

Throughout this paper, we assume that the shape of the domain $\Omega_m(t)$ smoothly changes in time in the following sense:

(A1) there exists a bounded domain $\Omega \subset \mathbb{R}^3$, with smooth boundary $\Gamma := \partial\Omega$, such that $Q_m \subset Q := (0,T) \times \Omega$. Moreover, there exists a function $\mathbf{y} \in \mathbf{C}^3(\overline{Q}) := C^3(\overline{Q})^3$, yielding a C^3-diffeomorphism $\mathbf{y}(t,\cdot) := (y_1(t,\cdot), y_2(t,\cdot), y_3(t,\cdot))$ from $\overline{\Omega}$ onto itself for all $t \in [0,T]$, such that

$$\mathbf{y}(t, \overline{\Omega_m(t)}) = \overline{\Omega_{m0}} \quad \text{for all } t \in [0,T],$$

namely $\mathbf{y}(0,\cdot) = \mathbf{I}$ is the identity on $\overline{\Omega_{m0}}$, where $\Omega_{m0} := \Omega_m(0)$.

This assumption implies our standpoint: although in research problems on the Czochralski method the material shape design, that is, the design of the interface between the material and the gas, is also an interesting problem, we will not consider it in this paper.

2.1. *Classical formulation*

In this subsection we introduce the classical formulation of the problem. We denote by $\mathbf{v} := \mathbf{v}(t,x)$ the velocity of the flow. In this paper, we derive the following equations in the liquid region Q_ℓ:

$$c_\ell \rho_\ell \left(\frac{\partial \theta}{\partial t} + \mathbf{v} \cdot \nabla\theta \right) - k_\ell \Delta\theta = f \quad \text{in } Q_\ell, \tag{4}$$

$$\rho_\ell \frac{\partial \mathbf{v}}{\partial t} + \rho_\ell (\mathbf{v} \cdot \nabla)\mathbf{v} - \mu_\ell \Delta \mathbf{v} = -\nabla p_\ell + \boldsymbol{g}_\ell(\theta) \quad \text{in } Q_\ell, \tag{5}$$

$$\text{div}\mathbf{v} = 0 \quad \text{in } Q_\ell. \tag{6}$$

The second part (5)-(6) is the incompressible Navier-Stokes equation governing the velocity of the material flow, where p_ℓ is a pressure, $\mu_\ell > 0$ stands for the first viscosity, $\boldsymbol{g}_\ell : \mathbb{R} \to \mathbb{R}^3$ is a given function, for example, the buoyancy force $\boldsymbol{g}_\ell(\theta) := (0, 0, g_0(1-\theta))$, with a positive constant g_0; in general, $\boldsymbol{g}_\ell$ is a nonlinear function. In the solid region Q_s we derive equations:

$$c_s\rho_s\left(\frac{\partial\theta}{\partial t} + \mathbf{v}\cdot\nabla\theta\right) - k_s\Delta\theta = f \quad \text{in } Q_s, \tag{7}$$

$$\mathbf{v} = \mathbf{v}_D \quad \text{in } Q_s, \tag{8}$$

where $\mathbf{v}_D := (v_{D1}, v_{D2}, v_{D3})$ is a designed vector on $\overline{Q}$. As a simple example, we have $\mathbf{v}_D(t,x) := [\partial\mathbf{x}/\partial t \circ \mathbf{y}](t,x) := \partial\mathbf{x}/\partial t(t, \mathbf{y}(t,x))$ for $(t,x) \in Q_s$, using the transformation $\mathbf{y}$ and its inverse $\mathbf{x} := \mathbf{y}^{-1}$. On the interface $S(t)$ we derive the Stefan condition

$$\theta = \theta_c, \quad \mathbf{v} = \mathbf{v}_D,$$

$$k_\ell\frac{\partial\theta}{\partial\mathbf{n}^+} + k_s\frac{\partial\theta}{\partial\mathbf{n}^-} = L(\mathbf{v}_D\cdot\mathbf{n}^+ - v_S) \quad \text{on } S := \bigcup_{t\in(0,T)}\{t\}\times S(t), \tag{9}$$

where θ_c is the critical temperature; without loss of generality, we set that $c_\ell\rho_\ell = c_s\rho_s = 1$ and the critical temperature θ_c is equal to 0. Actually,it is enough to take the new function $\tilde{\theta} := c_i\rho_i(\theta - \theta_c)$ in each domain $\Omega_\ell(t)$ and $\Omega_s(t)$, then $\tilde{\theta} = 0$ on $S(t)$, so that the continuity is kept on the boundary $S(t)$ between $\Omega_\ell(t)$ and $\Omega_s(t)$. Moreover, $\mathbf{n}^+ = \mathbf{n}^+(t,x) := (n_1(t,x)^+, n_2(t,x)^+, n_3(t,x)^+)$ is the 3-dimensional unit vector normal to $S(t)$, pointing to the solid region $\Omega_s(t)$,and $\mathbf{n}^- := -\mathbf{n}^+$; L is a positive constant which stands for the latent heat; $v_S := v_S(t,x)$ means the normal speed of $S(t)$. The boundary and initial conditions are of the forms, respectively:

$$k_i\frac{\partial\theta}{\partial\boldsymbol{\nu}} + n_0k_i\theta = h, \quad \mathbf{v} = \mathbf{v}_D \quad \text{on } \Sigma_{m,i} := \bigcup_{t\in(0,T)}\{t\}\times\Gamma_{m,i}(t), \tag{10}$$

$$\theta(0) = \theta_0, \quad S(0) = S_0, \quad \mathbf{v}(0) = \mathbf{v}_0 \quad \text{on } \Omega_{m0}, \tag{11}$$

where $\Gamma_{m,i}(t) := \Gamma_m(t)\cap\Gamma_i(t)$ for $i = \ell, s$; $\boldsymbol{\nu} = \boldsymbol{\nu}(t,x)$ is the 3-dimensional unit outward normal vector on $\Gamma_m(t)$ at $x \in \Gamma_m(t)$; n_0 is a positive constant, while h, θ_0 and $\mathbf{v}_0$ are given functions on Σ_m and Ω_{m0}, respectively. A natural assumption on $\mathbf{v}_D$ is that:

(A2) $\mathbf{v}_D \in \mathbf{C}^2(\overline{Q})$, and

$$\mathrm{div}\mathbf{v}_D(t,\cdot) = 0 \quad \text{in } \Omega_m(t) \quad \text{for all } t \in [0,T], \tag{12}$$

$$\mathbf{v}_D \cdot \boldsymbol{\nu} = v_{\Sigma_m} \quad \text{on } \Sigma_m, \tag{13}$$

where $v_{\Sigma_m}(t,\cdot)$ is the normal speed of $\Gamma_m(t)$, defined by $v_{\Sigma_m} := [\partial \mathbf{x}/\partial t \circ \mathbf{y}] \cdot \boldsymbol{\nu}$ on Σ_m.

The determination of $\mathbf{v}$ is an important question in our mathematical models of the Czochralski method. It seems that $\mathbf{v}$ mainly obeys the pulling velocity in the crystal, and it is a solution of the incompressible Navier-Stokes or simply of the Stokes equation in the melt.

2.2. *Weak formulation and existence result*

In this subsection we discuss about the solvability in the weak sense for the problem which was have just introduced. To do so, we give a weak variational formulation of the Stefan problem. This approach is called the enthalpy method for the Stefan problem. We consider a new parameter u, which is called enthalpy, and a function $\beta : \mathbb{R} \to \mathbb{R}$ defined by

$$u := \begin{cases} \theta & \text{if } \theta < 0, \\ r \in [0,L] & \text{if } \theta = 0, \\ \theta + L & \text{if } \theta > 0, \end{cases} \qquad \beta(r) := \begin{cases} k_s r & \text{if } r < 0, \\ 0 & \text{if } 0 \le r \le L, \\ k_\ell (r-L) & \text{if } r > L. \end{cases}$$

With these u and β, the Stefan problem {(4), (7), (9), (10), (11)} can be written as the following initial boundary value problem (SP):= {(14)-(16)} for a degenerate parabolic partial differential equation in the non-cylindrical domain Q_m:

$$\frac{\partial u}{\partial t} + \mathbf{v} \cdot \nabla u - \Delta \beta(u) = f \quad \text{in } Q_m, \tag{14}$$

$$\frac{\partial \beta(u)}{\partial \boldsymbol{\nu}} + n_0 \beta(u) = h \quad \text{on } \Sigma_m, \tag{15}$$

$$u(0) = u_0 := \theta_0 \mathcal{X}_{\Omega_s(0)} + (\theta_0 + L)\mathcal{X}_{\Omega_\ell(0)} \quad \text{in } \Omega_{m0}, \tag{16}$$

where $\mathcal{X}_{\Omega_i(0)}$ is the characteristic function of $\Omega_i(0)$ for $i = \ell, s$. Actually, multiplying (4) and (7) by any test function $\eta \in W := \{\eta \in H^1(Q_m); \eta(T) = 0\}$, integrating over Q_m and using the Stefan condition (9), along with the assumption (A1), we derive a variational identity, which for example we find in the paper of Fukao, Kenmochi and Pawłow [8]. One of the most important points is that the liquid region, in which the convection $\mathbf{v}$ is governed by the Navier-Stokes equation, is unknown. So, our first idea to treat this difficulty is the following reformulation of the original problem. Notice that both the liquid region Q_ℓ and solid region Q_s are unknown, and depend on the enthalpy u as follows

$$Q_\ell(u) := \left\{ (t,x) \in Q_m; u(t,x) > \frac{L}{2} \right\},$$

$$Q_s(u) := \left\{ (t,x) \in Q_m; u(t,x) < \frac{L}{2} \right\}.$$

Actually, by the definition of u they make sense as the liquid and solid regions, respectively. Then, we can write the Navier-Stokes part { (5), (6), (8)} in the following way (NS):= {(17), (18), (19), (10), (11)}

$$\frac{\partial \mathbf{v}}{\partial t} + (\mathbf{v} \cdot \nabla)\mathbf{v} - \nu_\ell \Delta \mathbf{v} = -\frac{1}{\rho_\ell}\nabla p_\ell + \frac{1}{\rho_\ell}\boldsymbol{g}_\ell(\beta(u)) \quad \text{in } Q_\ell(u), \tag{17}$$

$$\text{div}\mathbf{v} = 0 \quad \text{in } Q_\ell(u), \tag{18}$$

$$\mathbf{v} = \mathbf{v}_D \quad \text{in } Q_s(u), \tag{19}$$

where $\nu_\ell := \mu_\ell/\rho_\ell > 0$ is called the kinematic viscosity. In Fukao and Kenmochi [6], the existence of a weak solution of {(14)~(19)} is obtained. At first, for each $\varepsilon > 0$ we find the approximate solution of (SP) and (NS)$_\varepsilon$ by the regularization of the domains $Q_i(\rho_\varepsilon * u)$ for $i = \ell, s$, where $\rho_\varepsilon := \rho_\varepsilon(x)$ is a mollifier.

Theorem 2.1 (Fukao and Kenmochi [6]) *Assume (A1) and (A2) hold. Moreover, let $f \in L^\infty(Q_m)$, $h \in L^\infty(\Sigma_m)$, $u_0 \in L^\infty(\Omega_{m0})$ and $\mathbf{v}_0 \in \mathbf{L}^2(\Omega_{m0})$, with $\text{div}\mathbf{v}_0 = 0$ in Ω_{m0}. Then, for each $\varepsilon > 0$ there exists at least one pair of functions $\{u_\varepsilon, \mathbf{v}_\varepsilon\} \in L^\infty(Q_m) \times \mathbf{L}^2(Q_m)$ such that u_ε and $\mathbf{v}_\varepsilon$ satisfy (SP) and (NS)$_\varepsilon$ in the variational sense.*

As to the limit of the solutions $\{u_\varepsilon, \mathbf{v}_\varepsilon\}$ as $\varepsilon \downarrow 0$, the following result was obtained.

Theorem 2.2. (Fukao and Kenmochi [6]) *Under the same assumption of Theorem 2.1, let $\{u_\varepsilon, \mathbf{v}_\varepsilon\}_{\varepsilon>0}$ be the family of approximate solutions constructed by Theorem 2.1. Then, there exists a sequence $\{\varepsilon_n\}$ of positive numbers with $\varepsilon_n \to 0$ as $n \to +\infty$, such that*

$$u_{\varepsilon_n} \to u \quad \textit{weakly in } L^2(Q_m),$$

$$\mathbf{v}_{\varepsilon_n} - \mathbf{v}_D \to \mathbf{v} - \mathbf{v}_D \quad \textit{weakly in } L^2(0,T;\mathbf{H}^1_\sigma(\Omega)),$$

as $n \to +\infty$ and u satisfies (SP).

The space $\mathbf{H}^1_\sigma(\Omega)$ is the closure of $\boldsymbol{\mathcal{D}}_\sigma(\Omega)$ in spaces $\mathbf{H}^1(\Omega)$, where

$$\boldsymbol{\mathcal{D}}_\sigma(\Omega) := \left\{\mathbf{z} \in \mathbf{C}^\infty_0(\Omega); \mathrm{div}\mathbf{z} = 0 \text{ in } \Omega\right\}.$$

3. Transmission-Stefan problem

In this section, we survey an existence result by Fukao, Kenmochi and Pawlow, see [9], for the transmission-Stefan problem with a given convection. Roughly speaking, the problem is to find the solid-liquid interface accounting for the heat conduction from the crucible boundary through the gas region with the artificial convection.

Let $\Omega_m(t) = \Omega_\ell(t) \cup S(t) \cup \Omega_s(t) \subset \mathbb{R}^3$ be the material, and let the notation of the material domain be the same as in the above section. Moreover, let $\Omega \subset \mathbb{R}^3$ be the domain which is the crucible containing the material $\Omega_m(t)$ and the gas region $\Omega_g(t)$, that is, $\Omega = \Omega_m(t) \cup \Gamma_{gm}(t) \cup \Omega_g(t)$; $\Gamma_{gm}(t)$ is separated from $\Gamma_{gs}(t)$ and $\Gamma_{g\ell}(t)$, which are the interfaces between gas and solid, gas and liquid for each $t \in [0,T]$, respectively; $\Gamma := \partial\Omega$ is smooth and we set $\Gamma_i(t) := \Gamma \cap \partial\Omega_i(t)$ for $i = \ell, g$. The vector $\boldsymbol{\nu} := \boldsymbol{\nu}(x)$ is the unit 3-dimensional outward normal on Γ, and $\boldsymbol{\nu}^+ := \boldsymbol{\nu}^+(t,x)$ is the 3-dimensional unit normal to $\Gamma_{gm}(t)$ pointing to $\Omega_g(t)$; $\boldsymbol{\nu}^- := -\boldsymbol{\nu}^+$. Note here that our geometrical assumption ensures that the solid region is detached from the crucible boundary Γ. Moreover, we define the non-cylindrical lateral boundaries by

$$\Sigma_i := \bigcup_{t \in (0,T)} \{t\} \times \Gamma_i(t) \quad \text{for all } i = \ell, s, gm, g\ell, gs.$$

In this setting, we consider the transmission problem between the Stefan problem (4), with $\mathbf{v} = \mathbf{v}_{Dm}$ (where $\mathbf{v}_{Dm}$ is the designed vector on the material), (7), (10) and the heat equation in the gas region

$$c_g\rho_g\left(\frac{\partial\theta_g}{\partial t} + \mathbf{v}_{Dg}\cdot\nabla\theta_g\right) - k_g\Delta\theta_g = f \quad \text{in } Q_g := \bigcup_{t \in (0,T)} \{t\} \times \Omega_g(t), \quad (20)$$

where $\theta_g := \theta_g(t,x)$ is the temperature of the gas, $\mathbf{v}_{Dg} := \mathbf{v}_{Dg}(t,x)$ is the designed vector in the gas and c_g, ρ_g and k_g are positive constants. Moreover, on the gas-material boundary Σ_{gm} we require the following transmission condition:

$$\theta = \theta_g, \quad k_i \frac{\partial \theta}{\partial \boldsymbol{\nu}^+} + k_g \frac{\partial \theta_g}{\partial \boldsymbol{\nu}^-} = 0 \quad \text{on } \Sigma_{gi}, \quad \text{for } i = \ell, s. \tag{21}$$

The boundary condition on Σ is

$$k_\ell \frac{\partial \theta}{\partial \boldsymbol{\nu}} + n_0 \theta = h \quad \text{on } \Sigma_\ell, \quad k_g \frac{\partial \theta_g}{\partial \boldsymbol{\nu}} + n_0 \theta_g = h \quad \text{on } \Sigma_g, \tag{22}$$

and the initial conditions are prescribed as follows:

$$\theta(0) = \theta_0, \quad S(0) = S_0 \quad \text{on } \Omega_{m0}, \quad \theta_g(0) = \theta_0 \quad \text{on } \Omega_g(0). \tag{23}$$

We call the above transmission-Stefan problem Problem (TS) := { (4), (7), (10), (20)-(23)}, with $\mathbf{v} = \mathbf{v}_{Dm}$ in Q_m. Assume that designed vector fields $\mathbf{v}_{Dm}$ and $\mathbf{v}_{Dg}$ satisfy the following (A3):

(A3) $\mathbf{v}_{Dm} \in \mathbf{C}^2(\overline{Q_m})$, $\mathbf{v}_{Dg} \in \mathbf{C}^2(\overline{Q_g})$ and the following properties are satisfied:

$$\operatorname{div}\mathbf{v}_{Dm}(t, \cdot) = 0 \quad \text{in } \Omega_m(t),$$

$$\operatorname{div}\mathbf{v}_{Dg}(t, \cdot) = 0 \quad \text{in } \Omega_g(t) \quad \text{for all } t \in [0, T],$$

$$\mathbf{v}_{Dg} \cdot \boldsymbol{\nu}^+ = \mathbf{v}_{Dm} \cdot \boldsymbol{\nu}^+ = v_{\Sigma_m} \quad \text{on } \Sigma_{gm},$$

$$\mathbf{v}_{Dg} \cdot \boldsymbol{\nu} = \mathbf{v}_{Dm} \cdot \boldsymbol{\nu} = v_{\Sigma_m} := 0 \quad \text{on } \Sigma.$$

For simplicity, we again denote θ_g by θ and see it as a function on the cylindrical domain Q. Then, we have the following existence result for the transmission-Stefan problem (TS).

Theorem 3.1 (Fukao, Kenmochi and Pawłow [9]) *Assume that* (A1) *and* (A3) *hold. Let* $f \in L^2(Q)$, $h \in L^2(\Sigma)$ *and* $\theta_0 \in H^1(\Omega)$. *Then, there exists at least one function* $\theta \in L^2(Q)$ *such that* θ *satisfies (TS) in the variational sense.*

4. Mathematical modelling for the transmission problem with the Stefan and Navier-Stokes equations

In the final section, we introduce a mathematical model for the transmission-Stefan and Navier-Stokes problem. In this model, we consider two incompressible Navier-Stokes equations as the momentum balance equation of melt and gas. They are connected by the transmission condition on the gas-melt interface $\Gamma_{g\ell}(t)$. Mathematically, the difference of density and first viscosity between gas and melt results in the nonlinearity in the diffusion term.

We use the same geometrical assumptions as in the last section. Consider the transmission-Stefan problem (TS) and the transmission-Navier-Stokes equations (TN) as follows:

$$\frac{\partial \mathbf{v}}{\partial t} + (\mathbf{v} \cdot \nabla)\mathbf{v} - \nu_\ell \Delta \mathbf{v} = -\frac{1}{\rho_\ell} \nabla p_\ell + \frac{1}{\rho_\ell} \boldsymbol{g}_\ell(\theta) \quad \text{in } Q_\ell,$$

$$\operatorname{div} \mathbf{v} = 0 \quad \text{in } Q_\ell,$$

$$\mathbf{v} = \mathbf{v}_D \quad \text{on } Q_s,$$

$$\frac{\partial \mathbf{v}_g}{\partial t} + (\mathbf{v}_g \cdot \nabla)\mathbf{v}_g - \nu_g \Delta \mathbf{v}_g = -\frac{1}{\rho_g} \nabla p_g + \frac{1}{\rho_g} \boldsymbol{g}_g(\theta) \quad \text{in } Q_g, \tag{24}$$

$$\operatorname{div} \mathbf{v} = 0 \quad \text{in } Q_g, \tag{25}$$

where $\nu_g := \mu_g/\rho_g > 0$. Moreover, these equations are connected on the liquid-solid and gas-solid interface by a no-slip type transmission condition,

$$\mathbf{v} = \mathbf{v}_D \quad \text{on } S(t), \tag{26}$$

$$\mathbf{v}_g = \mathbf{v}_D \quad \text{on } \Gamma_{gs}(t). \tag{27}$$

On the other hand, on the gas-liquid interface, we connect them by a slip type transmission condition,

$$\mathbf{v}_g \cdot \boldsymbol{\nu}^- + \mathbf{v} \cdot \boldsymbol{\nu}^+ = 0 \quad \text{on } \Sigma_{g\ell}, \tag{28}$$

$$\nu_g \mathbf{D}(\mathbf{v}_g)\boldsymbol{\nu}^- + \nu_\ell \mathbf{D}(\mathbf{v})\boldsymbol{\nu}^+ = 0 \quad \text{on } \Sigma_{g\ell}, \tag{29}$$

where for each $\mathbf{v} := (v_1, v_2, v_3)$, $\mathbf{D}(\mathbf{v}) = (d_{ij})$ is the tensor defined by $d_{ij} := 1/2(\partial v_i/\partial x_j + \partial v_j/\partial v_i)$ for each $i, j = 1, 2, 3$.

Finally, we propose a weak variational form of (TN):= $\{$(5), (6), (8), (24)-(29)$\}$. Again, we denote $\mathbf{v}_g$ by $\mathbf{v}$ and see it as a function on the cylinder Q. It seems that the penalty method is useful. We introduce the vector valued function spaces $\mathbf{K}(t)$

$$\mathbf{K}(t) := \left\{ \mathbf{z} \in \mathbf{L}^2_\sigma(\Omega); \begin{array}{c} \mathbf{z}|_{\Omega_m(t)} \in \mathbf{H}^1_\sigma(\Omega_m(t)),\ \mathbf{z}|_{\Omega_g(t)} \in \mathbf{H}^1_\sigma(\Omega_g(t)), \\ \mathbf{z} \cdot \boldsymbol{\nu}^+ = \mathbf{z} \cdot \boldsymbol{\nu}^- \text{ on } \Gamma_{gm}(t) \end{array} \right\}.$$

Then, taking a test function $\boldsymbol{\eta} \in \mathbf{C}_0^\infty(\overline{Q})$ with $\boldsymbol{\eta}(T) = 0$ on Ω, we interpret (TS) in the distribution sense. As a first step, the following approximate problem is considered. This is the same approach as in the case of the result for the Stefan problem with convection, governed by the Navier-Stokes equations, in section 2, see Fukao and Kenmochi [6]. We extend u on the gas region by $u := \theta_g$ on $\Omega_g(t)$. For each $\delta > 0$, we find a solution $\{u_\delta, \theta_\delta, \mathbf{v}_\delta\}$ satisfying the system $\{(30), (31)\}$, namely

$$-\int_0^T (u_\delta, \eta')_{L^2(\Omega)} dt + \int_0^T \langle K(t;\theta_\delta), \eta\rangle dt - \int_0^T b(\mathbf{v}_\delta, \eta, u_\delta) dt$$
$$+n_0 \int_0^T (\theta_\delta, \eta)_{L^2(\Gamma)} dt = \int_0^T \langle f^*, \eta\rangle dt + (u_0, \eta(0))_{L^2(\Omega)} \quad \text{for all } \eta \in W, \tag{30}$$

where $W := \{\eta \in H^1(Q); \eta(T) = 0 \text{ on } \Omega\}$, $\eta' := \partial\eta/\partial t$,

$$u_\delta \in \alpha(\theta_\delta) := \begin{cases} \theta_\delta & \text{if } x \in \Omega_m(t),\ \theta(t,x) < 0, \\ [0, L] & \text{if } x \in \Omega_m(t),\ \theta(t,x) = 0, \\ \theta_\delta + L & \text{if } x \in \Omega_m(t),\ \theta(t,x) > 0, \\ \theta_\delta & \text{if } x \in \overline{\Omega_g(t)}, \end{cases} \quad \text{for all } (t,x) \in Q,$$

$$\langle K(t;z), \tilde{z}\rangle := \int_{\Omega_g(t)} k_g \nabla z \cdot \nabla \tilde{z} dx + \int_{\Omega_m(t)} k(z) \nabla z \cdot \nabla \tilde{z} dx,$$

for all $z, \tilde{z} \in H^1(\Omega)$ where $k(r) := k_s$ if $r < 0$ and $k(r) := k_\ell$ if $r > 0$ for all $r \in \mathbb{R}$, and

$$b(\mathbf{z}, \tilde{z}, \hat{z}) := \int_\Omega (\mathbf{z} \cdot \nabla \tilde{z}) \hat{z} dx \quad \text{for all } \mathbf{z} \in \mathbf{L}^2_\sigma(\Omega), \tilde{z} \in H^1(\Omega), \hat{z} \in L^2(\Omega),$$

$$\langle f^*, \tilde{z}\rangle := (f, \tilde{z})_{L^2(\Omega)} + (h, \tilde{z})_{L^2(\Gamma)} \quad \text{for all } \tilde{z} \in H^1(\Omega).$$

Equation (30) is coupled with

$$-\int_0^T (\mathbf{v}_\delta, \boldsymbol{\eta}')_{\mathbf{L}^2_\sigma(\Omega)} dt + \int_0^T \mathbf{a}(t; \mathbf{v}_\delta, \boldsymbol{\eta}) dt - \int_0^T \mathbf{b}(\mathbf{v}_\delta, \boldsymbol{\eta}, \mathbf{v}_\delta) dt$$
$$+\frac{1}{\delta} \int_0^T (\mathbf{p}(\theta_\delta, \mathbf{v}_\delta - \mathbf{v}_D), \boldsymbol{\eta})_{\mathbf{L}^2_\sigma(\Omega)} dt \tag{31}$$
$$= \int_0^T (\boldsymbol{g}(t,x;\theta_\delta), \boldsymbol{\eta})_{\mathbf{L}^2_\sigma(\Omega)} dt + (\mathbf{u}_0, \boldsymbol{\eta}(0))_{\mathbf{L}^2_\sigma(\Omega)} \quad \text{for all } \boldsymbol{\eta} \in \mathbf{W},$$

where $\mathbf{W} := \{\boldsymbol{\eta} \in \mathbf{C}_0^\infty(\overline{Q}); \boldsymbol{\eta}(T) = 0 \text{ on } \Omega\}$, $\boldsymbol{\eta}' := \partial\boldsymbol{\eta}/\partial t$,

$$\mathbf{a}(t; \mathbf{z}, \tilde{\mathbf{z}}) := \int_{\Omega_g(t)} 2\nu_g \mathbf{D}(\mathbf{z}) : \mathbf{D}(\tilde{\mathbf{z}}) dx + \int_{\Omega_m(t)} 2\nu_\ell \mathbf{D}(\mathbf{z}) : \mathbf{D}(\tilde{\mathbf{z}}) dx$$

$$\text{for all } \mathbf{z}, \tilde{\mathbf{z}} \in \mathbf{K}(t),$$

$$\begin{aligned}\mathbf{b}(\mathbf{z}, \tilde{\mathbf{z}}, \hat{\mathbf{z}}) &:= \sum_{j=1}^{3} b(\mathbf{z}, \tilde{z}_j, \hat{z}_j) \\ &= \int_\Omega \sum_{i,j=1}^{3} z_i \frac{\partial \tilde{z}_j}{\partial x_i} \hat{z}_j dx \quad \text{for all } \mathbf{z}, \hat{\mathbf{z}} \in \mathbf{L}^2_\sigma(\Omega) \text{ and } \tilde{\mathbf{z}} \in \mathbf{H}^1_\sigma(\Omega),\end{aligned}$$

and

$$\boldsymbol{g}(t, x; z) := \begin{cases} \dfrac{1}{\rho_\ell} \boldsymbol{g}_\ell(z) \text{ if } (t, x) \in Q_m, \\ \dfrac{1}{\rho_g} \boldsymbol{g}_g(z) \text{ if } (t, x) \in Q_g, \end{cases}$$

while the penalty term is defined by $\mathbf{p}(\theta_\delta, \mathbf{v}_\delta - \mathbf{v}_D) := [\theta_\delta]^-(\mathbf{v}_\delta - \mathbf{v}_D)$, where $[\theta_\delta]^- := -\min\{\theta_\delta, 0\}$. The limit of the approximate solution as $\delta \to 0$ is a candidate solution of the original problem, in which the limit coincides with $\mathbf{v}_D$ on the solid region. Actually, since $\mathbf{v}$ is divergence-free, the diffusion term in the Navier-Stokes equations is

$$\begin{aligned}&-\int_{Q_m} \nu_\ell \Delta \mathbf{v} \cdot \boldsymbol{\eta} dx \\ &= -\int_{Q_m} \sum_{i,j=1}^{3} 2\nu_\ell \frac{\partial}{\partial x_j} \frac{1}{2} \left(\frac{\partial v_i}{\partial x_j} + \frac{\partial v_j}{\partial x_i} \right) \eta_i dx \\ &= \int_{Q_m} 2\nu_\ell \sum_{i,j=1}^{3} \frac{1}{2} \left(\frac{\partial v_i}{\partial x_j} + \frac{\partial v_j}{\partial x_i} \right) \frac{\partial \eta_i}{\partial x_j} dx \\ &\quad - \int_0^T \int_{\Gamma_{gm}(t)} \nu_\ell \sum_{i,j=1}^{3} \left(\frac{\partial v_i}{\partial x_j} + \frac{\partial v_j}{\partial x_i} \right) \nu_j^+ \eta_i \, d\Gamma_{gm}(t) dt \\ &= \int_{Q_m} 2\nu_\ell \mathbf{D}(\mathbf{v}) : \mathbf{D}(\boldsymbol{\eta}) dx - \int_0^T \int_{\Gamma_{gm}(t)} 2\nu_\ell \mathbf{D}(\mathbf{v}) \boldsymbol{\nu}^+ \cdot \boldsymbol{\eta} \, d\Gamma_{gm}(t) dt.\end{aligned}$$

In the same way, we have

$$\begin{aligned}&-\int_{Q_g} \nu_g \Delta \mathbf{v}_g \cdot \boldsymbol{\eta} dx \\ &= \int_{Q_g} 2\nu_g \mathbf{D}(\mathbf{v}_g) : \mathbf{D}(\boldsymbol{\eta}) dx - \int_0^T \int_{\Gamma_{gm}(t)} 2\nu_g \mathbf{D}(\mathbf{v}_g) \boldsymbol{\nu}^- \cdot \boldsymbol{\eta} \, d\Gamma_{gm}(t) dt.\end{aligned}$$

Thus,

$$\int_0^T \mathbf{a}(t; \mathbf{v}, \boldsymbol{\eta})\, dt$$
$$= -\int_{Q_g} \nu_g \Delta \mathbf{v}_g \cdot \boldsymbol{\eta}\, dx - \int_{Q_m} \nu_\ell \Delta \mathbf{v} \cdot \boldsymbol{\eta}\, dx$$
$$+ \int_0^T \int_{\Gamma_{gm}(t)} 2\left(\nu_g \mathbf{D}(\mathbf{v}_g)\boldsymbol{\nu}^- + \nu_\ell \mathbf{D}(\mathbf{v})\boldsymbol{\nu}^+\right) \cdot \boldsymbol{\eta}\, d\Gamma_{gm}(t) dt.$$

This implies that if a function $\mathbf{v}$ satisfies the variational formulation (31) in which the Laplacian makes sense in $\mathbf{L}^2_\sigma(\Omega_m(t))$ and $\mathbf{L}^2_\sigma(\Omega_g(t))$, then $\mathbf{v}$ satisfies the slip type transmission condition (29) in the weak sense.

References

[1] M. Aso, T. Fukao and N. Kenmochi, A new class of doubly nonlinear evolution equations. *Taiwanese J. Math.* **8**, 103–124 (2004).

[2] J.R. Cannon, E. DiBenedetto and G.H. Knightly, The bidimensional Stefan problem with convection: The time dependent case. *Comm. Partial Differential Equations* **14**, 1549–1604 (1983).

[3] E. Casella and M. Giangi, An analytical and numerical study of the Stefan problem with convection by means of an enthalpy method. *Math. Methods Appl. Sci.* **24**, 623–639 (2001).

[4] A. Damlamian, Some results on the multi-phase Stefan problem. *Comm. Partial Differential Equations* **2**, 1017–1044 (1977).

[5] E. DiBenedetto and M. O'Leary, Three-dimensional conduction-convection problems with change of phase. *Arch. Ration. Mech. Anal.* **123**, 99–116 (1993).

[6] T. Fukao and N. Kenmochi, Stefan problems with convection governed by Navier-Stokes equation. *Adv. Math. Sci. Appl.* **15**, 29–48 (2005).

[7] T. Fukao and N. Kenmochi, Degenerate parabolic equations with convection in non-cylindrical domain. *Adv. Math. Sci. Appl.* **14**, 139–150 (2004).

[8] T. Fukao, N. Kenmochi and I. Pawłow, Stefan problems in non-cylindrical domains arising in Czochralski process of crystal growth. *Control Cybernet.* **32**, 201–221 (2003).

[9] T. Fukao, N. Kenmochi and I. Pawłow, Transmission-Stefan problems arising in Czochralski process of crystal growth. In: *Free boundary problems, Theory and applications* (Birkhäuser, Basel, 2003), p. 151.

[10] J. L. Lions, *Quelques méthodes de résolution des problèmes aux limites non linéaires.* Dunod Gauthier-Villars, Paris, 1968.

[11] I. Pawłow, Three-phase boundary Czochralski model. In: *Mathematical Aspects of Modeling Structure Formation Phenomena* (Gakkōtosho, Tokyo, 2002), p. 203.

[12] J.F. Rodrigues, Variational methods in the Stefan problem. In: *Phase Tran-*

sitions and Hysteresis (Montecatini Terme 1993), (Springer, Berlin, 1994), p. 147.

[13] J.F. Rodrigues and F. Yi, On a two-phase continuous casting Stefan problem with nonlinear flux. *European J. Appl. Math.* **1**, 259–278 (1990).

[14] V.N. Strarovoitov, On the Stefan problem with different phase densities. *Z. Angew. Math. Mech.* **80**, 103–111 (2000).

[15] R. Temam, *Navier-Stokes equations, Theory and numerical analysis, 3rd edition. Vol. 2.* North-Holland, Amsterdam-New York-Oxford, 1984.

[16] A. Visintin, *Models of phase transitions.* Birkhäuser, Boston, 1996.

MEMORY RELAXATION OF THE ONE-DIMENSIONAL CAHN-HILLIARD EQUATION

Stefania Gatti[1], Maurizio Grasselli[2],
Alain Miranville[3] and Vittorino Pata[4]

[1] *Dipartimento di Matematica, Università di Ferrara*
Via Machiavelli 35, I-44100 Ferrara, Italy
E-mail: s.gatti@economia.unife.it

[2] *Dipartimento di Matematica "F.Brioschi", Politecnico di Milano*
Via Bonardi 9, I-20133 Milano, Italy
E-mail: maugra@mate.polimi.it

[3] *Laboratoire de Mathématiques et Applications, Université de Poitiers*
Boulevard Marie et Pierre Curie - Téléport 2
F-86962 Chasseneuil Futuroscope Cedex, France
E-mail: miranv@math.univ-poitiers.fr

[4] *Dipartimento di Matematica "F.Brioschi", Politecnico di Milano*
Via Bonardi 9, I-20133 Milano, Italy
E-mail: pata@mate.polimi.it

We consider the memory relaxation of the one-dimensional Cahn-Hilliard equation, endowed with the no-flux boundary conditions. The resulting integrodifferential equation is characterized by a memory kernel, which is the rescaling of a given positive decreasing function. The Cahn-Hilliard equation is then viewed as the formal limit of the relaxed equation, when the scaling parameter (or relaxation time) ε tends to zero. In particular, if the memory kernel is the decreasing exponential, then the relaxed equation is equivalent to the standard hyperbolic relaxation. The main result of this paper is the existence of a family of robust exponential attractors for the one-parameter dissipative dynamical system generated by the relaxed equation. Such a family is stable with respect to the singular limit $\varepsilon \to 0$.

1. Introduction

1.1. *The model equation*

For $u = u(x,t) : (0,\ell) \times \mathbb{R} \to \mathbb{R}$, where ℓ is a fixed positive constant, we consider the initial and boundary value problem

$$\begin{cases} u_t - \displaystyle\int_0^\infty k_\varepsilon(s)\big(-u_{xx}(t-s) + \phi(u(t-s))\big)_{xx} ds = 0, & \text{in } (0,\ell) \times \mathbb{R}^+, \\ u_x(0,t) = u_x(\ell,t) = u_{xxx}(0,t) = u_{xxx}(\ell,t) = 0, & t \in \mathbb{R}, \\ u(x,t) = \psi(x,t), \quad x \in [0,\ell],\ t \leq 0. \end{cases} \tag{1}$$

Here, $\mathbb{R}^+ = (0,\infty)$, ϕ is a suitable nonlinear function (usually, the derivative of a double-well potential), ψ is a given datum satisfying the boundary conditions, and, for any *relaxation time* $\varepsilon \in (0,1]$,

$$k_\varepsilon(s) := \frac{1}{\varepsilon} k\Big(\frac{s}{\varepsilon}\Big)$$

is the rescaling of a sufficiently smooth, decreasing function $k : [0,\infty) \to [0,\infty)$, subject to the normalization conditions

$$\int_0^\infty k(s) ds = 1 \qquad \text{and} \qquad k(0) = 1.$$

Problem (1) is the *memory relaxation* of the well-known one-dimensional Cahn-Hilliard equation (see, e.g., [6] and references therein)

$$\begin{cases} u_t - \big(-u_{xx} + \phi(u)\big)_{xx} = 0, & \text{in } (0,\ell) \times \mathbb{R}^+, \\ u_x(0,t) = u_x(\ell,t) = u_{xxx}(0,t) = u_{xxx}(\ell,t) = 0, & t \geq 0, \\ u(x,0) = u_0(x), \quad x \in [0,\ell]. \end{cases} \tag{2}$$

With the particular choice $k(s) = e^{-s}$, one can (formally) check that, if u solves (1), then u is a solution of the so-called *hyperbolic relaxation* of the Cahn-Hilliard equation, namely,

$$\begin{cases} \varepsilon u_{tt} + u_t - \big(-u_{xx} + \phi(u)\big)_{xx} = 0, & \text{in } (0,\ell) \times \mathbb{R}^+, \\ u_x(0,t) = u_x(\ell,t) = u_{xxx}(0,t) = u_{xxx}(\ell,t) = 0, & t \geq 0, \\ u(x,0) = u_0(x), \quad x \in [0,\ell], \\ u_t(x,0) = u_1(x), \quad x \in [0,\ell], \end{cases} \tag{3}$$

with $u_0(x) := \psi(x,0)$ and

$$u_1(x) := \int_0^\infty k_\varepsilon(s)\big(-\psi_{xx}(x,-s) + \phi(\psi(x,-s))\big)_{xx} ds.$$

Conversely, suppose that u solves the above differential problem for some given initial data u_0 and u_1. Then, provided that u_1 has null average, it is possible to construct a solution of the memory relaxation problem (1), taking as past history a constant solution w to the equation

$$\big(- w_{xx} + \phi(w) \big)_{xx} = u_1,$$

endowed with no-flux boundary conditions, .

Problem (3) has recently been proposed in [7] as a possible model for rapid solidification processes (see also [8]), and it was first studied in the context of dynamical systems in [4]. Then, a deeper analysis was developed in [9] (see also [14, 15]).

Following a phenomenological approach, the hyperbolic relaxation of the Cahn-Hilliard equation can be deduced from the balance law

$$u_t + J_x = 0,$$

supposing that the diffusion flux J is delayed, namely,

$$J(t+\varepsilon) = -\big(- u_{xx} + \phi(u) \big)_x(t),$$

and taking $J + \varepsilon J_t$ as a first approximation of $J(t+\varepsilon)$. Hence, u_t must necessarily have null average. However, there are situations in which the memory effects cannot be modeled by just a single decreasing exponential kernel. Thus, more complicated integrodifferential equations, like the one of problem (1), need to be considered (see [12]).

1.2. *The singular limit*

The aim of this work is to analyze the long term behavior of (1) in the limit $\varepsilon \to 0$. In that case, k_ε converges, in the sense of distributions, to the Dirac mass at zero. Hence, the formal limit of (1) is the Cahn-Hilliard equation (2). In order to give a rigorous meaning to the aforementioned limit, we will show that both (1) and (2) generate dissipative dynamical systems, denoted by $S_\varepsilon(t)$, on suitable (and related) phase-spaces. Clearly, for $\varepsilon = 0$, we recover the limit problem (2). Then, we will demonstrate that, for every $\varepsilon \in [0,1]$, $S_\varepsilon(t)$ has "nice" asymptotic properties, guaranteed by the existence of the global attractor $\mathcal{A}_\varepsilon$ and of exponential attractors $\mathcal{E}_\varepsilon$ (see, e.g., [1, 11, 13] for more details on the subject). We recall that the global attractor is the smallest (for the inclusion) compact invariant set which attracts uniformly the bounded sets of initial data. On the other hand, an exponential attractor is a larger compact set, which is only semi-invariant, has finite fractal dimension and attracts exponentially fast the bounded sets of initial data (in contrast to the global attractor, which may attract the trajectories slowly). As a consequence, we can expect exponential attractors to be more robust under perturbations than global attractors.

Here, we will show that the family of the global attractors $\mathcal{A}_\varepsilon$ is upper semicontinuous at $\varepsilon = 0$, while the family of exponential attractors $\mathcal{E}_\varepsilon$ is continuous at $\varepsilon = 0$. These facts also ensure that the asymptotic dynamics of the two problems are close when ε is sufficiently small. Incidentally, we will also obtain the closeness of the trajectories of $S_\varepsilon(t)$ and $S_0(t)$ on finite-time intervals as $\varepsilon \to 0$.

The key tool for our scope is a recent result from [10]. In that paper, we considered the memory relaxation of an abstract first-order differential equation, and we provided a general theorem ensuring, under certain hypotheses, the convergence of the related attractors. Needless to say that such hypotheses are not automatically satisfied by every problem, and they have to be verified case by case, which, sometimes, might be far from being an obvious task. Indeed, the word "hypotheses" might be misleading, since these are actually results on the dynamical system that have an interest by themselves (e.g., the existence of absorbing sets, asymptotic compactness properties, and so on). Moreover, there is a further (often nontrivial) step to make, since, before checking the hypotheses, one has to formulate the problem in a suitable way, so that the abstract result is applicable.

In [10], we studied, as particular examples, the Allen-Cahn and the Cahn-Hilliard equations, the latter, though, endowed with the boundary conditions

$$u(0,t) = u(\ell,t) = u_{xx}(0,t) = u_{xx}(\ell,t) = 0, \qquad t \in \mathbb{R},$$

which are physically questionable for modeling phase separation processes, albeit technically much easier to handle. On the contrary, here we focus on the case of no-flux boundary conditions, which imply the mass conservation

$$\int_0^\ell u(x,t)dx = \int_0^\ell u(x,0)dx, \qquad \forall\, t \geq 0.$$

As we shall see, this feature brings to a careful formulation of the functional setting, which happens to be more delicate than the one used in the related application studied in [10].

Finally, it is worth to mention that, as soon as we obtain results for the memory relaxation (1), we automatically have the same results for the hyperbolic relaxation (3). This, again, follows from the abstract analysis developed in [10].

1.3. *The history space formulation*

We now translate problem (1) in the so-called "history space" setting (see [10] and references therein). First of all, we introduce the past history variable $\eta = \eta^t(x,s)$ by

$$\eta^t(x,s) := -\int_0^s \big(-u_{xx}(x,t-y) + \phi(u(x,t-y))\big)_{xx} dy,$$

for $t \geq 0$ and $s \in \mathbb{R}^+$. Notice that η fulfills the differential equation (all subscripts denote partial derivatives)

$$\eta_t + \eta_s = -\big(-u_{xx} + \phi(u)\big)_{xx}.$$

Then, setting

$$\mu_\varepsilon(s) := -k'_\varepsilon(s),$$

and integrating by parts in the convolution integral, we obtain the history space formulation of (1), namely,

$$\begin{cases} u_t + \displaystyle\int_0^\infty \mu_\varepsilon(s)\eta(s)ds = 0, & \text{in } (0,\ell)\times\mathbb{R}^+, \\ \eta_t + \eta_s = -\big(-u_{xx} + \phi(u)\big)_{xx}, & \text{in } (0,\ell)\times\mathbb{R}^+\times\mathbb{R}^+, \\ u_x(0,t) = u_x(\ell,t) = u_{xxx}(0,t) = u_{xxx}(\ell,t) = 0, & t\in\mathbb{R}, \\ \eta^t(x,0) = 0, \quad x\in[0,\ell],\ t\geq 0, \\ u(x,0) = u_0(x), \quad x\in[0,\ell], \\ \eta^0(x,s) = \eta_0(x,s), \quad x\in[0,\ell], s\in\mathbb{R}^+, \end{cases} \tag{4}$$

where $u_0(x) = \psi(x,0)$ and

$$\eta_0(x,s) = -\int_0^s \big(-\psi_{xx}(x,-y) + \phi(\psi(x,-y))\big)_{xx} dy.$$

Notice that the boundary conditions entail

$$\int_0^\ell \eta^t(x,s)dx = 0,$$

for any $t \geq 0$ and any $s \geq 0$.

1.4. *Assumptions on the memory kernel and on the nonlinearity*

Concerning the memory kernel, setting $\mu(s) = -k'(s)$, we assume μ nonnegative, $\mu \in W^{1,1}(\mathbb{R}^+)$, and

$$\mu'(s) + \delta\mu(s) \leq 0, \qquad \text{for a.e. } s\in\mathbb{R}^+, \tag{5}$$

for some $\delta > 0$. Due to the normalization conditions on k, we have

$$\int_0^\infty \mu_\varepsilon(s)ds = \frac{1}{\varepsilon} \qquad \text{and} \qquad \int_0^\infty s\mu_\varepsilon(s)ds = 1. \tag{6}$$

Regarding the nonlinearity ϕ, we will suppose $\phi \in C^3(\mathbb{R})$ and

$$\liminf_{|r|\to\infty} \phi'(r) > 0. \tag{7}$$

Clearly, this implies the existence of some $\theta > 0$ such that

$$\phi'(r) \geq -\theta, \qquad \forall r \in \mathbb{R}. \tag{8}$$

2. The dynamical system

In this section, we introduce a suitable functional setting in order to prove the generation of a strongly continuous semigroup, or dynamical system, on a proper phase-space.

Let $\mathbb{H} = L^2(0,\ell)$ be endowed with the usual scalar product $\langle\cdot,\cdot\rangle$ and the induced norm $\|\cdot\|$. We consider the linear nonnegative self-adjoint operator A on $\mathbb{H}$ defined by

$$A := \frac{d^4}{dx^4},$$

with

$$\mathcal{D}(A) := \big\{u \in H^4(0,\ell) \,:\, u_x(0) = u_x(\ell) = u_{xxx}(0) = u_{xxx}(\ell) = 0\big\}.$$

Observe that

$$A^{1/2} = -\frac{d^2}{dx^2}, \qquad \mathcal{D}(A^{1/2}) = \big\{u \in H^2(0,\ell) \,:\, u_x(0) = u_x(\ell) = 0\big\}.$$

For any $u \in \mathbb{H}$, we set

$$\bar{u} := \frac{1}{\ell}\langle u, 1\rangle.$$

Then, we introduce the closed subspace of $\mathbb{H}$

$$\mathbb{H}_0 := \big\{u \in \mathbb{H} \,:\, \bar{u} = 0\big\},$$

and, correspondingly, the projector $P : \mathbb{H} \to \mathbb{H}_0$ given by

$$Pu := u - \bar{u}.$$

It is not difficult to realize that

$$A = PA \qquad \text{and} \qquad A^{1/2} = PA^{1/2}.$$

Define then the self-adjoint strictly positive operator on $\mathbb{H}_0$

$$A_0 := A|_{\mathbb{H}_0},$$

and set

$$W^r := \mathcal{D}(A_0^{(1+2r)/4}), \qquad r \in \mathbb{R},$$

endowed with the scalar product

$$\langle w_1, w_2 \rangle_{W^r} := \langle A_0^{(1+2r)/4} w_1, A_0^{(1+2r)/4} w_2 \rangle.$$

Note that $W^{-1/2} = \mathbb{H}_0$. Besides, for any fixed $\alpha \in \mathbb{R}$, there holds

$$\langle A_0^\alpha w, A^{1/2} v \rangle = \langle A_0^{(2\alpha+1)/2} w, v \rangle, \quad \forall\, w \in W^{(4\alpha+1)/2},\ \forall\, v \in \mathcal{D}(A^{1/2}). \quad (9)$$

We denote by $\mathbb{K}$ the subspace of $\mathbb{H}$ consisting of all the constant functions. Then, for any $r \in [-\frac{1}{2}, \infty)$, we set

$$V^r := W^r \oplus \mathbb{K},$$

equipped with the scalar product

$$\langle v_1, v_2 \rangle_{V^r} = \langle P v_1, P v_2 \rangle_{W^r} + \langle \bar{v}_1, \bar{v}_2 \rangle.$$

Clearly, $V^{-1/2} = \mathbb{H}$. We also need to use the weighted L^2-space

$$\mathcal{M}_\varepsilon^r := L^2_{\mu_\varepsilon}(\mathbb{R}^+; W^{r-1}),$$

endowed with the scalar product

$$\langle \eta_1, \eta_2 \rangle_{\mathcal{M}_\varepsilon^r} := \int_0^\infty \mu_\varepsilon(s) \langle \eta_1(s), \eta_2(s) \rangle_{W^{r-1}} ds.$$

On account of the mass conservation, we introduce, for each $r \geq -\frac{1}{2}$ and any fixed $\gamma \geq 0$, the closed subsets of the Hilbert space V^r

$$V_\gamma^r := \{ v \in V^r \,:\, |\bar{v}| \leq \gamma \}.$$

Moreover, for $\varepsilon \in [0,1]$ and $r \geq -\frac{1}{2}$, we define the Banach spaces

$$\mathcal{H}_\varepsilon^r := \begin{cases} V^r \times \mathcal{M}_\varepsilon^r, & \text{if } \varepsilon > 0, \\ V^r, & \text{if } \varepsilon = 0, \end{cases}$$

and their related closed subsets

$$\mathcal{H}_{\varepsilon,\gamma}^r := \begin{cases} V_\gamma^r \times \mathcal{M}_\varepsilon^r, & \text{if } \varepsilon > 0, \\ V_\gamma^r, & \text{if } \varepsilon = 0. \end{cases}$$

Observe that $\mathcal{H}_{\varepsilon,\gamma}^r$ is a complete metric space, whose metric is induced by the norm in $\mathcal{H}_\varepsilon^r$. When $\varepsilon = 0$, we agree to interpret the pair (u, η) just as

u. Accordingly, the norm reduces to the first component only. We shall also make use of the *lifting map* $\mathbb{L}_\varepsilon : \mathcal{H}^0_{0,\gamma} \to \mathcal{H}^0_{\varepsilon,\gamma}$, and of the *projection maps* $\mathbb{P}$ and $\mathbb{Q}_\varepsilon$ on $\mathcal{H}^0_{\varepsilon,\gamma}$, given by

$$\mathbb{L}_\varepsilon u := (u, 0), \qquad \mathbb{P}(u,\eta) := u, \qquad \mathbb{Q}_\varepsilon(u,\eta) := \eta.$$

To write down the problem in a convenient abstract form, we introduce the nonlinear operator $B : V^0 \to W^{-1}$

$$B[v] := A^{1/2}\phi(v), \qquad \forall v \in V^0.$$

Note that, for every $R \geq 0$, there exists $C = C(R) \geq 0$ such that

$$\sup_{\|u_i\|_{V^1} \leq R} \|B[u_1] - B[u_2]\|_{W^{-1}} \leq C\|u_1 - u_2\|_{V^0}, \tag{10}$$

$$\sup_{\|u\|_{V^1} \leq R} \|B[u]\|_{W^0} \leq C. \tag{11}$$

We also need to define the linear operator $T_\varepsilon = -\partial_s$ with domain

$$\mathcal{D}(T_\varepsilon) := \{\eta \in \mathcal{M}^0_\varepsilon : \eta_s \in \mathcal{M}^0_\varepsilon,\, \eta(0) = 0\}.$$

Then, we can introduce the abstract formulation of (4) in the history space setting

$$\begin{cases} u_t + \int_0^\infty \mu_\varepsilon(s)\eta(s)ds = 0, \\ \eta_t = T_\varepsilon \eta + A_0 P u + B[u], \\ u(0) = u_0, \\ \eta^0 = \eta_0. \end{cases} \tag{12}$$

Remark 1: According to our notation, (12) also includes the limiting case, corresponding to $\varepsilon = 0$, provided that we properly interpret the term $\int_0^\infty \mu_\varepsilon(s)\eta(s)ds$ as $A_0 P u + B[u]$.

Given $z_0 = (u_0, \eta_0) \in \mathcal{H}^0_{\varepsilon,\gamma}$ ($z_0 = u_0$ for $\varepsilon = 0$), let us denote by

$$S_\varepsilon(t) z_0 = \begin{cases} (u(t), \eta^t), & \text{if } \varepsilon > 0, \\ u(t), & \text{if } \varepsilon = 0 \end{cases}$$

the solution at time t of problem (12), with initial datum z_0.

Let $\gamma \geq 0$ be fixed. Thanks to the assumptions on μ and ϕ, there holds

Proposition 2: *For any $\varepsilon \in [0,1]$ $S_\varepsilon(t)$ is a strongly continuous semigroup on $\mathcal{H}^0_{\varepsilon,\gamma}$. For any $R \geq 0$, there exists $K = K(R) \geq 0$ such that, for any two initial data $z_1, z_2 \in \mathcal{H}^0_{\varepsilon,\gamma}$ with $\|z_i\|_{\mathcal{H}^0_\varepsilon} \leq R$, there holds*

$$\|S_\varepsilon(t)z_1 - S_\varepsilon(t)z_2\|_{\mathcal{H}^0_\varepsilon} \leq K e^{Kt} \|z_1 - z_2\|_{\mathcal{H}^0_\varepsilon}.$$

The proof of Proposition 2 parallels the corresponding one for the hyperbolic relaxation of the Cahn-Hilliard equation, suitably adapted to the past history setting (cf. [9]). However, comparing the formulation that we have just introduced with the one used in Sec. 7 of [10], we realize that new technical difficulties, related with the fact that the spatial average of u is now conserved, may arise when dealing with the longterm behavior.

Remark 3: Observe that, if $\eta \in \mathcal{M}_\varepsilon^r$, then, thanks to (6), the first equation of (12) yields the inequality

$$\|u_t\|^2_{W^{r-1}} \leq \frac{1}{\varepsilon}\|\eta\|^2_{\mathcal{M}_\varepsilon^r}. \tag{13}$$

3. The main result

We now state the main result of this paper, whose proof is postponed to the next section.

Theorem 4: For every $\varepsilon \in [0,1]$, there exists a set $\mathcal{E}_\varepsilon \subset \mathcal{H}^0_{\varepsilon,\gamma}$ (called exponential attractor), compact in $\mathcal{H}^0_\varepsilon$ and bounded in $\mathcal{H}^1_\varepsilon$, which satisfies the following conditions.

(i) The set $\mathcal{E}_\varepsilon$ is positively invariant for $S_\varepsilon(t)$, that is, $S_\varepsilon(t)\mathcal{E}_\varepsilon \subset \mathcal{E}_\varepsilon$ for every $t \geq 0$.

(ii) There exist $\omega > 0$ and a positive, increasing function M such that, for every $R \geq 0$, there holds

$$\text{dist}_{\mathcal{H}^0_\varepsilon}(S_\varepsilon(t)\mathcal{B}^0_\varepsilon(R), \mathcal{E}_\varepsilon) \leq M(R)e^{-\omega t}.$$

(iii) The fractal dimension of $\mathcal{E}_\varepsilon$ in $\mathcal{H}^0_\varepsilon$ is uniformly bounded with respect to ε.

(iv) There exist $c \geq 0$ and $\tau \in (0, \frac{1}{16}]$ such that

$$\text{dist}^{\text{sym}}_{\mathcal{H}^0_\varepsilon}(\mathcal{E}_\varepsilon, \mathbb{L}_\varepsilon\mathcal{E}_0) \leq c\varepsilon^\tau.$$

Both c and τ can be explicitly calculated.

With the usual notation, dist and dist^{sym} denote the Hausdorff semidistance and the symmetric distance, respectively. Observe that (iv) can be equivalently written as

$$\text{dist}^{\text{sym}}_{V^0}(\mathbb{P}\mathcal{E}_\varepsilon, \mathcal{E}_0) + \sup_{z\in\mathcal{E}_\varepsilon} \|\mathbb{Q}_\varepsilon z\|_{\mathcal{M}^0_\varepsilon} \leq c\varepsilon^\tau.$$

This means that the projection of the perturbed attractor is close to the attractor of the limit problem, while the norm of the history component squeezes to zero as $\varepsilon \to 0$, in an explicitly controlled way.

Here are some noteworthy consequences of Theorem 4 (cf. [10]).

Corollary 5: *The semigroup $S_\varepsilon(t)$ acting on the phase-space $\mathcal{H}^0_{\varepsilon,\gamma}$ possesses a connected global attractor $\mathcal{A}_\varepsilon \subset \mathcal{E}_\varepsilon$. In particular, the fractal dimension of $\mathcal{A}_\varepsilon$ in $\mathcal{H}^0_\varepsilon$ is uniformly bounded with respect to ε.*

Corollary 6: *The semigroup $S_\varepsilon(t)$ uniquely extends to a strongly continuous group of operators $\{\widetilde{S_\varepsilon}(t)\}_{t\in\mathbb{R}}$ on $\mathcal{A}_\varepsilon$. Moreover, the family $\mathcal{A}_\varepsilon$ is upper semicontinuous at $\varepsilon = 0$, that is,*

$$\lim_{\varepsilon\to 0}\Big[\mathrm{dist}_{\mathcal{H}^0_\varepsilon}(\mathcal{A}_\varepsilon, \mathbb{L}_\varepsilon \mathcal{A}_0)\Big] = 0.$$

Corollary 7: *For every given $T > 0$ and $R \geq 0$, there exist $K_T \geq 0$ and a positive constant C_R such that*

$$\|S_\varepsilon(t)z - \mathbb{L}_\varepsilon S_0(t)\mathbb{P}z\|_{\mathcal{H}^0_\varepsilon} \leq \|\mathbb{Q}_\varepsilon z\|_{\mathcal{M}^0_\varepsilon} e^{-\frac{\delta t}{4\varepsilon}} + K_T C_R \sqrt[16]{\varepsilon},$$

for every $t \in [0,T]$ and every $z \in \mathcal{H}^1_{\varepsilon,\gamma}$ with $\|z\|_{\mathcal{H}^1_\varepsilon} \leq R$.

4. Sketch of the proof of Theorem 4

On account of (10)-(11) and Proposition 2, Theorem 4 is a consequence of the following suitable adaptation of Theorem 4.2 in [10].

Lemma 8: *The thesis of Theorem 4 follows provided that the following assumptions hold.*

(H1) *For $i = 0, 1$, there exists $R_i > 0$ such that $\mathcal{B}^i_\varepsilon(R_i)$ is an absorbing set for $S_\varepsilon(t)$ on $\mathcal{H}^i_{\varepsilon,\gamma}$, uniformly with respect to ε. Namely, given any $R \geq 0$, there exists a time $t_i \geq 0$, depending only on R, such that*

$$S_\varepsilon(t)\mathcal{B}^i_\varepsilon(R) \subset \mathcal{B}^i_\varepsilon(R_i), \qquad \forall t \geq t_i.$$

Moreover, for every $R \geq 0$ there exists $C_i = C_i(R) \geq 0$ such that, for any $z \in \mathcal{H}^i_{\varepsilon,\gamma}$,

$$\sup_{\|z\|_{\mathcal{H}^i_\varepsilon} \leq R} \|S_\varepsilon(t)z\|_{\mathcal{H}^i_\varepsilon} \leq C_i, \qquad \forall t \geq 0.$$

(H2) *There exists $R_2 > 0$ such that $\mathcal{B}^1_\varepsilon(R_2)$ satisfies*

$$\mathrm{dist}_{\mathcal{H}^0_\varepsilon}(S_\varepsilon(t)\mathcal{B}^0_\varepsilon(R_0), \mathcal{B}^1_\varepsilon(R_2)) \leq Me^{-\kappa t},$$

for some $M \geq 0$ and $\kappa > 0$.

(H3) *Given any $R > 0$, there exist $\lambda, \Lambda : [0, \infty) \to \mathbb{R}^+$ (possibly depending on R), with $\lambda(t) < \frac{1}{2}$ for t large enough, such that, for every $z_1, z_2 \in \mathcal{B}^1_\varepsilon(R)$, the map $S_\varepsilon(t)$ admits the decomposition*

$$S_\varepsilon(t)z_1 - S_\varepsilon(t)z_2 = L_\varepsilon(t)(z_1, z_2) + N_\varepsilon(t)(z_1, z_2),$$

for some $L_\varepsilon(t)$ and $N_\varepsilon(t)$ satisfying

$$\|L_\varepsilon(t)(z_1, z_2)\|_{\mathcal{H}^0_\varepsilon} \leq \lambda(t)\|z_1 - z_2\|_{\mathcal{H}^0_\varepsilon},$$

$$\|N_\varepsilon(t)(z_1, z_2)\|_{\mathcal{H}^{1/2}_\varepsilon} \leq \Lambda(t)\|z_1 - z_2\|_{\mathcal{H}^0_\varepsilon}.$$

Besides, $\tilde{\eta}^t = \mathbb{Q}_\varepsilon N_\varepsilon(t)(z_1, z_2)$ solves the Cauchy problem

$$\begin{cases} \tilde{\eta}_t = T_\varepsilon \tilde{\eta} + \tilde{g}, & t > 0, \\ \tilde{\eta}^0 = 0, \end{cases}$$

for some $\tilde{g}$ satisfying, for all $T > 0$,

$$\|\tilde{g}(t)\|_{W^{-3/2}} \leq \Lambda(T)\|z_1 - z_2\|_{\mathcal{H}^0_\varepsilon}, \qquad \forall t \in (0, T).$$

(H4) *For every $R \geq 0$, there exists $C = C(R) \geq 0$ such that, for all $T > 0$,*

$$\sup_{\|u\|_{V^1} \leq R} \int_0^T \|\partial_t B[S_0(y)u]\|^2_{W^{-2}}\, dy \leq C(1 + T).$$

Therefore, the proof of this theorem consists in showing that conditions (H1)-(H3) hold true, since (H4) easily follows from (H1) for $i = 1$ and the limiting equation, i.e., the standard Cahn-Hilliard equation. But, indeed, once the formulation (12) is obtained (and this is actually the distinctive feature of the present note), we can proceed along the same lines as in [10] to verify (H1)-(H3), with some minor changes due to the fact that the average of u is now conserved. For the reader's convenience, we write in detail the proof of (H1) for $i = 0$ (that is, the existence of an absorbing set), which was omitted in [10], referring to [10] for the remaining steps.

We introduce the energy functional

$$E(u, \eta) = \|Pu\|^2_{W^0} + \|\eta\|^2_{\mathcal{M}^0_\varepsilon} - 2\alpha\varepsilon\langle Pu, \eta\rangle_{\mathcal{M}^0_\varepsilon} + 2\langle\Phi(u), 1\rangle,$$

for any $z = (u, \eta) \in \mathcal{H}^0_{\varepsilon,\gamma}$. Here, $\Phi(r) = \int_0^r \phi(\rho)d\rho$, $\alpha > 0$ is a parameter to be chosen small enough so that the ensuing estimates are fulfilled. Owing to (6)-(7), there exists $\nu \in (0,1)$ such that

$$2\langle Pu, \phi(u) - \phi(\bar{u})\rangle \geq -2(1-\nu)\|Pu\|^2_{W^0} - c,$$
$$2\langle \Phi(u), 1\rangle \geq -(1-\nu)\|Pu\|^2_{W^0} - c,$$
$$2\alpha\varepsilon|\langle Pu, \eta\rangle_{\mathcal{M}^0_\varepsilon}| \leq \frac{\nu}{2}\|Pu\|^2_{W^0} + \frac{1}{2}\|\eta\|^2_{\mathcal{M}^0_\varepsilon}.$$

Here and in the sequel, $c \geq 0$ denotes a generic constant independent of ε. Then, from the embedding $W^0 \hookrightarrow L^\infty(0,\ell)$ we conclude that

$$\frac{\nu}{2}\|z\|^2_{\mathcal{H}^0_\varepsilon} - c \leq E(z) \qquad \text{and} \qquad E(z) \leq M(\|z\|_{\mathcal{H}^0_\varepsilon}), \tag{14}$$

where M is a nondecreasing and nonnegative continuous function. We now fix $z \in \mathcal{H}^0_{\varepsilon,\gamma}$ with $\|z\|_{\mathcal{H}^0_\varepsilon} \leq R$, and we denote $(u(t), \eta^t) = S_\varepsilon(t)z$. Multiplying the first equation of (12) by $A_0Pu + A^{1/2}\phi(u)$ in W^{-1} and by $-\alpha\varepsilon\eta$ in $\mathcal{M}^0_\varepsilon$, and the second one by η in $\mathcal{M}^0_\varepsilon$, we obtain that $E = E(S_\varepsilon(t)z)$ satisfies

$$\begin{aligned}
&\frac{dE}{dt} + 2\alpha\|Pu\|^2_{W^0} - \int_0^\infty \mu_\varepsilon'(s)\|\eta(s)\|^2_{W^{-1}}ds + 2\alpha\langle Pu, \phi(u) - \phi(\bar{u})\rangle \\
&= -2\alpha\langle Pu, \phi(\bar{u})\rangle + 2\alpha\varepsilon\left\|\int_0^\infty \mu_\varepsilon(s)\eta(s)ds\right\|^2_{W^{-1}} \\
&\quad - 2\alpha\varepsilon\int_0^\infty \mu_\varepsilon'(s)\langle Pu, \eta(s)\rangle_{W^{-1}}ds.
\end{aligned}$$

The terms on the right-hand side can be controlled by

$$\begin{aligned}
&- 2\alpha\langle Pu, \phi(\bar{u})\rangle + 2\alpha\varepsilon\left\|\int_0^\infty \mu_\varepsilon(s)\eta(s)ds\right\|^2_{W^{-1}} \\
&\leq \frac{\alpha\nu}{2}\|Pu\|^2_{W^0} + \frac{\delta}{2\varepsilon}\|\eta\|^2_{\mathcal{M}^0_\varepsilon} + \alpha c,
\end{aligned}$$

and

$$\begin{aligned}
&- 2\alpha\varepsilon\int_0^\infty \mu_\varepsilon'(s)\langle Pu, \eta(s)\rangle_{W^{-1}}ds \\
&\leq \frac{\alpha\nu}{2}\|Pu\|^2_{W^0} + \frac{1}{2}\int_0^\infty \mu_\varepsilon'(s)\|\eta(s)\|^2_{W^{-1}}ds.
\end{aligned}$$

Owing to (5) and the above inequalities, we obtain

$$\frac{d}{dt}E(S_\varepsilon(t)z) + \alpha\nu\|S_\varepsilon(t)z\|^2_{\mathcal{H}^0_\varepsilon} \leq c.$$

In view of (14), from Lemma 2.7 of [2] we deduce that there exists $t_0 = t_0(R)$ such that

$$E(S_\varepsilon(t)z) \leq \sup_{\zeta \in \mathcal{H}^0_{\varepsilon,\gamma}} \{E(\zeta) : \|\zeta\|^2_{\mathcal{H}^0_\varepsilon} \leq c\}, \qquad \forall t \geq t_0.$$

Using again (14), we get the existence of an absorbing set $\mathcal{B}_0(R_0)$ in $\mathcal{H}^0_{\varepsilon,\gamma}$ for $S_\varepsilon(t)$.

Acknowledgments

This work was partially supported by the Italian MIUR PRIN Research Project *Aspetti Teorici e Applicativi di Equazioni a Derivate Parziali* and by the Italian MIUR FIRB Research Project *Analisi di Equazioni a Derivate Parziali, Lineari e Non Lineari: Aspetti Metodologici, Modellistica, Applicazioni.*

References

[1] A.V. Babin and M.I. Vishik, *Attractors of evolution equations.* North-Holland, Amsterdam, 1992.

[2] V. Belleri and V. Pata, Attractors for semilinear strongly damped wave equations on $\mathbb{R}^3$. *Discrete Contin. Dynam. Systems* **7**, 719–735 (2001).

[3] M. Conti, V. Pata and M. Squassina, Singular limit of differential systems with memory. *Indiana Univ. Math. J.* (to appear).

[4] A. Debussche, A singular perturbation of the Cahn-Hilliard equation. *Asymptotic Anal.* **4**, 161–185 (1991).

[5] P. Fabrie, C. Galusinski, A. Miranville and S. Zelik, Uniform exponential attractors for a singularly perturbed damped wave equation. *Discrete Contin. Dynam. Systems* **10**, 211–238 (2004).

[6] P.C. Fife, Models for phase separation and their mathematics. *Electron. J. Differential Equations* **48**, 1–26 (2000).

[7] P. Galenko, Phase-field models with relaxation of the diffusion flux in nonequilibrium solidification of a binary system. *Phys. Lett. A* **287**, 190–197 (2001).

[8] P. Galenko and D. Jou, Diffuse-interface model for rapid phase transformations in nonequilibrium systems. *Phys. Rev. E* (to appear).

[9] S. Gatti, M. Grasselli, A. Miranville and V. Pata, On the hyperbolic relaxation of the one-dimensional Cahn-Hilliard equation. *J. Math. Anal. Appl.* (to appear).

[10] S. Gatti, M. Grasselli, A. Miranville and V. Pata, Memory relaxation of first order evolution equations. *Nonlinearity* (to appear).

[11] J. Hale, *Asymptotic behavior of dissipative systems.* Amer. Math. Soc., Providence, 1988.

[12] J. Jäckle, Models of glass transition. *Rep. Prog. Phys.* **49**, 171–231 (1986).

[13] R. Temam, *Infinite-dimensional dynamical systems in mechanics and physics*. Springer, New York, 1988.

[14] S. Zheng and A. Milani, Global Attractors for singular perturbations of the Cahn-Hilliard equations. *J. Differential Equations* **209**, 101–139 (2005).

[15] S. Zheng and A. Milani, Exponential attractors and inertial manifolds for singular perturbations of the Cahn-Hilliard equations. *Nonlinear Anal.* **57**, 843–877 (2004).

MATHEMATICAL MODELS FOR PHASE TRANSITION IN MATERIALS WITH THERMAL MEMORY

Giorgio Gentili[1] and Claudio Giorgi[2]

[1] *Deceased on December 2, 2000*

[2] *Dipartimento di Matematica*
Università di Brescia
Via Valotti 9, I-25133 Brescia, Italy
E-mail: claudio.giorgi@ing.unibs.it

Thermally induced phase transitions in materials with thermal memory are modelled by means of an order parameter, or phase–field, that changes smoothly in a fixed region. In this paper, two thermodynamic approaches are developed by regarding the phase–field evolution equation as a supplementary balance law. In the first one, the kinetic equation is regarded as the balance of some microscopic or accretive forces, and the first law of thermodynamics is modified taking into account their mechanical power. The second approach takes any mechanical meaning off the phase–field evolution, and modifies the second law of thermodynamics through some entropy extra–flux. Restrictions due to thermodynamics are derived and a general form of the kinetic equation is obtained in both cases. A simplified model ruling the system evolution around the transition temperature is derived according to the first approach.

1. Introduction

Many efforts have been devoted to formulate a *"thermodynamically consistent"* non–isothermal phase–field model, especially in connection with thermally–induced reversible phase transitions. In particular, we mention the works by Penrose & Fife [17, 18] Alt & Pawlow [3] Fried & Gurtin [10, 14] and Frémond [9] For a critical survey see also [5]. All of them look for a modelling approach consistent with the second law of thermodynamics for non–isothermal processes and, at the same time, such that the *standard model* (see Caginalp [6] for instance) is recovered after linearization with respect to the temperature around its transition value.

Mainly, Penrose & Fife [17, 18] consider a scheme for the non–conserved phase change completely based on relaxation laws with entropy or free energy as potential. Consistency with thermodynamics is meant as the requirement that the entropy (free energy) functional cannot decrease (increase) along solution paths. Alt & Pawlow [3] proposed a similar approach to model the conserved phase separation.

On the contrary, both Fried & Gurtin and Frémond models are framed within modern continuum thermodynamics. Fried & Gurtin [10] make a systematic use of general balance laws, and exploit the key idea that the phase evolution is governed by the balance of *accretive forces*, which expend mechanical power. Compatibility with thermodynamics is imposed through the classical entropy inequality and exploiting its consequences. Frémond, cf. [9], develops a similar approach by having recourse to a principle of virtual power and including the power of *interior microforces.* Instead of imposing the entropy inequality, compatibility with thermodynamics is satisfied by introducing a pseudo–potential of dissipation.

The crucial point that distinguishes the latter approach from the former one is the different expression they take into account for the first law of thermodynamics. Assuming, as usual, a negligible contribution of the macroscopic stress power, in the papers by Caginalp and Penrose & Fife the energy balance equation only involves the heat flux, namely

$$\rho\, \dot{e} = -\nabla \cdot \boldsymbol{q} + \rho\, r\,. \tag{1}$$

On the contrary, in the theories proposed by Fried & Gurtin and Frémond the energy balance involves an *extra energy flux* $\boldsymbol{h}$, which accounts for the power of the *accretive stresses* (Fried & Gurtin), or the *microscopic forces* (Frémond), namely

$$\rho\, \dot{e} = -\nabla \cdot (\boldsymbol{q} - \boldsymbol{h}) + \rho\, r\,, \tag{2}$$

According to [10], this suggestion traces back to S.-K.Chan. After linearization with respect to the temperature in a neighborhood of its transition value, the classical form of the energy balance is recovered (see 6).

The first aim of the present paper is to compare the Fried & Gurtin's approach with a different one, based on the introduction of an *extra entropy flux* $\boldsymbol{k}$, due to the nonlocal character of the phase–field evolution, into the Clausius–Duhem entropy inequality

$$\rho\, \dot{\eta} + \nabla \cdot \left(\frac{\boldsymbol{q}}{\theta} + \boldsymbol{k} \right) - \frac{\rho\, r}{\theta} \geq 0\,. \tag{3}$$

Essentially, $\boldsymbol{k}$ accounts for the entropy production of the phase change process. It is worth noting that, by virtue of the external boundary condition $\boldsymbol{k} \cdot \boldsymbol{n} = 0$, the entropy inequality takes its classical integral form on the whole body. On the other hand, the energy balance law is unaffected by *extra energy fluxes* and takes the local form (1). Combining (1) and (3) we derive the *dissipation inequality*

$$-\rho(\dot{\psi} + \eta\,\dot{\theta}) - \frac{1}{\theta}\boldsymbol{q} \cdot \nabla\theta + \theta\nabla \cdot \boldsymbol{k} \geq 0\,, \tag{4}$$

which differs from the one obtained by Fried & Gurtin in that the term $\theta\nabla \cdot \boldsymbol{k}$ replaces $\nabla \cdot \boldsymbol{h}$. As in the Fried & Gurtin's approach, the evolution law for the phase–field (the so called *kinetic* or *Ginzburg–Landau equation*) is introduced by assuming a *supplementary balance law*, but in a dynamical form ($\nu \neq 0$)

$$\nu\dot{\varphi} = \nabla \cdot \boldsymbol{\xi} - \pi\,,$$

and (4) is exploited to restrict the choice of the constitutive functions ψ, η, $\boldsymbol{q}$, $\boldsymbol{k}$, $\boldsymbol{\xi}$, π, ν, etc.

An alternative model based on the same procedure, but treating the phase field as an internal variable, is developed in [8]. The authors prove that the entropy extra–flux approach is able to account for both conserved and non conserved phase–field, first and second order phase changes. In addition, the models proposed by Penrose & Fife and Alt & Pawlow are recovered as special cases (see Sect.5 of [8]). The results in [8], as well as the present paper, can be viewed as a contribution to give a rigorous statement of their theories in the framework of continuum thermodynamics.

The second aim of this work is to construct a general theory for phase transitions of the first and second order in materials with memory. This can be done using the previous analysis as a basis and allowing all constitutive functionals to depend on the so called "summed past histories" of temperature and temperature gradient, respectively. It is worth noting that a lot of papers were devoted to study well–posedness and long–time behaviour for phase–field systems with memory, see e.g. [1, 2, 4, 13, 12]. In spite of this, a thermodynamic consistent model for these materials is not at hand.

A general theory for temperature–induced phase transitions in materials with thermal memory provides a proper starting point to model different materials. An example is a fluid at rest whose liquid–solid first–order transition occurs at a very low temperature (liquid helium, for instance). If this is the case, a finite speed of propagation for thermal waves is observed (the so called *second sound* effect) and the heat conduction Fourier law

$$\boldsymbol{q} = -k\nabla\theta \tag{5}$$

fails to be valid. In particular, this phenomenon can be accounted for by allowing the heat flux vector $\boldsymbol{q}$ to depend on the past history of $\nabla\theta$, see [15, 7]. Another example is the second–order glass transition in liquids of high viscosity (glycerol, for instance). In fact, as pointed out by Jackle [16], liquids of this kind exhibit both thermal and mechanical memory effects at temperatures near the glass transition value. However, a proper description of this class of materials needs a model in which deformations and mass transfer are allowed.

This paper is organized as follows. In Sec. 2, the basic notation and general assumptions are introduced. Starting from the Fried & Gurtin's approach, in which the energy balance is modified as in (2), in Secs. 3 and 4 we develop the full phase–field model with thermal memory. Sec. 5 is devoted to obtain a different model by means of the modified inequality (3), in which an entropy extra–flux appears. Finally, in 6 we adapt an approximation procedure due to Fried & Gurtin [10], in order to write down a set of simpler equations ruling the system evolution around the transition temperature.

2. Notations and basic assumptions

In a bounded region $\mathcal{B}$ we consider a continuous body which consists of a two-phases material (liquid–solid, for instance), and we assume that its phase change depends on temperature. Let $\theta(\boldsymbol{x}, t)$ be the *absolute temperature* and $\varphi(\boldsymbol{x}, t)$ the *order parameter*, describing the phase of the material in a point $\boldsymbol{x} \in \mathcal{B}$ at time t.

As usual, we assume $\varphi(\boldsymbol{x}, t) = -1$ when the material element in $\boldsymbol{x}$ is in the (more disordered) liquid phase, whereas $\varphi(\boldsymbol{x}, t) = 1$ when the material element in $\boldsymbol{x}$ is in the (more ordered) solid phase. Intermediate values of the order parameter, $-1 < \varphi(\boldsymbol{x}, t) < 1$, represent situations in which the material element in $\boldsymbol{x}$ is composed of a phase mixture. Let θ_c be the *transition temperature*, so that in stationary conditions we have the liquid phase $\varphi = -1$ over the transition temperature, $\theta > \theta_c$, and the solid one $\varphi = 1$ below it, $\theta < \theta_c$.

According to the phase–field description, two pure phases ($\varphi = +1, -1$) are not separated by a sharp interface, but there exists a bulk thin region separating them where $|\varphi| < 1$, the so–called *mushy region*. For the sake of simplicity, at any fixed time t this *interfacial region* is viewed as the union of an uncountable family of smooth surfaces Σ_c, namely

$$\Sigma_c = \{\boldsymbol{x} : \varphi(\boldsymbol{x}, t) = c\} , \qquad |c| < 1 \tag{6}$$

each of them moving with a velocity

$$\boldsymbol{v}_\varphi = h\boldsymbol{n}_\varphi \qquad \text{where} \quad h = \dot{\varphi} \quad \text{and} \quad \boldsymbol{n}_\varphi = \frac{\boldsymbol{p}}{|\boldsymbol{p}|}, \ \boldsymbol{p} = \nabla\varphi ,$$

where the superposed dot denotes derivation with respect to time along the trajectory of motion. It is worth noting that

Remark 1: If the material body $\mathcal{B}$ can undergo deformations, then

$$h = \partial_t\varphi + \boldsymbol{v} \cdot \boldsymbol{p} \quad \text{and} \quad \dot{\boldsymbol{p}} = \nabla h - \boldsymbol{L}^T\boldsymbol{p} .$$

Otherwise, in a rigid body $h = \partial_t\varphi$ and $\dot{\boldsymbol{p}} = \nabla h$.

The quantity $\boldsymbol{n}_\varphi = \boldsymbol{p}/|\boldsymbol{p}|$ locally denotes the orientation of the surfaces (6), whereas $|\nabla\varphi|$ is locally related to their closeness and globally provides a measure of the thickness of the interfacial zone. On the other hand, $h = \dot{\varphi}$ and $\boldsymbol{f} = \dot{\boldsymbol{p}}$ describe the variation of these features of the interfacial zone in time. In fact, the normal velocity is given by $\boldsymbol{v}_\varphi = h\boldsymbol{n}_\varphi$, whereas orientation and thickness changes are accounted for by $\boldsymbol{f} = \dot{\boldsymbol{p}} = |\nabla\varphi|\dot{\boldsymbol{n}}_\varphi + \frac{d}{dt}|\nabla\varphi|\boldsymbol{n}_\varphi$.

For later convenience, we take the following

Assumption 2: Mass diffusion and deformation are negligible if compared with phase transition and heat conduction effects. As a consequence, the body looks like a rigid one, and at each point $\boldsymbol{x}$ of the interfacial surface Σ_c the rate of change of the phase is proportional to the normal velocity $\boldsymbol{v}_\varphi$ of that surface.

Since we are interested in materials with thermal memory, we introduce the causal function θ^t, called *past history of the temperature* up to t at $\boldsymbol{x}$, defined by

$$\theta^t(\boldsymbol{x}, s) = \theta(\boldsymbol{x}, t - s) \qquad s > 0.$$

Accordingly, we define the *past history of the temperature gradient*,

$$\boldsymbol{G}^t(\boldsymbol{x}, s) = \boldsymbol{G}(\boldsymbol{x}, t - s) \qquad s > 0 \quad \text{where} \quad \boldsymbol{G} = \nabla\theta ,$$

and the *summed past histories* of temperature and temperature gradient, respectively,

$$\zeta^t(\boldsymbol{x}, \tau) = \int_0^\tau \theta(\boldsymbol{x}, t-s)\, ds , \qquad \boldsymbol{g}^t(\boldsymbol{x}, \tau) = \int_0^\tau \boldsymbol{G}(\boldsymbol{x}, t-s)\, ds , \quad t > \tau \geq 0 .$$

It is worth noting that

$$\dot{\zeta}^t(s) = \theta(t) - \theta^t(s) \quad \text{and} \quad \dot{\boldsymbol{g}}^t(s) = \boldsymbol{G}(t) - \boldsymbol{G}^t(s) \,. \tag{7}$$

Given a summable, positive and decreasing scalar function α on $\mathbb{R}^+ = (0, +\infty)$ and a real Hilbert space X, let $L^2_\alpha(\mathbb{R}^+, X)$ be the Hilbert space of X-valued functions on $\mathbb{R}^+$ endowed with the inner product

$$\langle f, g \rangle_{L^2_\alpha} = \int_0^\infty \alpha(s) \langle f(s), g(s) \rangle_X \, ds \,.$$

Definition 3: (Fading Memory). Let λ and μ be two positive scalar functions, called *influence functions*, defined and decreasing on $\mathbb{R}^+$, vanishing at infinity and such that

$$\int_0^\infty \lambda(s) s^2 \, ds = \lambda_0 > 0 \,, \qquad \int_0^\infty \mu(s) s^2 \, ds = \mu_0 > 0 \,.$$

Let $\zeta^t(\boldsymbol{x}, \cdot)$ and $\boldsymbol{g}^t(\boldsymbol{x}, \cdot)$ belong to $\mathcal{H}^+ = L^2_\lambda(\mathbb{R}^+, \mathbb{R}^+)$ and $\mathbb{H} = L^2_\mu(\mathbb{R}^+, \mathbb{R}^3)$, respectively. Then, a constitutive functional $\mathcal{F}(\zeta^t, \boldsymbol{g}^t)$ is said to fulfil the "fading memory property" if it is continuous in $\mathcal{H}^+ \times \mathbb{H}$.

3. Phase–field models with a modified energy balance

According to Fried & Gurtin [10] and Frémond [9], the presence of two different phases can yield some forces and stresses on the interfacial region. Such quantities are named "accretive forces" by Fried & Gurtin, since they induce the accretion of one pure phase at the expense of the other one, and obey a suitable balance equation. Frémond called them "microforces", in that they account for microscopical movements and actions, and their balance is derived from the principle of virtual power.

In any case, since the order parameter φ is a scalar quantity, the balance must be set in a two–dimensional domain and "forces" must be restricted accordingly. Essentially, the transition only occurs on surfaces. We can show this as follows. Let $\mathcal{P} \subset \mathcal{B}$ any portion of the material body and assume a balance law of the *accretive (micro)forces*, namely

$$\int_{\mathcal{P}} \rho \boldsymbol{\pi} \, dv = \int_{\partial\mathcal{P}} \Xi \boldsymbol{n} \, da + \int_{\mathcal{P}} \rho \boldsymbol{\gamma} \, dv \tag{8}$$

where ρ is the constant mass density, Ξ is a symmetric tensor called *accretive stress*, $\boldsymbol{\pi}$ and $\boldsymbol{\gamma}$ are vectors, called respectively *internal accretive force* and *external accretive supply* per unit mass, and $\boldsymbol{n}$ is the outward normal to the boundary $\partial\mathcal{P}$ of $\mathcal{P}$. Of course, we are forced to assume that the stress Ξ in (8) is identically vanishing away from the transition zone. On the other hand, we have actions (represented by $\boldsymbol{\pi}$) that drive the phase transition even if no interfacial region is present (uniform melting or freezing). In this connection, many authors assume that no contribution from the outside of the body can affect the internal accretion of the phase, see [9]. This in turn implies

$$\int_{\partial \mathcal{B}} \boldsymbol{\Xi} \boldsymbol{n} \, da + \int_{\mathcal{B}} \rho \boldsymbol{\gamma} \, dv = \mathbf{0}. \tag{9}$$

In order to obtain a scalar balance law by projection, we make use of Assumption 2. Therefore, the only meaningful component of (8) is the one along the normal $\boldsymbol{n}_\varphi$. Introducing the projections $\boldsymbol{\xi} = \boldsymbol{\Xi} \cdot \boldsymbol{n}_\varphi$, $\pi = \boldsymbol{\pi} \cdot \boldsymbol{n}_\varphi$ and $\gamma = \boldsymbol{\gamma} \cdot \boldsymbol{n}_\varphi$, the balance equation (8) becomes

$$\int_{\mathcal{P}} \rho \, \pi \, dv = \int_{\partial \mathcal{P}} \boldsymbol{\xi} \cdot \boldsymbol{n} \, da + \int_{\mathcal{P}} \rho \, \gamma \, dv. \tag{10}$$

In addition, condition (9) takes the form

$$\int_{\partial \mathcal{B}} \boldsymbol{\xi} \cdot \boldsymbol{n} \, da + \int_{\mathcal{B}} \rho \, \gamma \, dv = \mathbf{0}.$$

which is usually satisfied by assuming

$$\boldsymbol{\xi} \cdot \boldsymbol{n} = 0 \quad \text{on } \partial \mathcal{B} \qquad \text{and} \qquad \gamma = 0 \quad \text{in } \mathcal{B}\,.$$

The main assumption of the energy approach is the modification of the *energy balance law*. In view of Assumption 2, it can be stated as follows

$$\begin{aligned} \frac{d}{dt} \int_{\mathcal{P}} \rho \, e \, dv &= - \int_{\partial \mathcal{P}} \boldsymbol{q} \cdot \boldsymbol{n} \, da + \int_{\mathcal{P}} \rho \, r \, dv + \int_{\partial \mathcal{P}} (\boldsymbol{\Xi} \boldsymbol{v}_\varphi) \cdot \boldsymbol{n} \, da + \int_{\mathcal{P}} \rho \, \boldsymbol{\gamma} \cdot \boldsymbol{v}_\varphi \, dv \\ &= - \int_{\partial \mathcal{P}} \boldsymbol{q} \cdot \boldsymbol{n} \, da + \int_{\mathcal{P}} \rho \, r \, dv + \int_{\partial \mathcal{P}} \dot{\varphi} \boldsymbol{\xi} \cdot \boldsymbol{n} \, da + \int_{\mathcal{P}} \rho \, \dot{\varphi} \gamma \, dv, \end{aligned} \tag{11}$$

where e is the internal energy density, $\boldsymbol{q}$ the heat flux and r the heat source density. Note that the first two addenda on the right-hand side of $(11)_2$ account for the thermal power, whereas the last two addenda represent the power expended by the accretive forces. When $\gamma = 0$, they reduce to the *energy extra-flux* $\boldsymbol{h} = \dot{\varphi} \boldsymbol{\xi}$. Finally we introduce the second law of thermodynamics in the classical form of the *Clausius–Duhem entropy inequality*

$$\frac{d}{dt} \int_{\mathcal{P}} \rho \, \eta \, dv \geq - \int_{\partial \mathcal{P}} \frac{\boldsymbol{q} \cdot \boldsymbol{n}}{\theta} \, da + \int_{\mathcal{P}} \frac{\rho \, r}{\theta} \, dv, \tag{12}$$

being η the entropy and θ the absolute temperature. Relations (11), (10) and (12) may be written in local form

$$\begin{aligned} \rho\,\dot{e} &= -\nabla\cdot\boldsymbol{q} + \rho\,r + \nabla\cdot(\dot{\varphi}\boldsymbol{\xi}) \\ \rho\,\pi &= \nabla\cdot\boldsymbol{\xi} \\ \rho\,\dot{\eta} &\geq -\nabla\cdot\left(\frac{\boldsymbol{q}}{\theta}\right) + \frac{\rho\,r}{\theta}. \end{aligned} \tag{13}$$

By a suitable combination, $(13)_1$ yields an alternative form of the energy balance

$$\rho\,\dot{e} = -\nabla\cdot\boldsymbol{q} + \rho\,r + \boldsymbol{\xi}\cdot\dot{\boldsymbol{p}} + \rho\,\pi\dot{\varphi}\,, \tag{14}$$

and $(13)_3$ leads to the so called *dissipation inequality*

$$\rho\,(\dot{\psi} + \eta\dot{\theta}) \leq \boldsymbol{\xi}\cdot\dot{\boldsymbol{p}} + \rho\,\pi\dot{\varphi} - \frac{1}{\theta}\boldsymbol{q}\cdot\boldsymbol{G}\,, \tag{15}$$

where $\boldsymbol{G} = \nabla\theta$ and

$$\psi = e - \theta\eta \tag{16}$$

is the Helmholtz free energy density per unit mass.

4. A phase–field model with thermal memory

Here, we develop a model accounting for memory thermal effect in a rigid heat conductor as suggested in [11]. In order to fix the minimum set of independent variables which the constitutive equations may depend on (often called *material state*), we separately consider the phenomena of heat conduction and of phase transition. For simplicity, we assume $\rho = 1$, that is we are considering phase transitions at constant density.

According to the theory developed by Gurtin & Pipkin, see [15], for a (single phase) rigid heat conductor with memory, all the constitutive quantities at time t can be determined, once we fix the set of the independent variables $(\theta(t), \zeta^t, \boldsymbol{g}^t)$ at any point $\boldsymbol{x} \in \mathcal{B}$.

On the other hand, if we consider a two-phase material under isothermal conditions, the constitutive equations are required to depend on the phase φ and on some features of the interface zone, summed up by $\boldsymbol{p} = \nabla\varphi$. In addition, taking into account dynamic dissipation, they are also allowed to depend on $h = \dot{\varphi}$ and $\boldsymbol{f} = \dot{\boldsymbol{p}} = \nabla\dot{\varphi}$. Thus, the widest set of independent variables for such a material is $(\varphi(t), \boldsymbol{p}(t), h(t), \boldsymbol{f}(t))$.

In the general case of a two-phase, rigid heat conductor, we assume that at any fixed point $\boldsymbol{x} \in \mathcal{B}$ all the constitutive quantities η, ψ, e, π, $\boldsymbol{q}$ and $\boldsymbol{\xi}$

at time t depend on the *material state*

$$\omega(t) = (\varphi(t), \boldsymbol{p}(t), h(t), \boldsymbol{f}(t), \theta(t), \zeta^t, \boldsymbol{g}^t)$$

and that, for all time t, $\omega(t)$ belongs to the state space

$$\mathcal{Z} = I \times \mathbb{R}^3 \times \mathbb{R} \times \mathbb{R}^3 \times \mathbb{R}^+ \times \mathcal{H}^+ \times \mathbb{H},$$

where $I = [-1, +1]$. It is worth noting that this modelization mainly differs from that proposed in [10] in that the actual value $\boldsymbol{G}(t)$ of the temperature gradient does not enter the state of the material element. In the next subsections, the dissipation inequality (15) is exploited to find thermodynamic restrictions on the constitutive relations, and the general evolution equations of the model is established. Finally, in 6 a simpler model is obtained by linearization with respect to the temperature near its transition value.

4.1. *Thermodynamic restrictions*

Consider a two–phase rigid heat conductor with memory, whose general constitutive equations are functions of the state ω. Its dynamics is ruled by equations (13) and (14). After rewriting the dissipation inequality (15) in the form

$$\rho(\dot{\psi} + \eta\dot{\theta}) - \boldsymbol{\xi} \cdot \boldsymbol{f} - \rho\pi h + \frac{1}{\theta}\boldsymbol{q} \cdot \boldsymbol{G} \leq 0\,, \tag{17}$$

we need to calculate the time derivative of the free energy function

$$\dot{\psi}(t) = \frac{d}{dt}\tilde{\psi}(\omega(t)) = \delta_\omega\tilde{\psi}(\omega|\dot{\omega}(t))\,.$$

Here, $\delta_\omega\tilde{\psi}$ denotes the Fréchet differential of $\tilde{\psi}$ on $\mathcal{Z}$, namely

$$\begin{aligned}\delta_\omega\tilde{\psi}(\omega|\dot{\omega}(t)) = {} & \partial_\varphi\tilde{\psi}\, h(t) + \partial_{\boldsymbol{p}}\tilde{\psi} \cdot \boldsymbol{f}(t) + \partial_h\tilde{\psi}\, \dot{h}(t) + \partial_{\mathrm{f}}\tilde{\psi} \cdot \dot{\boldsymbol{f}}(t) \\ & + \partial_\theta\tilde{\psi}\, \dot{\theta}(t) + \delta_1\tilde{\psi}(\omega|\dot{\zeta}^t) + \delta_2\tilde{\psi}(\omega|\, \dot{\boldsymbol{g}}^t),\end{aligned}$$

where $\delta_1\tilde{\psi}$ and $\delta_2\tilde{\psi}$ represent the Fréchet differentials of $\tilde{\psi}$ on $\mathcal{H}^+$ and $\mathbb{H}$, respectively. Because of the linearity of δ, from (7) we have the splitting

$$\begin{aligned}\delta_1\tilde{\psi}(\omega|\dot{\zeta}^t) &= J_\theta\tilde{\psi}(\omega)\theta(t) - \delta_1\tilde{\psi}(\omega|\theta^t)\,, \\ \delta_2\tilde{\psi}(\omega|\, \dot{\boldsymbol{g}}^t) &= \boldsymbol{J}_{\mathrm{G}}\ \tilde{\psi}(\omega) \cdot \boldsymbol{G}(t) - \delta_2\tilde{\psi}(\omega|\boldsymbol{G}^t)\,,\end{aligned}$$

where $J_\theta\tilde{\psi}$ and $\boldsymbol{J}_{\mathrm{G}}\ \tilde{\psi}$ are state functions such that

$$J_\theta\tilde{\psi}(\omega)\theta(t) = \delta_1\tilde{\psi}(\omega|\theta(t))\,, \qquad \boldsymbol{J}_{\mathrm{G}}\ \tilde{\psi}(\omega) \cdot \boldsymbol{G}(t) = \delta_2\tilde{\psi}(\omega|\boldsymbol{G}(t))\,. \tag{18}$$

Letting, for the sake of simplicity,

$$\delta\tilde{\psi}(\omega|\theta^t, \boldsymbol{G}^t) := \delta_1\tilde{\psi}(\omega|\theta^t) + \delta_2\tilde{\psi}(\omega|\boldsymbol{G}^t),$$

we finally obtain

$$\dot{\psi}(t) = \partial_\varphi \tilde{\psi}(\omega) h(t) + \partial_{\boldsymbol{p}} \tilde{\psi}(\omega) \cdot \boldsymbol{f}(t) + \partial_h \tilde{\psi}(\omega) \dot{h}(t) + \partial_{\mathfrak{f}} \tilde{\psi}(\omega) \cdot \dot{\boldsymbol{f}}(t)$$
$$+ \partial_\theta \tilde{\psi}(\omega) \dot{\theta}(t) + J_\theta \tilde{\psi}(\omega) \theta(t) + \boldsymbol{J}_{\mathsf{G}}\ \tilde{\psi}(\omega) \cdot \boldsymbol{G}(t) - \delta \tilde{\psi}(\omega | \theta^t, \boldsymbol{G}^t) \, .$$

As a consequence, the dissipation inequality (17) takes the form

$$(\partial_\varphi \psi - \pi)\, h(t) + (\partial_{\boldsymbol{p}} \psi - \boldsymbol{\xi}) \cdot \boldsymbol{f}(t) + \partial_h \psi \dot{h}(t) + \partial_{\mathfrak{f}} \tilde{\psi}(\omega) \cdot \dot{\boldsymbol{f}}(t) \tag{19}$$
$$+ (\partial_\theta \psi + \eta)\, \dot{\theta}(t) + J_\theta \psi\, \theta(t) + \frac{1}{\theta} (\boldsymbol{q} + \theta \boldsymbol{J}_{\mathsf{G}}\ \psi) \cdot \boldsymbol{G}(t) - \delta \tilde{\psi}(\omega | \theta^t, \boldsymbol{G}^t) \leq 0 \, .$$

On the other hand, taking into account the balance laws (13) and (14) and their boundary conditions, it is apparent that the quantities $\dot{\boldsymbol{f}}$, $\dot{h}$, $\dot{\theta}$ and $\boldsymbol{G}$ at $(\boldsymbol{x}, t)$ may be chosen arbitrarily, for any fixed ω at $(\boldsymbol{x}, t)$, since they do not enter the definition of ω and each of them is independent of the others. Accordingly, (19) implies

$$\partial_{\mathfrak{f}} \psi = \boldsymbol{0} \, , \quad \partial_h \psi = 0 \, , \quad \partial_\theta \psi = -\eta \, , \quad \boldsymbol{J}_{\mathsf{G}}\ \psi = -\boldsymbol{q}/\theta \, . \tag{20}$$

Letting

$$\sigma(t) := (\varphi(t), \boldsymbol{p}(t), \theta(t), \zeta^t, \boldsymbol{g}^t) \quad \text{and} \quad \mathcal{S} := I \times \mathbb{R}^3 \times \mathbb{R}^+ \times \mathcal{H}^+ \times \mathbb{H}$$

so that $\omega = (\sigma, h, \boldsymbol{f})$ and $\mathcal{Z} = \mathcal{S} \times \mathbb{R} \times \mathbb{R}^3$, we have

$$\psi = \hat{\psi}(\sigma), \qquad \eta = \hat{\eta}(\sigma) = -\partial_\theta \hat{\psi}(\sigma), \qquad \boldsymbol{q} = \hat{\boldsymbol{q}}(\sigma) = -\theta \boldsymbol{J}_{\mathsf{G}}\ \hat{\psi}(\sigma),$$
$$\pi = \hat{\pi}(\sigma, h, \boldsymbol{f}), \qquad \boldsymbol{\xi} = \hat{\boldsymbol{\xi}}(\sigma, h, \boldsymbol{f}) \, .$$

Moreover, taking into account (16), it follows

$$e = \hat{e}(\sigma) = \hat{\psi}(\sigma) - \theta \partial_\theta \hat{\psi}(\sigma) = -\theta^2 \partial_\theta \left(\frac{\hat{\psi}(\sigma)}{\theta} \right), \tag{21}$$

and (19) provides the *reduced dissipation inequality*

$$\Big(\partial_\varphi \hat{\psi}(\sigma) - \hat{\pi}(\sigma, h, \boldsymbol{f})\Big)\, h(t) + \Big(\partial_{\boldsymbol{p}} \hat{\psi}(\sigma) - \hat{\boldsymbol{\xi}}(\sigma, h, \boldsymbol{f})\Big) \cdot \boldsymbol{f}(t) \tag{22}$$
$$+ J_\theta \hat{\psi}(\sigma)\, \theta(t) - \delta \hat{\psi}(\sigma | \theta^t, \boldsymbol{G}^t) \leq 0$$

Since h and $\boldsymbol{f}$ do not enter the definition of σ, for any fixed value of $\sigma(t)$ we can choose $h(t) = 0$ and $\boldsymbol{f}(t) = \boldsymbol{0}$. As an obvious consequence, for all $\sigma(t) \in \mathcal{S}$ we obtain

$$J_\theta \hat{\psi}(\sigma(t)) \theta(t) - \delta \hat{\psi}(\sigma(t) | \theta^t, \boldsymbol{G}^t) \leq 0 \, .$$

Assumption 4: Among all possible choices compatible with (22), we assume that each of the constitutive functions $\hat{\pi}$ and $\hat{\boldsymbol{\xi}}$ can be split into the sum of two terms, a "stationary" part, $\pi^{(s)}$ and $\boldsymbol{\xi}^{(s)}$, and a "dynamic" part, $\pi^{(d)}$ and $\boldsymbol{\xi}^{(d)}$, where

$$\hat{\pi}^{(s)}(\sigma) = \hat{\pi}(\sigma, 0, \mathbf{0}) = \partial_\varphi \hat{\psi}(\sigma)\,, \qquad \hat{\boldsymbol{\xi}}^{(s)}(\sigma) = \hat{\boldsymbol{\xi}}(\sigma, 0, \mathbf{0}) = \partial_{\boldsymbol{p}} \hat{\psi}(\sigma)$$

$$\hat{\pi}^{(d)}(\sigma, h, \boldsymbol{f}) = \beta(\sigma, h, \boldsymbol{f})h\,, \qquad \hat{\boldsymbol{\xi}}^{(d)}(\sigma, h, \boldsymbol{f}) = \boldsymbol{B}(\sigma, h, \boldsymbol{f})\boldsymbol{f}\,.$$

Here, β is a non negative scalar function and $\boldsymbol{B}$ is a positive semidefinite tensor-valued function.

From this Assumption it follows

$$\hat{\pi}(\sigma, h, \boldsymbol{f}) = \partial_\varphi \hat{\psi}(\sigma) + \beta(\sigma, h, \boldsymbol{f})h, \tag{23}$$

$$\hat{\boldsymbol{\xi}}(\sigma, h, \boldsymbol{f}) = \partial_{\boldsymbol{p}} \hat{\psi}(\sigma) + \boldsymbol{B}(\sigma, h, \boldsymbol{f})\boldsymbol{f}, \tag{24}$$

and (22) is satisfied since

$$(\partial_\varphi \psi - \pi)\, h = -\beta h^2 \leq 0\,, \qquad (\partial_{\boldsymbol{p}} \psi - \boldsymbol{\xi}) \cdot \boldsymbol{f} = -\boldsymbol{B}\boldsymbol{f} \cdot \boldsymbol{f} \leq 0\,.$$

In view of (16) and Assumption 4, the time derivatives of ψ and e may be rewritten as

$$\dot{\psi} = \pi^{(s)}\, h + \boldsymbol{\xi}^{(s)} \cdot \boldsymbol{f} - \eta\dot{\theta} - \frac{1}{\theta}\boldsymbol{q} \cdot \boldsymbol{G} + J_\theta \psi\, \theta - \delta\psi\,,$$

$$\dot{e} = \pi^{(s)}\, h + \boldsymbol{\xi}^{(s)} \cdot \boldsymbol{f} - \frac{1}{\theta}\boldsymbol{q} \cdot \boldsymbol{G} + (J_\theta \psi + \dot{\eta})\theta - \delta\psi.$$

The comparison with (14) yields

$$\dot{\eta} = -\nabla \cdot \left(\frac{\boldsymbol{q}}{\theta}\right) + \frac{1}{\theta}\left[\beta\, h^2 + \boldsymbol{B}\boldsymbol{f} \cdot \boldsymbol{f} + r + \delta\psi - \theta\, J_\theta \psi\right].$$

4.2. *Evolution equations*

Taking into account $(13)_2$ and (14) and letting $\rho = 1$, the evolution of the system is ruled by

$$\dot{e} = -\nabla \cdot \boldsymbol{q} + r + \boldsymbol{\xi} \cdot \dot{\boldsymbol{p}} + \pi\dot{\varphi},$$

$$0 = \nabla \cdot \boldsymbol{\xi} - \pi\,.$$

First, we observe that

$$\dot{e}(t) = \partial_\varphi \hat{e}(\sigma)h(t) + \partial_{\boldsymbol{p}} \hat{e}(\sigma) \cdot \boldsymbol{f}(t) + \partial_\theta \hat{e}(\sigma)\dot{\theta}(t) + J_\theta \hat{e}(\sigma)\theta(t) + \boldsymbol{J}_{\mathsf{G}}\, \hat{e}(\sigma) \cdot \boldsymbol{G}(t) - \delta\hat{e}(\sigma|\theta^t, \boldsymbol{G}^t)\,.$$

In view of (16), we have

$$\begin{aligned}\partial_\theta e &= \partial_\theta \psi + \partial_\theta(\theta\eta) = \theta\partial_\theta\eta = -\theta\partial^2_{\theta\theta}\psi,\\ \partial_\varphi e &= \partial_\varphi\psi + \theta\partial_\varphi\eta = \pi^{(s)} - \theta\partial^2_{\theta\varphi}\psi,\\ \partial_{\boldsymbol{p}} e &= \partial_{\boldsymbol{p}}\psi + \theta\partial_{\boldsymbol{p}}\eta = \boldsymbol{\xi}^{(s)} - \theta\partial^2_{\theta\boldsymbol{p}}\psi,\\ J_\theta e &= J_\theta\psi + \theta J_\theta\eta = -\theta^2\partial_\theta(J_\theta\psi/\theta),\\ \boldsymbol{J}_{\mathrm{G}}\ e &= \boldsymbol{J}_{\mathrm{G}}\ \psi + \theta\boldsymbol{J}_{\mathrm{G}}\ \eta = \boldsymbol{J}_{\mathrm{G}}\ \psi - \theta\partial_\theta\boldsymbol{J}_{\mathrm{G}}\ \psi,\end{aligned} \tag{25}$$

where the specific heat $c(\sigma) = \partial_\theta\hat{e}(\sigma)$ is assumed to be positive valued, as usual. This, in turn, implies the concavity of ψ with respect to θ, namely $\partial^2_{\theta\theta}\psi < 0$. According to previous relations, the evolution equations take the form

$$\begin{aligned} c\dot{\theta} &= (\theta\,\partial^2_{\theta\boldsymbol{p}}\psi + \boldsymbol{\xi}^{(d)})\cdot\dot{\boldsymbol{p}} + (\theta\,\partial^2_{\theta\varphi}\psi + \pi^{(d)})\dot{\varphi}\\ &\quad +\theta\,\partial_\theta\boldsymbol{J}_{\mathrm{G}}\ \psi\cdot\boldsymbol{G} + \theta\nabla\cdot(\boldsymbol{J}_{\mathrm{G}}\ \psi) - D_\theta\psi + r,\\ 0 &= \nabla\cdot(\partial_{\boldsymbol{p}}\psi + \boldsymbol{\xi}^{(d)}) - \partial_\varphi\psi - \pi^{(d)},\end{aligned}$$

and (23)–(24) yield

$$\begin{aligned} 0 &= c\dot{\theta} - (\theta\partial^2_{\theta\boldsymbol{p}}\psi + \boldsymbol{B}\,\dot{\boldsymbol{p}})\cdot\dot{\boldsymbol{p}} - (\theta\partial^2_{\theta\varphi}\psi + \beta\dot{\varphi})\dot{\varphi} \\ &\quad -\theta\,\partial_\theta\boldsymbol{J}_{\mathrm{G}}\ \psi\cdot\boldsymbol{G} - \theta\nabla\cdot(\boldsymbol{J}_{\mathrm{G}}\ \psi) + D_\theta\psi - r,\end{aligned} \tag{26}$$

$$0 = \beta\dot{\varphi} + \nabla\cdot(\boldsymbol{B}\,\dot{\boldsymbol{p}}) - \nabla\cdot\partial_{\boldsymbol{p}}\psi + \partial_\varphi\psi, \tag{27}$$

where

$$D_\theta\psi = \theta J_\theta e - \delta e = \theta^2[\partial_\theta(\delta\psi/\theta) - \theta\partial_\theta(J_\theta\psi/\theta)]\,.$$

Note that these equations depend on the constitutive functions ψ, β and $\boldsymbol{B}$, only. Following the approach proposed by Frémond, functions $\hat{\pi}^{(d)}$ and $\hat{\boldsymbol{\xi}}^{(d)}$ can be derived from a *pseudopotential of dissipation*, so that both β and $\boldsymbol{B}$ can be obtained from it, cf. [9].

Remark 5: The general form of equation (27) reduces to the kinetic equation of the standard model when $\boldsymbol{B} = \boldsymbol{0}$, $\beta = 1$ and $\partial_{\boldsymbol{p}}\psi = \alpha\,\boldsymbol{p}$. On the contrary, the classical form of the energy balance can be obtained from (26) only if some approximation with respect to θ is performed.

Because of its length and technicality, this approximation procedure is developed in Section 6.

5. Phase–field models with an entropy extra–flux

Here, we follow a different approach, see [8], according to which the phase change cannot be traced back to the action of microforces. Thus, the latter give no contribution to the energy balance. The material is modelled as a continuum without any internal structure, and the phase–field scalar variable $\varphi(\boldsymbol{x},t)$ summarizes the order degree of the material in $\boldsymbol{x}$ at time t. It is named *order parameter*, and obeys the local balance law

$$\nu\dot{\varphi} = \nabla\cdot\boldsymbol{\xi} - \pi, \tag{28}$$

where π and $\boldsymbol{\xi}$ are constitutive functions representing the *internal supply of order* and the *specific order flux vector*, respectively. The negative sign in front to π is merely taken in order to compare the results of Secs. 4 and 5. No external source supplies order from the outside. Accordingly, the total flux of order into the body vanishes identically, namely

$$\boldsymbol{\xi}\cdot\boldsymbol{n} = 0 \quad \text{on } \partial\mathcal{B}\,.$$

Any change of the order parameter is assumed to give no contribution to the balance equation for internal energy. Then, the latter is taken in the classical form of continuum mechanics when deformations are negligible, namely

$$\rho\,\dot{e} = -\nabla\cdot\boldsymbol{q} + \rho\,r, \tag{29}$$

where ρ is the mass density.

To account for the thermodynamic effects due to changes in the order parameter, the Clausius–Duhem inequality is modified by adding an *entropy extra–flux* $\boldsymbol{k}$, namely

$$\rho\,\dot{\eta} + \nabla\cdot\left(\frac{\boldsymbol{q}}{\theta} + \boldsymbol{k}\right) - \frac{\rho\,r}{\theta} \geq 0. \tag{30}$$

Letting $\psi = e - \theta\eta$, from (29) and (30) we obtain the dissipation inequality

$$-\rho(\dot{\psi} + \eta\,\dot{\theta}) - \frac{1}{\theta}\boldsymbol{q}\cdot\nabla\theta + \theta\nabla\cdot\boldsymbol{k} \geq 0\,.$$

The extra–flux $\boldsymbol{k}$ is regarded as an unknown. Nevertheless, we are forced to assume that it vanishes along all processes for which no phase change occurs ($\dot{\varphi} = \nabla\varphi = 0$), in order to recover the classical local form of (30). To this aim, we state the following constitutive assumption:

Assumption 6: The entropy extra flux is proportional to the *specific order flux vector* and to the time rate variation of the phase. Namely,

$$\boldsymbol{k}(\boldsymbol{x},t) = \dot{\varphi}(\boldsymbol{x},t)\boldsymbol{\xi}(\boldsymbol{x},t) \tag{31}$$

where we let the constant of proportionality equal to 1.

As a consequence, $\boldsymbol{k}\cdot\boldsymbol{n}=0$ on $\partial\mathcal{B}$ and this, in turn, implies that the second law (30) for the whole body takes the standard form

$$\frac{d}{dt}\int_{\mathcal{B}}\rho\,\eta\,dv \geq -\int_{\partial\mathcal{B}}\frac{\boldsymbol{q}}{\theta}\cdot\boldsymbol{n}\,da+\int_{\mathcal{B}}\frac{\rho\,r}{\theta}\,dv\,.$$

By virtue of (28) and (31), the dissipation inequality reads

$$\rho(\dot{\psi}-\eta\,\dot{\theta})+\frac{1}{\theta}\boldsymbol{q}\cdot\nabla\theta-\theta\pi\dot{\varphi}-\theta\nu\dot{\varphi}^2-\theta\boldsymbol{\xi}\cdot\dot{\boldsymbol{p}}\leq 0\,. \tag{32}$$

The general case of a deformable two-phase heat conductor of Fourier (non hereditary) type has been proposed in [8]. Here, we develop a model accounting for memory thermal effects in a rigid heat conductor. We consider separately the phenomena of hereditary heat conduction and phase transition in order to fix the minimum set of independent variables which represent the state of the material at $(\boldsymbol{x},t)$. The main difference with respect to the previous section is that the phase-field variable φ is considered here as a balanced quantity, so that its time derivative $h=\dot{\varphi}$ and the time derivative of its gradient $\boldsymbol{f}=\dot{\boldsymbol{p}}$ cannot enter the set of the state variables.

Accordingly, we set $\rho=1$ and assume that at any fixed point $\boldsymbol{x}\in\mathcal{B}$ all the constitutive quantities η, ψ, e, π, ν, $\boldsymbol{q}$ and $\boldsymbol{\xi}$ at time t depend on the *material state*

$$\sigma(t):=(\varphi(t),\boldsymbol{p}(t),\theta(t),\zeta^t,\boldsymbol{g}^t)$$

and, for all time t, $\sigma(t)$ belongs to the *state space*

$$\mathcal{Z}:=I\times\mathbb{R}^3\times\mathbb{R}^+\times\mathcal{H}^+\times\mathbb{H}$$

where $I=[-1,+1]$.

5.1. *Thermodynamic restrictions and evolution equations*

The dissipation inequality predicted by the entropy extra–flux approach is exploited here to restrict the constitutive relations and derive the evolution equations of the model.

Theorem 7: If (28), (29), (30) and (31) are assumed, then the constitutive relations fulfil

$$\partial_{\varphi}\psi=\theta\,\pi\,,\quad \partial_{\boldsymbol{p}}\psi=\theta\,\boldsymbol{\xi}\,,\quad \partial_{\theta}\psi=-\eta\,,\quad \boldsymbol{J}_{\mathsf{G}}\,\psi=-\boldsymbol{q}/\theta, \tag{33}$$

$$J_{\theta}\psi\,\theta(t)-\delta\psi(\sigma|\theta^t,\boldsymbol{G}^t)\leq 0\quad\text{and}\quad \nu\geq 0\,. \tag{34}$$

Proof: Let ψ, η, π, $\boldsymbol{\xi}$, $\boldsymbol{q}$ and ν be defined on $\mathcal{S}$. Substituting the constitutive relations into the dissipation inequality (32) and exploiting of the time derivative of $\hat{\psi}(\sigma)$, we obtain

$$\begin{aligned}(\partial_\theta\psi+\eta)\,\dot{\theta}(t)+(\partial_\varphi\psi-\theta\,\pi)\,\dot{\varphi}(t)+(\partial_{\boldsymbol{p}}\psi-\theta\,\boldsymbol{\xi})\cdot\dot{\boldsymbol{p}}(t)\\ +\frac{1}{\theta}\,(\boldsymbol{q}+\theta\boldsymbol{J}_{\mathrm{G}}\;\psi)\cdot\boldsymbol{G}(t)+J_\theta\psi\,\theta(t)-\delta\psi(\sigma|\theta^t,\boldsymbol{G}^t)-\theta\nu\dot{\varphi}^2(t)\le 0\end{aligned} \tag{35}$$

with the same notation as in the previous Section. On the other hand, taking into account the balance laws (28) and (29) and their boundary conditions, it is apparent that the quantities $\dot{\boldsymbol{p}}$, $\dot{\varphi}$, $\dot{\theta}$ and $\boldsymbol{G}$ at $(\boldsymbol{x},t)$ are independent of each other and may be chosen arbitrarily, for any fixed σ at $(\boldsymbol{x},t)$. Hence, from (35) we obtain the thermodynamic restrictions (33) and the reduced inequality

$$J_\theta\psi(\sigma)\,\theta(t)-\delta\psi(\sigma|\theta^t,\boldsymbol{G}^t)-\theta\nu(\sigma)\dot{\varphi}^2(t)\le 0\,. \tag{36}$$

Observing that $\dot{\varphi}(t)$ is independent of $\sigma(t)$, the first inequality in (34) is a consequence of (36) when $\dot{\varphi}(t)=0$. The second one can be proved by contradiction. In fact, if $\nu(\sigma)<0$ for some $\sigma\in\mathcal{S}$, we could choose $|\dot{\varphi}(t)|$ as large as we need to violate inequality (36). □

It is worth noting that, here, π and $\boldsymbol{\xi}$ play the role of $\pi^{(s)}$ and $\boldsymbol{\xi}^{(s)}$ in the previous section. In addition, we have

$$\begin{gathered}-\partial_\varphi\hat{\eta}=\hat{\pi}+\theta\,\partial_\theta\hat{\pi}\,,\qquad -\partial_{\boldsymbol{p}}\hat{\eta}=\hat{\boldsymbol{\xi}}+\theta\,\partial_\theta\hat{\boldsymbol{\xi}}\,,\qquad -\theta^2\boldsymbol{J}_{\mathrm{G}}\;\hat{\eta}=\hat{\boldsymbol{q}}-\theta\partial_\theta\hat{\boldsymbol{q}}\,,\\ \partial_{\boldsymbol{p}}\hat{\pi}=\partial_\varphi\hat{\boldsymbol{\xi}}\,,\qquad \partial_\varphi\hat{\boldsymbol{q}}=-\theta\boldsymbol{J}_{\mathrm{G}}\;\hat{\pi}\,,\qquad \partial_{\boldsymbol{p}}\hat{\boldsymbol{q}}=-\theta\boldsymbol{J}_{\mathrm{G}}\;\hat{\boldsymbol{\xi}}\,.\end{gathered}$$

The expressions of $\dot{\psi}$ and $\dot{e}$ can be rewritten as follows

$$\begin{aligned}\dot{\psi}&=\theta\,(\pi\,\dot{\varphi}+\boldsymbol{\xi}\cdot\dot{\boldsymbol{p}}+J_\theta\psi)-\eta\dot{\theta}-\frac{1}{\theta}\boldsymbol{q}\cdot\boldsymbol{G}-\delta\psi\,,\\ \dot{e}&=\theta\,(\pi\,\dot{\varphi}+\boldsymbol{\xi}\cdot\dot{\boldsymbol{p}}+J_\theta\psi+\dot{\eta})-\frac{1}{\theta}\boldsymbol{q}\cdot\boldsymbol{G}-\delta\psi\,,\end{aligned}$$

and $\dot{\eta}$ can be obtained from comparing the latter with (29)

$$\dot{\eta}=-\nabla\cdot\left(\frac{\boldsymbol{q}}{\theta}\right)+\frac{1}{\theta}\,[r+\delta\psi]-J_\theta\psi-\pi\,\dot{\varphi}-\boldsymbol{\xi}\,\dot{\boldsymbol{p}}.$$

Finally, inserting (20) into the balance laws (28) and (29), we easily obtain the evolution equations of the system, namely

$$\nu\dot{\varphi}=\nabla\cdot(\partial_{\boldsymbol{p}}\psi/\theta)-\partial_\varphi\psi/\theta, \tag{37}$$

$$\dot{e}=\nabla\cdot(\theta\boldsymbol{J}_{\mathrm{G}}\;\psi)+r. \tag{38}$$

Since $e = -\theta^2 \partial_\theta(\psi/\theta)$, this model is only characterized by two constitutive functions, i.e., ψ and ν. The explicit form of the time derivative of e reads

$$\begin{aligned}\dot{e}(t) &= \partial_\varphi \hat{e}(\sigma)\dot{\varphi}(t) + \partial_{\boldsymbol{p}}\hat{e}(\sigma)\cdot\dot{\boldsymbol{p}}\,(t) + \partial_\theta \hat{e}(\sigma)\dot{\theta}(t) \\ &\quad + J_\theta \hat{e}(\sigma)\theta(t) + \boldsymbol{J}_{\mathrm{G}}\ \hat{e}(\sigma)\cdot\boldsymbol{G}(t) - \delta\hat{e}(\sigma|\theta^t,\boldsymbol{G}^t)\ ,\end{aligned}$$

where

$$\delta\hat{e}(\sigma|\theta^t,\boldsymbol{G}^t) = \delta_1\hat{e}(\sigma|\theta^t) + \delta_2\hat{e}(\sigma|\boldsymbol{G}^t)\,.$$

Thus, taking into account the relations

$$\begin{aligned}\partial_\varphi e &= \partial_\varphi\psi + \theta\partial_\varphi\eta = -\theta^2\partial_\theta(\partial_\varphi\psi/\theta),\\ \partial_{\boldsymbol{p}} e &= \partial_{\boldsymbol{p}}\psi + \theta\partial_{\boldsymbol{p}}\eta = -\theta^2\partial_\theta(\partial_{\boldsymbol{p}}\psi/\theta),\\ J_\theta e &= J_\theta\psi + \theta J_\theta\eta = -\theta^2\partial_\theta(J_\theta\psi/\theta),\end{aligned}$$

the energy balance (38) takes the form

$$\begin{aligned}c\dot{\theta} &= \theta^2\partial_\theta(\partial_{\boldsymbol{p}}\psi/\theta)\cdot\dot{\boldsymbol{p}} + \theta^2\partial_\theta(\partial_\varphi\psi/\theta)\dot{\varphi} + \theta^2 D_\theta\psi \\ &\quad + \theta\,\partial_\theta\boldsymbol{J}_{\mathrm{G}}\ \psi\cdot\boldsymbol{G} + \theta\nabla\cdot(\boldsymbol{J}_{\mathrm{G}}\ \psi) + r\ ,\end{aligned} \tag{39}$$

where $c = -\theta\partial^2_{\theta\theta}\psi$ is the specific heat and

$$D_\theta\psi = \theta\partial_\theta(J_\partial\psi/\theta) - \partial_\theta(\delta\psi/\theta)\ .$$

Equations (37) and (39) completely rule the evolution of the system.

5.2. *Quasilinear theory*

Henceforth, we assume a special form of the constitutive relations in order to extend the Penrose & Fife model and describe first or second order phase transitions with thermal memory.

Let e depend at most linearly on the thermal part of the state $(\theta, \zeta^t, \boldsymbol{g}^t)$, namely

$$\begin{aligned}e(\sigma) &= c(\varphi,\boldsymbol{p})\theta + \int_0^\infty \alpha(\varphi,\boldsymbol{p},s)\zeta^t(s)\,ds \\ &\quad + \int_0^\infty \boldsymbol{\kappa}(\varphi,\boldsymbol{p},s)\cdot\boldsymbol{g}^t(s)\,ds + E(\varphi,\boldsymbol{p})\ ,\end{aligned} \tag{40}$$

and let $\boldsymbol{q}$ be a linear functional of $\boldsymbol{g}^t$, vanishing at $\boldsymbol{g}^t = \boldsymbol{0}$, namely

$$\boldsymbol{q}(\sigma) = \int_0^\infty \boldsymbol{K}(\theta,\varphi,\boldsymbol{p},s)\boldsymbol{g}^t(s)\,ds\ , \tag{41}$$

where $\boldsymbol{K}$ is a tensor-valued function. Finally, let π and $\boldsymbol{\xi}$ be independent of the thermal histories $(\zeta^t, \boldsymbol{g}^t)$, namely

$$\pi(\sigma) = P(\theta, \varphi, \boldsymbol{p}) \ , \qquad \boldsymbol{\xi}(\sigma) = \boldsymbol{X}(\theta, \varphi, \boldsymbol{p}) \ . \tag{42}$$

By virtue of thermodynamic restrictions (20), we prove the following result.

Theorem 8: Let the material be homogeneous and isotropic, possibly with the exception of the interface between phases. Then, the general form of the constitutive relations is

$$e = c(\varphi, \boldsymbol{p})\theta + \int_0^\infty \alpha(s)\zeta^t(s)\, ds + E(\varphi, \boldsymbol{p}), \tag{43}$$

$$\boldsymbol{q} = -\theta^2 \int_0^\infty K(s)\boldsymbol{g}^t(s)\, ds, \tag{44}$$

$$\pi = (1 - \log\theta)\ \partial_\varphi c(\varphi, \boldsymbol{p}) + \frac{1}{\theta}\, \partial_\varphi E(\varphi, \boldsymbol{p}) + \partial_\varphi F(\varphi, \boldsymbol{p}) \ , \tag{45}$$

$$\boldsymbol{\xi} = (1 - \log\theta)\ \partial_{\boldsymbol{p}} c(\varphi, \boldsymbol{p}) + \frac{1}{\theta}\, \partial_{\boldsymbol{p}} E(\varphi, \boldsymbol{p}) + \partial_{\boldsymbol{p}} F(\varphi, \boldsymbol{p}) \ , \tag{46}$$

where c, E and F are suitable smooth functions.

Proof: Because of the isotropy of the material, any scalar constitutive function cannot linearly depend on a vector variable, as well as any vector valued function cannot linearly depend on a scalar variable. Hence, constitutive relations fulfil

$$\boldsymbol{\kappa} = \boldsymbol{0} \quad \text{and} \quad \boldsymbol{K} = K\, \boldsymbol{I} \ .$$

Moreover, (16) and $(20)_3$ yield $\psi - \theta\, \partial_\theta \psi = e$, which can be solved with respect to ψ. Assuming $\psi|_{\theta=0} = e|_{\theta=0}$, from (40) we obtain

$$\psi(\sigma) = \theta(1 - \log\theta)\, c(\varphi, \boldsymbol{p}) + \int_0^\infty \alpha(\varphi, \boldsymbol{p}, s)\zeta^t(s)\, ds \tag{47}$$
$$+ E(\varphi, \boldsymbol{p}) + \theta\, \mathcal{J}(\varphi, \boldsymbol{p}, \zeta^t, \boldsymbol{g}^t),$$

where $\mathcal{J}$ is independent of e. Taking into account the cross relations

$$\partial_\varphi e = -\theta^2 \partial_\theta \pi \ , \quad \partial_{\boldsymbol{p}} e = -\theta^2 \partial_\theta \boldsymbol{\xi} \ ,$$

we argue that $\partial_\varphi \alpha = 0$ and $\partial_{\boldsymbol{p}}\, \alpha = \boldsymbol{0}$. Thus (43) easily follows.

Using (41) and (47), the relation $(20)_4$ takes the form

$$\int_0^\infty \boldsymbol{K}(\theta, \varphi, \boldsymbol{p}, s)\boldsymbol{g}^t(s)\, ds = -\theta^2 \boldsymbol{J}_{\mathsf{G}}\ \mathcal{J}(\varphi, \boldsymbol{p}, \zeta^t, \boldsymbol{g}^t),$$

which, by virtue of the isotropy condition, yields

$$\begin{aligned}\boldsymbol{K}(\theta,\varphi,\boldsymbol{p},s) &= -\theta^2 K(\varphi,\boldsymbol{p},s)\,\boldsymbol{I}\,,\\ \mathcal{J}(\varphi,\boldsymbol{p},\boldsymbol{g}^t) &= \frac{1}{2}\int_0^\infty K(\varphi,\boldsymbol{p},s)\boldsymbol{g}^t(s)\cdot\boldsymbol{g}^t(s)\,ds + F(\varphi,\boldsymbol{p})\,,\end{aligned}$$

where F is a regular function independent of e and $\boldsymbol{q}$. Then, owing to (42) and the cross relations

$$\partial_\varphi \boldsymbol{q} = -\theta\,\boldsymbol{J}_{\mathrm{G}}\;\pi, \quad \partial_{\boldsymbol{p}}\boldsymbol{q} = -\theta\,\boldsymbol{J}_{\mathrm{G}}\;\boldsymbol{\xi}\,,$$

we argue that $\partial_{\boldsymbol{p}}k = \mathbf{0}$, $\partial_\varphi k = 0$, and this, in turn, implies (44).

Finally, summarizing previous results, we have

$$\begin{aligned}\psi(\sigma) = \theta(1-\log\theta)\,c(\varphi,\boldsymbol{p}) &+ \int_0^\infty \alpha(s)\zeta^t(s)\,ds\\ &+E(\varphi,\boldsymbol{p}) + \theta F(\varphi,\boldsymbol{p}) + \frac{1}{2}\theta\int_0^\infty K(s)\boldsymbol{g}^t(s)\cdot\boldsymbol{g}^t(s)\,ds\,.\end{aligned} \tag{48}$$

As a consequence, (45) and (46) follow in view of $(20)_{1,2}$. □

Assuming the specific heat to be constant, $c(\varphi,\boldsymbol{p}) = C_0$, and

$$E(\varphi,\boldsymbol{p}) = -A\varphi^2 - \frac{L_0}{2}\varphi - C_0\theta_c\,, \quad F(\varphi,\boldsymbol{p}) = \frac{L_0}{2\theta_c}\varphi + C_0\log\theta_c + \frac{1}{4}(\varphi^2-1)^2 + \frac{\kappa_1}{2}\boldsymbol{p}$$

where θ_c is the transition temperature and L_0 the latent heat, then the free and internal energy functionals take the form

$$\begin{aligned}\hat\psi(\sigma) &= \Big(\frac{\theta}{\theta_c}-1\Big)\Big(\frac{L_0}{2}\varphi + A\varphi^2\Big) + C_0\,\theta\Big(1-\log\frac{\theta}{\theta_c}\Big)\\ &\quad+\frac{\theta}{4}(\varphi^2-1)^2 + \int_0^\infty \alpha(s)\zeta^t(s)\,ds\\ &\quad+\frac{\kappa_1}{2}\,\theta\boldsymbol{p}^2 + \frac{1}{2}\theta\int_0^\infty K(s)\boldsymbol{g}^t(s)\cdot\boldsymbol{g}^t(s)\,ds - C_0\theta_c\,,\\ \hat e(\sigma) &= C_0(\theta-\theta_c) - A\varphi^2 - \frac{L_0}{2}\varphi + \int_0^\infty \alpha(s)\zeta^t(s)\,ds\,.\end{aligned} \tag{49}$$

Accordingly,

$$\hat\pi(\theta,\varphi) = \Big(\frac{L_0}{2} + 2A\varphi\Big)\Big(\frac{1}{\theta_c} - \frac{1}{\theta}\Big) + \varphi(\varphi^2-1)\,, \qquad \hat{\boldsymbol{\xi}}(\boldsymbol{p}) = \kappa_1\boldsymbol{p}\,, \tag{50}$$

so that, inserting (49) and (44) into (37) and (38), the kinetic and energy balance equations take the form

$$\begin{aligned}\nu\dot{\varphi} &= \Big(\frac{L_0}{2} + 2A\varphi\Big)\Big(\frac{1}{\theta} - \frac{1}{\theta_c}\Big) + \varphi(1-\varphi^2) + \kappa_1\Delta\varphi\,, \\ C_0\dot{\theta} &= \Big(\frac{L_0}{2} + 2A\varphi\Big)\dot{\varphi} - \alpha_0\theta + \int_0^\infty \alpha(s)\theta^t(s)\,ds \\ &\quad + \nabla\cdot\theta^2\int_0^\infty k(s)\boldsymbol{G}^t(s)\,ds,\end{aligned} \tag{51}$$

where $\alpha_0 = \int_0^\infty \alpha(s)\,ds$ and $k(s) = \int_s^\infty K(\tau)\,d\tau$.

Remark 9: Equation $(51)_1$ has the same form of the kinetic equation of the Penrose & Fife model (see eqn.(8) of [18]). In particular, when $L_0 = 0$ the model describe a second–order transition, whereas $L_0 \neq 0$ and $A = L_0/4$ accounts for a liquid–solid first–order transition.

Unfortunately, the classical form of the energy balance, as it appears into the Penrose & Fife model, cannot be recovered from $(51)_2$. Indeed, by choosing $\alpha_0 = \alpha(s) = 0$ and replacing the memory kernel $k(s)$ with the distribution $k\,\delta_0$ (δ_0 being the Dirac mass at zero), instead of (5) a non standard Fourier law for the heat flux is obtained, namely

$$\boldsymbol{q} = -k\,\theta^2\nabla\theta\,.$$

6. Appendix: quasilinear approximation

Starting from the approach developed in Sec. 4, our goal here is the derivation of quasilinear evolution equations and their comparison with (51). To this aim, we parallel the procedure proposed by Fried & Gurtin in [10]. Let $u \in U := (-\infty, 1)$ denote the dimensionless variable

$$u = \frac{\theta - \theta_c}{\theta}, \tag{52}$$

such that $u = 0$ at the transition temperature θ_c. Introducing the Gibbs function

$$y = e - \theta_c\eta = \psi + (\theta - \theta_c)\eta\,,$$

the local form of the dissipation inequality reads ($\rho = 1$)

$$\dot{y} \leq -\nabla\cdot(u\boldsymbol{q}) + ur + \boldsymbol{\xi}\cdot\dot{\boldsymbol{p}} + \pi\dot{\varphi}\,. \tag{53}$$

Moreover, exploiting the dependence of y on θ through u

$$\partial_\theta y = (\theta - \theta_c)\partial_\theta\eta,$$

and recalling the expression $(25)_1$ of the specific heat c, we have

$$\partial^2_{\theta\theta} y = \partial_\theta \eta + (\theta - \theta_c)\partial^2_{\theta\theta}\eta = \frac{c}{\theta} + (\theta - \theta_c)\partial^2_{\theta\theta}\eta \,,$$

so that y has a local minimum with respect to temperature at θ_c because of the positivity of c, namely

$$\partial_\theta y = 0 \quad \text{and} \quad \partial^2_{\theta\theta} y > 0 \qquad \text{at } \theta = \theta_c \,. \tag{54}$$

Following [10], we introduce the function $\psi^* = y - ue$, which fulfils the relation

$$\psi^* = (1-u)\psi = \frac{\theta_c}{\theta}\psi \,. \tag{55}$$

It is named *rescaled free energy*, see [3]. Using (52) and (21), it follows

$$e = -\theta^2 \frac{\partial}{\partial\theta}\left(\frac{\psi}{\theta}\right) = -\frac{\partial\psi^*}{\partial u} \,,$$

and, by virtue of (14), the inequality (53) becomes

$$\dot{\psi}^* + e\,\dot{u} \le -\boldsymbol{q}\cdot\boldsymbol{G}_u + (1-u)\left[\boldsymbol{\xi}\cdot\dot{\boldsymbol{p}} + \dot{\varphi}\pi\right] \,, \tag{56}$$

where $\boldsymbol{G}_u = \nabla u$. In the sequel, we scrutinise this inequality for small values of u, that is, in a neighbourhood of the transition temperature.

Let consider the phase–field evolution problem near equilibrium, when only a small variation of the temperature from the critical value θ_c occurs. More precisely, we assume

$$\sup_{t\in\mathbb{R}} |u(t)| < \delta \,, \qquad \sup_{t\in\mathbb{R}} \|\boldsymbol{G}_u(t)\| < \delta \,,$$

and we have the following approximations

$$\theta = \theta_c(1+u) + O(\delta^2) \qquad \boldsymbol{G} = \frac{\theta_c}{(1-u)^2}\boldsymbol{G}_u = \theta_c \boldsymbol{G}_u + O(\delta^2) \,.$$

Accordingly, we introduce the summed past histories $\bar{u}^t$ and $\boldsymbol{\Gamma}^t_u$

$$\bar{u}^t(\tau) = \int_0^\tau u(t-s)\,ds \,, \qquad \boldsymbol{\Gamma}^t_u(\tau) = \int_0^\tau \boldsymbol{G}_u(t-s)\,ds \,,$$

belonging to $\mathcal{H}_u = L^2_\lambda(\mathbb{R}^+, U)$ and $\mathbb{H} = L^2_\mu(\mathbb{R}^+, \mathbb{R}^3)$, respectively. In this approximation scheme, the thermal variables in the state, namely $(\theta(t), \zeta^t, \boldsymbol{g}^t)$, can be substituted by $(u(t), \bar{u}^t, \boldsymbol{\Gamma}^t_u)$, whose norm in $\mathbb{R}\times\mathcal{H}_u\times\mathbb{H}$ is of order $O(\delta)$, namely

$$\sup_{t\in\mathbb{R}} |u(t)|^2 + \|\bar{u}^t\|^2_{\mathcal{H}_u} + \|\boldsymbol{\Gamma}^t_u\|^2_{\mathbb{H}} \le C\delta^2 \,. \tag{57}$$

In fact, owing to Definition 3,

$$\|\bar{u}^t\|_{\mathcal{H}_u} = \left(\int_0^\infty \lambda(s)|\bar{u}^t(s)|^2 ds\right)^{\frac{1}{2}} \leq \left(\int_0^\infty \lambda(s) s^2 ds\right)^{\frac{1}{2}} \delta = \sqrt{\lambda_0}\delta\,,$$

$$\|\boldsymbol{\Gamma}_u^t\|_{\mathbb{H}} = \left(\int_0^\infty \mu(s)|\boldsymbol{\Gamma}_u^t(s)|^2 ds\right)^{\frac{1}{2}} \leq \delta \left(\int_0^\infty \mu(s) s^2 ds\right)^{\frac{1}{2}} = \sqrt{\mu_0}\delta\,.$$

Proposition 10: *Assuming that the heat source r is of the same order $O(\delta)$ as the temperature variation, we can neglect the accretive terms in the energy balance (14), which thus yields its classical expression*

$$\dot{e} = -\nabla \cdot \boldsymbol{q} + r\,. \tag{58}$$

On the other hand, neglecting terms of order $o(\delta^2)$, (56) gives

$$\dot{\psi}^* + \dot{u}\, e + \boldsymbol{q} \cdot \boldsymbol{G}_u - \boldsymbol{\xi} \cdot \dot{\boldsymbol{p}} - \dot{\varphi}\pi \leq 0\,. \tag{59}$$

Proof: The procedure is exactly the same as in [10]. It is easy to check that the quantities e and $\boldsymbol{q}$ are of order $O(\delta)$. On the contrary, denoting with $y_c = y\,|_{\theta=\theta_c}$, the variation of the Gibbs function $y - y_c$ is of order $O(\delta^2)$, because of (54). Then, observing inequality (53), it turns out that the accretive term $\boldsymbol{\xi} \cdot \dot{\boldsymbol{p}} - \dot{\varphi}\,\pi$ is of order $O(\delta^2)$, too, and (58) follows from (14). Accordingly, (59) follows from (56). □

Starting from the linearized relations (58), (59) and the accretive balance law $(13)_2$, we are looking for the thermodynamic restrictions. Letting

$$\sigma_u = (\varphi, \boldsymbol{p}, u(t), \bar{u}^t, \boldsymbol{\Gamma}_u^t) \qquad \text{and} \qquad \omega_u = (\sigma_u, h, \boldsymbol{f})\,,$$

by the same arguments of Sec. 4 we have

$$e = \hat{e}(\sigma_u)\,, \quad \boldsymbol{q} = \hat{\boldsymbol{q}}(\sigma_u)\,, \quad \eta = \hat{\eta}(\sigma_u)\,, \quad \psi^* = \hat{\psi}^*(\sigma_u)$$

$$\boldsymbol{\xi} = \hat{\boldsymbol{\xi}}^{(s)}(\sigma_u) + \hat{\boldsymbol{\xi}}^{(d)}(\omega_u)\,, \quad \pi = \hat{\pi}^{(s)}(\sigma_u) + \hat{\pi}^{(d)}(\omega_u)\,.$$

Observe that the time derivative of ψ^* is given by

$$\begin{aligned}\dot{\psi}^*(t) = &\partial_\varphi\hat{\psi}^*(\sigma_u)h(t) + \partial_{\boldsymbol{p}}\hat{\psi}^*(\sigma_u)\cdot\boldsymbol{f}(t) + \partial_h\hat{\psi}^*(\sigma_u)\cdot\dot{h}(t) + \partial_{\mathrm{f}}\hat{\psi}^*(\sigma_u)\cdot\dot{\boldsymbol{f}}(t)\\ &+\partial_u\hat{\psi}^*(\sigma_u)\dot{u} + J_u\hat{\psi}^*(\sigma_u)\,u(t) + \boldsymbol{J}_{\mathrm{G}_u}\hat{\psi}^*(\sigma_u)\cdot\boldsymbol{G}_u(t) - \delta\hat{\psi}^*(\sigma_u|u^t, \boldsymbol{G}_u^t)\,,\end{aligned}$$

where J_u and $\boldsymbol{J}_{\mathrm{G}_u}$ have the analogous meaning of J_θ and $\boldsymbol{J}_{\mathrm{G}}$ in (18), respectively. As a consequence, (59) yields

$$\begin{aligned}&(\partial_\varphi\psi^* - \pi)\,h + (\partial_{\boldsymbol{p}}\psi^* - \boldsymbol{\xi})\cdot\boldsymbol{f} + \partial_h\psi^*\,\dot{h} + \partial_{\mathrm{f}}\psi^*\cdot\dot{\boldsymbol{f}}\\ &+(\partial_u\psi^* + e)\dot{u} + \left(\boldsymbol{q} + \boldsymbol{J}_{\mathrm{G}_u}\hat{\psi}^*\right)\cdot\boldsymbol{G}_u + J_u\psi^* u - \delta\psi^* \leq 0\,,\end{aligned} \tag{60}$$

and the balance laws (13) and (58) may be rewritten as

$$\partial_\varphi \boldsymbol{\xi} \cdot \boldsymbol{p} + \partial_{\boldsymbol{p}} \boldsymbol{\xi} \cdot \nabla \boldsymbol{p} + \partial_h \boldsymbol{\xi} \cdot \dot{\boldsymbol{p}} + \partial_{\mathbb{f}} \boldsymbol{\xi} \cdot \nabla \boldsymbol{f} + \boldsymbol{\xi}_u \cdot \boldsymbol{g}_u + \delta \boldsymbol{\xi} - \pi = 0$$

$$\partial_\varphi e\, h + \partial_{\boldsymbol{p}} e \cdot \boldsymbol{f} + \partial_h e \cdot \dot{h} + \partial_{\mathbb{f}} e \cdot \dot{\boldsymbol{f}} + \partial_u e\, \dot{u} + J_u e\, u + \boldsymbol{J}_{\mathsf{G}_u} e \cdot \boldsymbol{G}_u$$
$$-\delta e + \partial_\varphi \boldsymbol{q} \cdot \boldsymbol{p} + \partial_{\boldsymbol{p}} \boldsymbol{q} \cdot \nabla \boldsymbol{p} + \partial_h \boldsymbol{q} \cdot \boldsymbol{f} + \partial_{\mathbb{f}} \boldsymbol{q} \cdot \nabla \boldsymbol{f} + \boldsymbol{q}_u \cdot \boldsymbol{G}_u + \delta \tilde{\boldsymbol{q}} = r\,.$$

Since the quantities $\dot{\boldsymbol{f}}$, $\dot{h}$, $\dot{u}$ and $\boldsymbol{G}_u$ may be chosen arbitrarily, in that they do not enter the definition of ω_u, (60) implies $\psi^* = \hat{\psi}^*(\sigma_u)$ and

$$e = \hat{e}(\sigma_u) = -\partial_u \hat{\psi}^*(\sigma_u), \tag{61}$$

$$\boldsymbol{q} = \hat{\boldsymbol{q}}(\sigma_u) = -\boldsymbol{J}_{\mathsf{G}_u} \hat{\psi}(\sigma_u)\,. \tag{62}$$

Substituting (61) and (62) into (60), we obtain

$$(\partial_\varphi \psi^* - \pi)\, h + (\partial_{\boldsymbol{p}} \psi^* - \boldsymbol{\xi}) \cdot \boldsymbol{f} + J_u \psi^* u - \delta \psi^* \leq 0\,.$$

Since ψ^* is independent of h and $\boldsymbol{f}$, the reduced inequality follows when $h = 0$, $\boldsymbol{f} = \boldsymbol{0}$:

$$J_u \psi^*(\sigma_u) u - \delta \psi^*(\sigma_u) \leq 0\,. \tag{63}$$

Finally, in view of Assumption 4 we have

$$\hat{\pi}^{(s)}(\sigma_u) = \partial_\varphi \hat{\psi}^*(\sigma_u) \qquad \hat{\boldsymbol{\xi}}^{(s)}(\sigma_u) = \partial_{\boldsymbol{p}} \hat{\psi}^*(\sigma_u)\,, \tag{64}$$

$$\hat{\pi}^{(d)}(\sigma_u, h, \boldsymbol{f}) = \beta(\sigma_u, h, \boldsymbol{f}) h \qquad \hat{\boldsymbol{\xi}}^{(d)}(\sigma_u, h, \boldsymbol{f}) = \boldsymbol{B}(\sigma_u, h, \boldsymbol{f}) \boldsymbol{f}\,, \tag{65}$$

where β is non negative and $\boldsymbol{B}$ is positive semidefinite.

In order to obtain a *quasilinear* theory, according to (57) we consider constitutive relations for e, $\boldsymbol{q}$, $\boldsymbol{\xi}$ and π which depend at most linearly on the thermal part of the state $(u, \bar{u}^t, \boldsymbol{\Gamma}_u^t)$, namely

$$e(\sigma_u) = c^{(e)}(\varphi, \boldsymbol{p}) u + \mathcal{F}^{(e)}(\varphi, \boldsymbol{p}\,|\,\bar{u}^t) + \mathcal{G}^{(e)}(\varphi, \boldsymbol{p}\,|\,\boldsymbol{\Gamma}_u^t) + E(\varphi, \boldsymbol{p})$$
$$\boldsymbol{q}(\sigma_u) = \boldsymbol{c}^{(\mathbb{q})}(\varphi, \boldsymbol{p}) u + \boldsymbol{\mathcal{F}}^{(\mathbb{q})}(\varphi, \boldsymbol{p}\,|\,\bar{u}^t) + \boldsymbol{\mathcal{G}}^{(\mathbb{q})}(\varphi, \boldsymbol{p}\,|\,\boldsymbol{\Gamma}_u^t) + \boldsymbol{Q}(\varphi, \boldsymbol{p})$$
$$\boldsymbol{\xi}^{(s)}(\sigma_u) = \boldsymbol{c}^{(\boldsymbol{\xi})}(\varphi, \boldsymbol{p}) u + \boldsymbol{\mathcal{F}}^{(\boldsymbol{\xi})}(\varphi, \boldsymbol{p}\,|\,\bar{u}^t) + \boldsymbol{\mathcal{G}}^{(\boldsymbol{\xi})}(\varphi, \boldsymbol{p}\,|\,\boldsymbol{\Gamma}_u^t) + \boldsymbol{X}(\varphi, \boldsymbol{p})$$
$$\pi^{(s)}(\sigma_u) = c^{(\pi)}(\varphi, \boldsymbol{p}) u + \mathcal{F}^{(\pi)}(\varphi, \boldsymbol{p}\,|\,\bar{u}^t) + \mathcal{G}^{(\pi)}(\varphi, \boldsymbol{p}\,|\,\boldsymbol{\Gamma}_u^t) + P(\varphi, \boldsymbol{p})\,.$$

Terms $\mathcal{F}$, $\boldsymbol{\mathcal{F}}$ and $\mathcal{G}$, $\boldsymbol{\mathcal{G}}$ are linear functionals on $\mathcal{H}_u$ and $\mathbb{H}$, respectively, and take the form

$$\boldsymbol{\mathcal{F}}^{(\mathbb{I})} = \int_0^\infty \boldsymbol{\alpha}^{(\mathbb{I})}(\varphi, \boldsymbol{p}, s) \bar{u}^t(s)\, ds\,, \quad \boldsymbol{\mathcal{G}}^{(\mathbb{I})} = \int_0^\infty \boldsymbol{K}^{(\mathbb{I})}(\varphi, \boldsymbol{p}, s) \boldsymbol{\Gamma}_u^t(s)\, ds\,,$$

$$\mathcal{F}^{(m)} = \int_0^\infty \alpha^{(m)}(\varphi, \boldsymbol{p}, s) \bar{u}^t(s)\, ds\,, \quad \mathcal{G}^{(m)} = \int_0^\infty \boldsymbol{\kappa}^{(m)}(\varphi, \boldsymbol{p}, s) \cdot \boldsymbol{\Gamma}_u^t(s)\, ds\,,$$

where $\boldsymbol{l} = \boldsymbol{q}, \boldsymbol{\xi}$ and $m = e, \pi$. Here $\boldsymbol{K}^{(\boldsymbol{l})}$ is tensor valued, whereas $\boldsymbol{\alpha}^{(\boldsymbol{l})}$ and $\boldsymbol{\kappa}^{(m)}$ are vector valued. Taking into account thermodynamic restrictions (61) (62) and relations (64) (65), we prove the following result.

Theorem 11: Assume the material to be homogeneous and isotropic, possibly with the exception of the interface between phases. In addition, let the heat flux $\boldsymbol{q}(t)$ vanish when $\boldsymbol{G}_u(s) = \mathbf{0}$ for all $s < t$. Then, the general form of the constitutive relations is

$$e = c_0 u + E(\varphi) + \int_0^\infty \alpha(s)\bar{u}^t(s)\, ds, \tag{66}$$

$$\boldsymbol{q} = \int_0^\infty K_{\mathrm{q}}(s)\boldsymbol{\Gamma}_u^t(s)\, ds, \tag{67}$$

$$\boldsymbol{\xi}^{(s)} = \partial_{\boldsymbol{p}} R(\varphi, \boldsymbol{p}) + \int_0^\infty K_{\boldsymbol{\xi}}(s)\boldsymbol{\Gamma}_u^t(s)\, ds, \tag{68}$$

$$\pi^{(s)} = -E'(\varphi)u + \partial_\varphi R(\varphi, \boldsymbol{p}) + \int_0^\infty \partial_\varphi F(\varphi, s)\bar{u}^t(s)\, ds, \tag{69}$$

where R and F are suitable smooth functions.

Proof: By virtue of the isotropy of the material, the constitutive relations fulfil

$$\mathcal{G}^{(e)} = \mathcal{G}^{(\pi)} = 0\,, \quad \boldsymbol{c}^{(\mathrm{q})} = \boldsymbol{c}^{(\boldsymbol{\xi})} = \mathbf{0}\,, \quad \boldsymbol{\mathcal{F}}^{(\mathrm{q})} = \boldsymbol{\mathcal{F}}^{(\boldsymbol{\xi})} = \mathbf{0}\,, \tag{70}$$

$$\boldsymbol{K}^{(\mathrm{q})} = K_{\mathrm{q}}\boldsymbol{I}\,, \quad \boldsymbol{K}^{(\boldsymbol{\xi})} = K_{\boldsymbol{\xi}}\,\boldsymbol{I}\,, \tag{71}$$

and, from the vanishing of $\boldsymbol{q}$ when $\boldsymbol{\Gamma}_u^t = \boldsymbol{G}_u^t = 0$, it follows

$$\boldsymbol{Q}(\varphi, \boldsymbol{p}) = \mathbf{0}\,. \tag{72}$$

Moreover, in view of (61), (62), and (64), we have

$$\partial_\varphi e = -\partial_u \pi^{(s)} = -c^{(\pi)}(\varphi, \boldsymbol{p})\,, \quad \partial_{\boldsymbol{p}} e = -\partial_u \boldsymbol{\xi}^{(s)} = \mathbf{0}\,, \tag{73}$$

$$\partial_\varphi \boldsymbol{\xi}^{(s)} = \partial_{\boldsymbol{p}} \pi^{(s)}\,, \tag{74}$$

$$\partial_\varphi \boldsymbol{q} = -\boldsymbol{J}_{\mathrm{G}_u}\pi^{(s)} = \mathbf{0}\,, \quad \partial_{\boldsymbol{p}} \boldsymbol{q} = -\boldsymbol{J}_{\mathrm{G}_u}\boldsymbol{\xi}^{(s)} = -\boldsymbol{J}_{\mathrm{G}_u}\boldsymbol{\mathcal{G}}^{(\boldsymbol{\xi})}\,. \tag{75}$$

From (73) we obtain

$$\partial_\varphi c^{(e)} = \partial_\varphi \mathcal{F}^{(e)} = 0, \qquad \partial_{\boldsymbol{p}} c^{(e)} = \partial_{\boldsymbol{p}}\, \mathcal{F}^{(e)} = \mathbf{0},$$

and (66) follows by letting $c_0 = c^{(e)}$ and $\alpha = \alpha^{(e)}$. Because of the definition of $\boldsymbol{\mathcal{G}}^{(\boldsymbol{\xi})}$ and (71),

$$\boldsymbol{J}_{\mathrm{G}_u}\boldsymbol{\mathcal{G}}^{(\boldsymbol{\xi})} = \int_0^\infty K_{\boldsymbol{\xi}}(\varphi, \boldsymbol{p}, s)\, ds\, \boldsymbol{I}\,,$$

and from $(75)_2$ we argue that $\partial_{\boldsymbol{p}}\boldsymbol{q}$ at most depends on $(\varphi, \boldsymbol{p})$. As a consequence, $\partial_{\boldsymbol{p}}\boldsymbol{\mathcal{G}}^{(\mathrm{q})} = \mathbf{0}$ and $(75)_1$ yields $\partial_\varphi \boldsymbol{\mathcal{G}}^{(\mathrm{q})} = \mathbf{0}$. Summarizing and remembering (70), (71) and (72), we obtain (67). By virtue of $(75)_2$ and (67), $\boldsymbol{\mathcal{G}}^{(\boldsymbol{\xi})}$ is independent of φ and $\boldsymbol{p}$, and

$$\boldsymbol{J}_{\mathrm{G}_u}\boldsymbol{\mathcal{G}}^{(\boldsymbol{\xi})} = \int_0^\infty K_{\boldsymbol{\xi}}(s)\,ds\,\boldsymbol{I} = \mathbf{0}\,.$$

From (70) and (74) we get

$$\partial_\varphi \boldsymbol{X}(\varphi, \boldsymbol{p}) = \partial_{\boldsymbol{p}} c^{(\pi)}(\varphi, \boldsymbol{p})u + \partial_{\boldsymbol{p}}\mathcal{F}^{(\pi)}(\varphi, \boldsymbol{p}\,|\bar{u}^t) + \partial_{\boldsymbol{p}} P(\varphi, \boldsymbol{p})\,,$$

so that $\partial_{\boldsymbol{p}} c^{(\pi)} = \partial_{\boldsymbol{p}}\mathcal{F}^{(\pi)} = \mathbf{0}$. Moreover, (73) yields $c^{(\pi)} = -E'(\varphi)$, and (68)–(69) follow by letting

$$\alpha^{(\pi)}(\varphi, s) = \partial_\varphi F(\varphi, s)\,, \quad \boldsymbol{X}(\varphi, \boldsymbol{p}) = \partial_{\boldsymbol{p}} R(\varphi, \boldsymbol{p})\,, \quad P(\varphi, \boldsymbol{p}) = \partial_\varphi R(\varphi, \boldsymbol{p})\,,$$

where F and R are regular functions. □

In view of (66)–(69) the functional ψ^* takes the form

$$\begin{aligned}\psi^*(\sigma_u) = &-\frac{1}{2}c_0 u^2 - u\int_0^\infty \alpha(s)\bar{u}^t(s)\,ds - uE(\varphi)\\ &-\frac{1}{2}\int_0^\infty K_{\mathrm{q}}(s)\boldsymbol{\Gamma}_u^t(s)\cdot\boldsymbol{\Gamma}_u^t(s)\,ds + R(\varphi, \boldsymbol{p})\\ &+\boldsymbol{p}\cdot\int_0^\infty K_{\boldsymbol{\xi}}(s)\boldsymbol{\Gamma}_u^t(s)\,ds + \int_0^\infty F(\varphi, s)\bar{u}^t(s)\,ds + \Psi^*[\bar{u}^t]\end{aligned} \tag{76}$$

where $\int_0^\infty K_{\boldsymbol{\xi}}(s)\,ds = 0$. Here, $\Psi^*[\bar{u}^t]$ is any linear functional on $\mathcal{H}_u$ fulfilling the reduced inequality (63).

A simpler model is obtained when the accretive quantities are independent of thermal histories, namely when functions $K_{\boldsymbol{\xi}}$ and F vanish identically.

Corollary 12: *Under the hypotheses of Theorem 11, assuming that $\hat{\pi}$ and $\hat{\boldsymbol{\xi}}$ are independent of $\bar{u}^t$ and $\boldsymbol{\Gamma}_u^t$, and letting $R(\varphi, \boldsymbol{p})$ be a quadratic function of $\boldsymbol{p}$, the evolution equations of the system (26) and (27) take the simple form*

$$c_0\dot{u} + E'\dot{\varphi} = -\alpha_0 u + \int_0^\infty \alpha(s)u^t(s)\,ds + \nabla\cdot\int_0^\infty k(s)\nabla u^t(s)\,ds, \tag{77}$$

$$\beta\dot{\varphi} = E'u - G' + \nabla\cdot\boldsymbol{\Lambda}\nabla\varphi + \nabla\cdot\boldsymbol{B}\,\nabla\dot{\varphi} - \frac{1}{2}\nabla\varphi\cdot\boldsymbol{\Lambda}'\nabla\varphi, \tag{78}$$

where $\alpha_0 = \int_0^\infty \alpha(s)\,ds$ and $k(s) = \int_s^\infty K_{\mathrm{q}}(\tau)\,d\tau$.

Proof: In such a case, $K_{\boldsymbol{\xi}} = F = 0$ and (67) becomes

$$\boldsymbol{q}\left(\boldsymbol{\Gamma}_u^t\right) = \int_0^\infty K_{\mathrm{q}}(s)\boldsymbol{\Gamma}_u^t(s)\,ds\;, \tag{79}$$

whereas (68) and (69) take the form

$$\begin{aligned}\pi^{(s)}(\varphi,\boldsymbol{p}) &= -E'(\varphi)u + \partial_\varphi R(\varphi,\boldsymbol{p}),\\ \boldsymbol{\xi}^{(s)}(\varphi,\boldsymbol{p}) &= \partial_{\boldsymbol{p}} R(\varphi,\boldsymbol{p}).\end{aligned}$$

In particular, when $R(\varphi,\boldsymbol{p})$ is a quadratic function of $\boldsymbol{p}$, namely

$$R(\varphi,\boldsymbol{p}) = G(\varphi) + \frac{1}{2}\boldsymbol{p}\cdot\boldsymbol{\Lambda}(\varphi)\boldsymbol{p}$$

with $\boldsymbol{\Lambda}(\varphi)$ symmetric, we obtain

$$\pi^{(s)}(\varphi,\boldsymbol{p}) = -E'(\varphi)u + G'(\varphi) + \frac{1}{2}\boldsymbol{p}\cdot\boldsymbol{\Lambda}'(\varphi)\boldsymbol{p}, \tag{80}$$

$$\boldsymbol{\xi}^{(s)}(\varphi,\boldsymbol{p}) = \boldsymbol{\Lambda}(\varphi)\boldsymbol{p}\;. \tag{81}$$

Accordingly, the rescaled free energy (76) is given by

$$\psi^*(\sigma_u) = -E(\varphi)u + G(\varphi) + \frac{1}{2}\boldsymbol{p}\cdot\boldsymbol{\Lambda}(\varphi)\boldsymbol{p} + \Psi^*(u,\bar{u}^t,\boldsymbol{\Gamma}_u^t), \tag{82}$$

where

$$\Psi^*(u,\bar{u}^t,\boldsymbol{\Gamma}_u^t) = -\frac{1}{2}c_0u^2 - u\int_0^\infty \alpha(s)\bar{u}^t(s)\,ds - \frac{1}{2}\int_0^\infty K_{\mathrm{q}}(s)\boldsymbol{\Gamma}_u^t(s)\cdot\boldsymbol{\Gamma}_u^t(s)\,ds$$

contains all terms of order $O(\delta^2)$. Finally, in view of Proposition 10, the thesis follows. □

Well-posedness and long time behaviour for some special form of the system (77)-(78), with proper initial and boundary conditions, have been investigated in [13, 12].

Remark 13: The general form of equation (78) reduces to the kinetic equation of the standard model when $\boldsymbol{B} = \boldsymbol{0}$, $\beta = 1$ and $\boldsymbol{\Lambda}(\varphi) = \lambda_0\boldsymbol{I}$. In particular, $E(\varphi)$ and $G(\varphi)$ stand for the latent heat function and the double–well potential, respectively. The classical form of the energy balance (as it appears into the standard phase–field model), along with the Fourier law for the heat flux can be recovered from (77) by choosing $\alpha_0 = \alpha(s) = 0$ and replacing the memory kernel $k(s)$ with the distribution $k\,\delta_0$ (δ_0 being the Dirac mass at zero).

References

[1] S. Aizicovici and E. Feireisl, Long-time stabilization of solutions to a phase-field model with memory. *J. Evol. Equ.* **1**, 69–84 (2001).

[2] S. Aizicovici and H. Petzeltova, Asymptotic behavior of solutions of a conserved phase-field system with memory. *J. Integral Equations Appl.* **15**, 217–240 (2003).

[3] H.W. Alt and I. Pawlow, A mathematical model of dynamics of non–isothermal phase separation. *Phys. D* **59**, 389–416 (1992).

[4] E. Bonetti, P. Colli and M. Fremond, A phase field model with thermal memory governed by the entropy balance. *Math. Models Methods Appl. Sci.* **13**, 1565–1588 (2003).

[5] M. Brokate and J. Sprekels, *Hysteresis and Phase Transitions.* Springer, New York, 1996.

[6] G. Caginalp, An analysis of a phase field model of a free boundary. *Arch. Rational Mech. Anal.* **92**, 205–245 (1986).

[7] B.D. Coleman, M. Fabrizio and D.R. Owen, Thermodynamics and the constitutive relations for second sound in crystals. In: *New perspectives in Thermodynamics.* Ed. J. Serrin (Springer-Verlag, Berlin-Heidelberg, 1986.).

[8] M. Fabrizio, C. Giorgi and A. Morro, A Thermodynamic approach to non–isothermal phase–field evolution in continuum physics. *Submitted*, 2005.

[9] M. Frémond, *Non-smooth Thermomechanics.* Springer-Verlag, Berlin, 2002.

[10] E. Fried and M.E. Gurtin, Continuum theory of thermally induced phase transition based on an order parameter. *Phys. D* **68**, 326–343 (1993).

[11] G. Gentili and C. Giorgi, A mathematical model for phase transitions at low temperature. *Quad. Sem. Mat. Brescia* n. 35/2000.

[12] C. Giorgi, M. Grasselli and V. Pata, Uniform attractors for a phase-field model with memory and quadratic nonlinearity. *Indiana Univ. Math. J.* **48**, 1395–1445 (1999).

[13] C. Giorgi, M. Grasselli and V. Pata, Well-posedness and longtime behavior of the phase-field model with memory in a history space setting. *Quart. Appl. Math.* **59**, 701–736 (2001).

[14] M.E. Gurtin, Generalized Ginzburg-Landau and Cahn-Hilliard equations based on a microforce balance. *Phys. D* **92**, 178–192 (1996).

[15] M.E. Gurtin and A.C. Pipkin, A general theory of heat conduction with finite wave speeds. *Arch. Rational Mech. Anal.* **31**, 113–126 (1968).

[16] J. Jackle, Heat conduction and relaxation in liquids of high vioscosity. *Phys. A* **162**, 377–404 (1990).

[17] O. Penrose and P.C. Fife, Thermodynamically consistent models of phase–field type for the kinetics of phase transitions. *Phys. D* **43**, 44–62 (1990).

[18] O. Penrose and P.C. Fife, On the relation between the standard phase–field model and "a thermodynamically consistent" phase–field model. *Phys. D* **69**, 107–113 (1993).

HYSTERESIS IN A FIRST ORDER HYPERBOLIC EQUATION

Jana Kopfová

Mathematical Institute of the Silesian University at Opava
Na Rybníčku 1, 746 01 Opava, Czech Republic
E-mail: Jana.Kopfova@math.slu.cz

We study a hyperbolic equation of first order with a hysteresis nonlinearity. For the integral solution of this equation we derive an entropy condition of the type introduced by Kružkov.

1. Introduction

In this paper, we study the quasilinear hyperbolic equation with hysteresis

$$\frac{\partial}{\partial t}(u+w) + \sum_{j=1}^{N} \frac{\partial}{\partial x_j}(b_j u) + cu = f, \tag{1}$$

where $w = \mathcal{F}(u)$ is a generalized play or a Prandtl-Ishlinskii hysteresis operator of play type; both of them can possibly be discontinuous. Notice that the hysteresis brings nonlinearity to the equation. Equation (1) arises as a generic model for transport and adsorption of chemical concentration in chemical and geological engineering, see [9].

The quasilinear hyperbolic equation of first order

$$u_t + [\phi(u)]_x = 0, \qquad u(0) = u_0, \tag{2}$$

where the nonlinearity is caused by the function ϕ, was studied in a number of papers. It is well-known that even for ϕ and u_0 smooth, (2) exhibits singularities in a finite time. To be able to continue the solution, one has to pass to the concept of weak solutions, where discontinuities are allowed. Weak solutions are in general not uniquely determined by the data, and further, physically motivated conditions have to be prescribed. The simplest one is an entropy condition, stating that the entropy of the system must be decreasing, generalized by Olejnik [7] (for $N = 1$). S.N.Kružkov [6] gave a different definition of a generalized solution of (2), which contains a condition characterizing the admissible discontinuities of solutions. Kružkov showed that such a solution is unique. Note that this solution notion is equivalent, in the case $N = 1$, to the entropy solution from [7]. Inspired by Kružkovs work, Crandall [1] shows that the unique integral solution of condition derived by Kružkov. In the first section we give a brief overview of their results.

There is an example of P. Kordulová [3], showing that for the quasilinear hyperbolic equation with hysteresis (1), coupled with a generalized play operator whose hysteresis boundary curves are specially chosen, the situation is similar to the quasilinear hyperbolic equation without hysteresis (2). The quasilinear hyperbolic equation with hysteresis was also recently studied in [8], where for a linear play operator and a simpler form of the equation a differentiable solution is obtained using the nonlinear semigroup theory in L^2, and also in [12], where the operator considered is the relay operator.

It was expected, see [11], that the integral solution of (1), for which existence was proved in [11] using the nonlinear semigroup theory, and which is unique by construction, fulfills a condition of the type introduced by Kružkov. To derive such an entropy condition for the integral solution of (1) with hysteresis was posed as an open problem in [11]. The nonlinear semigroup theory enables us to deal with continuous and discontinuous hysteresis as well. These also include completed delayed relay operators. The solution of this problem was presented by the author in [2], but the conditions for the hysteresis operator were quite restricted and did not include the example from [3].

The main result of this paper is a theorem which shows that an integral solution of (1) defined by the semigroup theory satisfies an entropy condition introduced by Kružkov. This is proved without the assumption of the symmetry of hysteresis boundary curves and also, contrary to [2], for a non-zero right-hand side term f.

It remains an interesting open problem to see if the presence of hysteresis, whose hysteresis loop is strictly convex, in the hyperbolic equation of first order prevents the formation of shocks, as it was proved by P. Krejčí [4] for a hyperbolic equation of second order with hysteresis. This question is recently under study.

2. Hysteresis

Let b_j and c be given smooth functions, Ω an open bounded subset of R^n of Lipschitz class.

We consider the equation (1) and couple it with the hysteresis relation

$$w(x,t) = [\mathcal{E}(u(x,.), w_0(x)] (t) \qquad \text{in } [0,T], \text{ a.e. in } \Omega. \tag{3}$$

Here $\mathcal{E}$ is a multivalued functional, $w_0(x)$ a given initial value. Its values do not only depend on the current value of $u(.,t)$ at $t > 0$, but also on the past history $u(.,s), 0 < s < t$. We consider at first $\mathcal{E}$ to be a generalized play operator.

We set $\tilde{\mathbb{R}} := [-\infty, +\infty], \tilde{\mathbb{R}}^+ := [0, +\infty], \tilde{\mathbb{R}}^- := [-\infty, 0]$, $Q := \Omega \times [0,T]$. Let

$$\gamma_r, \gamma_l : R \to \tilde{\mathbb{R}} \text{ be maximal monotone (possibly multivalued) functions,}$$
$$\text{and} \qquad \inf \gamma_r(u) \le \sup \gamma_l(u) \qquad \forall u \in \mathbb{R}. \tag{4}$$

Now, given $w_0 \in \mathbb{R}$, we construct the hysteresis operator $\mathcal{E}(\cdot, w_0)$ as follows. Let u be any continuous, piecewise linear function on $\mathbb{R}^+$ such that u is linear on $[t_{i-1}, t_i]$ for $i = 1, 2, ...$ We then define $w := \mathcal{E}(u, w_0) : \mathbb{R}^+ \to \mathbb{R}$ by

$$w(t) := \begin{cases} \min\{\gamma_l(u(0)), \max\{\gamma_r(u(0)), w_0\}\} & \text{if } t = 0, \\ \min\{\gamma_l(u(t)), \max\{\gamma_r(u(t)), w(t_{i-1})\}\} & \text{if } t \in (t_{i-1}, t_i], i = 1, 2, ... \end{cases}$$

Note that $w(0) = w_0$ only if $\gamma_r(u(0)) \le w_0 \le \gamma_l(u(0))$. The hysteresis relation is assumed to hold pointwise in space :

$$w(x,t) = [\mathcal{E}(u(x,.), w_0(x))](t) \qquad \text{in } [0,T], \text{ a.e. in } \Omega. \tag{5}$$

As proved in Visintin [11], Section III.2, $\mathcal{E}(\cdot, w_0)$ can be extended to $C^0([0,T])$ by continuity. This operator is called a generalized play, see Figure 8.1.

To define the generalized Prandtl-Ishlinskii operator of play type, let us assume that we are given a measure space $(\mathcal{P}, \mathcal{A}, \mu)$, where μ is a finite Borel measure. For μ-almost any $\rho \in \mathcal{P}$, let $(\gamma_{\rho l}, \gamma_{\rho r})$ be a pair of functions $\gamma_{\rho l}, \gamma_{\rho r} : \mathbb{R} \to \mathbb{R}$, satisfying (4), and for each $\rho \in \mathcal{P}$ let $w_{\rho 0} \in \mathbb{R}$ be a given initial value. Let $\mathcal{E}_\rho(\cdot, w_{\rho 0})$ be the generalized play operator corresponding to the couple $(\gamma_{\rho l}, \gamma_{\rho r})$. Then, the operator defined as

$$\tilde{\mathcal{E}}_\mu \left(\tilde{u}, \{w_{\rho 0}\}_{\rho \in \mathcal{P}}\right) := \int_{\mathcal{P}} \mathcal{E}_\rho \left(\tilde{u}, w_{\rho 0}\right) d\mu(\rho)$$

is called generalized Prandtl-Ishlinskii operator of play type.

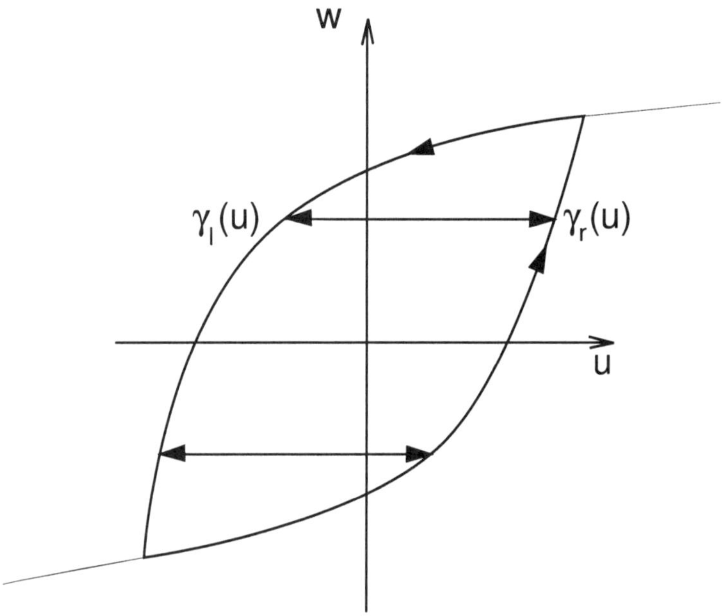

Fig. 8.1. The generalized play.

3. Semigroup approach

The system (1) with (3), where $\mathcal{E}$ is the generalized play operator, is formally equivalent to

$$\begin{cases} \frac{\partial u}{\partial t} + \xi + \sum_{j=1}^{N} \frac{\partial}{\partial x_j}(b_j u) + cu = f & \text{in } Q, \\ \frac{\partial w}{\partial t} - \xi = 0 & \text{in } Q, \\ \xi \in \phi(u, w) & \text{in } Q, \end{cases} \tag{6}$$

where

$$\phi(u,w) = \begin{cases} +\infty & \text{if } w < \inf \gamma_r(u), \\ \tilde{\mathbb{R}}^+ & \text{if } w \in \gamma_r(u) \backslash \gamma_l(u), \\ \{0\} & \text{if } \sup \gamma_r(u) < w < \inf \gamma_l(u), \\ \tilde{\mathbb{R}}^- & \text{if } w \in \gamma_l(u) \backslash \gamma_r(u), \\ -\infty & \text{if } w > \sup \gamma_l(u), \\ \tilde{\mathbb{R}} & \text{if } w \in \gamma_l(u) \cap \gamma_r(u). \end{cases} \tag{7}$$

To simplify the discussion, we assume that $\{b_j \in C^1(\overline{\Omega})\}_{j=1,\dots,N}$, $\sum_{j=1}^N b_j \nu_j = 0$ a.e. on $\partial\Omega$ and $c \in L^\infty(\Omega)$, where $\overrightarrow{\nu}$ denotes a field normal to $\partial\Omega$.

By introducing the following operators

$$
\begin{aligned}
D(A) &:= \{U := (u,w) \in \mathbb{R}^2 : \inf \gamma_r(u) \le w \le \sup \gamma_l(u)\},\\
A(U) &:= \{(\xi, -\xi) : \xi \in \phi(U) \cap \mathbb{R}\} \qquad \forall U \in D(A),\\
B(u) &:= \sum_{j=1}^N \frac{\partial}{\partial x_j}(b_j u) + cu,\\
R(U) &:= (B(u), 0),
\end{aligned}
$$

and by setting

$$U := (u,w), \qquad U_0 := (u_0, w_0), \qquad F := (f, 0),$$

the Cauchy problem for the system (6) can be written in the form

$$
\begin{gathered}
\frac{\partial U}{\partial t} + A(U) + R(U) \ni F \qquad \text{in } Q,\\
U(0) = U_0.
\end{gathered}
\tag{8}
$$

This approach can be easily extended to the case in which $\mathcal{E}$ is replaced by a generalized Prandtl-Ishlinskii operator of play type.

Then, we have the following theorem [11]:

Theorem 1: Let Ω be an open subset of $\mathbb{R}^N$ $(N \ge 1)$ of Lipschitz class. Let $L^1(\Omega; \mathbb{R}^2)$ be endowed with the norm

$$\|U\|_{L^1(\Omega;\mathbb{R}^2)} := \int_\Omega (|u(x)| + |w(x)|)\, dx$$

$$\forall U = (u,w) \in L^1(\Omega; \mathbb{R}^2).$$

Define the operator R as

$$R(U) := (Bu, 0) \qquad \forall U \in D(R) := \{U \in L^1(\Omega;\mathbb{R}^2) : Bu \in L^1(\Omega)\}.$$

Let A be defined for

$$\begin{cases} \gamma_l, \gamma_r \text{ maximal monotone (possibly multivalued) functions:}\\ \mathbb{R} \to P(\widetilde{\mathbb{R}}) \text{ (the power set), such that } \inf \gamma_r(u) \le \sup \gamma_l(u), \qquad \forall u \in \mathbb{R}. \end{cases}$$

Also, assume that γ_l, γ_r are affinely bounded, that is, there exist constants $C_1, C_2 > 0$, such that $\forall v \in \mathbb{R}, \forall z \in \gamma_h(v)$

$$|z| \le C_1|v| + C_2 \quad (h = l, r).$$

Take any $U_0 := (u_0, w_0) \in L^1(\Omega; \mathbb{R}^2)$, such that $U_0 \in D(A)$ a.e. in Ω, and any $f \in L^1(\Omega \times (0,T))$. Then, the Cauchy problem (8) has one and only one integral solution $U : [0,T] \to L^1(\Omega, \mathbb{R}^2)$, which depends continuously on the data u_0, w_0, f. Moreover, if $f \in BV(0,T; L^1(\Omega))$ and $Ru_0 \in L^1(\Omega)$, then U is Lipschitz continuous.

A similar statement is true for a generalized Prandtl-Ishlinskii operator of play type.

4. Main result

Let Λ denote the hysteresis region, i.e. the subset of R^2 of admissible pairs (u, w) such that $\inf \gamma_r(u) \le w \le \sup \gamma_l(u)$.

Theorem 2: Let the assumptions of Theorem 1 hold. Let $A_0 U := A(U) + R(U)$ on $D(A_0)$, and let $S(t) = (S_1(t), S_2(t))$ be the corresponding semigroup of contractions.

Let $v \in \overline{D(A)}$ and $t \ge 0$. Then, if $v = (v_1, v_2) \in L^\infty(\Omega) \times L^\infty(\Omega)$,

$$\int_0^T \int_\Omega |S_1(t)v_1(x) - k|\psi_t(x,t)dxdt + \int_0^T \int_\Omega |S_2(t)v_2(x) - l|\psi_t(x,t)dxdt$$

$$+ \int_0^T \int_\Omega \left\{ \sum_{j=1}^N b_j |S_1(t)v_1(x) - k| \frac{\partial}{\partial x_j}\psi(x,t) - c|S_1(t)v_1(x) - k|\psi(x,t) \right.$$

$$\left. - [\mathrm{sign}(S_1(t)v_1(x) - k)] \left[k \left(\sum_{j=1}^N \frac{\partial}{\partial x_j} b_j + c \right) - f(x,t) \right] \psi(x,t) \right\} dxdt \ge 0$$

for every $\psi(x,t) \in C_0^\infty((0,T) \times \Omega)$ such that $\psi \ge 0$ and every $(k,l) \in \Lambda$ and $T > 0$.

Remark 3: The hysteresis operator can be also discontinuous, e.g. the relay operator.

Proof: Let $v \in \overline{D(A_0)} \cap L^\infty(\Omega, \mathbb{R}^2)$ and $u_\epsilon(t)$, $w_\epsilon(t)$ satisfy:

$$\left.\begin{array}{r} \frac{u_\epsilon(t) - u_\epsilon(t-\epsilon)}{\epsilon} + \xi + \sum_{j=1}^N \frac{\partial}{\partial x_j}(b_j u_\epsilon(t)) + c u_\epsilon(t) = f \\ \frac{w_\epsilon(t) - w_\epsilon(t-\epsilon)}{\epsilon} - \xi = 0 \end{array}\right\} \text{ for } t \ge 0, \tag{9}$$

$$\begin{pmatrix} u_\epsilon(t) \\ w_\epsilon(t) \end{pmatrix} = v \qquad \text{for } t < 0. \tag{10}$$

If $k \in \mathbb{R}$ is any constant, then we have

$$-\left(\sum_{j=1}^{N} \frac{\partial}{\partial x_j} b_j k + ck\right) = -k\left(\sum_{j=1}^{N} \frac{\partial}{\partial x_j} b_j + c\right). \tag{11}$$

We get from the second equation in (9) that

$$\xi = \frac{w_\epsilon(t) - w_\epsilon(t-\epsilon)}{\epsilon},$$

which we can put into the first equation in (9). Adding the resulting equation with (11) gives us

$$\begin{aligned}\frac{u_\epsilon(t) - u_\epsilon(t-\epsilon)}{\epsilon} + \frac{w_\epsilon(t) - w_\epsilon(t-\epsilon)}{\epsilon} + \sum_{j=1}^{N} \frac{\partial}{\partial x_j}[b_j(u_\epsilon(t) - k)] \\ +[c(u_\epsilon(t) - k)] + k\left(\sum_{j=1}^{N} \frac{\partial}{\partial x_j} b_j + c\right) = f.\end{aligned} \tag{12}$$

Let $u_\epsilon(x,t) := u_\epsilon(t)(x)$ and $w_\epsilon(x,t) := w_\epsilon(t)(x)$. Multiply the last equation by $-[\text{sign}(u_\epsilon(x,t) - k)]\psi(x,t)$ and integrate over Ω to get:

$$\begin{aligned}\int_\Omega \psi(x,t)\Bigg\{ & [\text{sign}(u_\epsilon(x,t) - k)] \\ & \times\left[\frac{u_\epsilon(x,t-\epsilon) - u_\epsilon(x,t)}{\epsilon} + \frac{w_\epsilon(x,t-\epsilon) - w_\epsilon(x,t)}{\epsilon}\right] \\ -[\text{sign}(u_\epsilon(x,t) - k)] & \left[\sum_{j=1}^{N} \frac{\partial}{\partial x_j}[b_j(u_\epsilon(x,t) - k)] + [c(u_\epsilon(x,t) - k)]\right. \\ & \left. +k\left(\sum_{j=1}^{N} \frac{\partial}{\partial x_j} b_j + c\right) - f(x,t)\right]\Bigg\} dx = 0\end{aligned} \tag{13}$$

Let $h_\epsilon(x,t) := (u_\epsilon(x,t) - k)[\text{sign } (u_\epsilon(x,t) - k)] = |u_\epsilon(x,t) - k|$. Recall that

$$\begin{aligned}&(u_\epsilon(x,t-\epsilon) - u_\epsilon(x,t))[\text{sign } (u_\epsilon(x,t) - k)] \\ &= (u_\epsilon(x,t-\epsilon) - k)[\text{sign } (u_\epsilon(x,t) - k)] - (u_\epsilon(x,t) - k)[\text{sign } (u_\epsilon(x,t) - k)] \\ &\le h_\epsilon(x,t-\epsilon) - h_\epsilon(x,t).\end{aligned} \tag{14}$$

Also, $\forall\,(k,l) \in \Lambda$

$$(w_\epsilon(x,t-\epsilon) - w_\epsilon(x,t))[\text{sign }(u_\epsilon(x,t) - k)] \\ \leq (w_\epsilon(x,t-\epsilon) - w_\epsilon(x,t))[\text{sign }(w_\epsilon(x,t) - l)]. \quad (15)$$

This last inequality is true because of the following argument. The only way it could fail would be if either:

$$\text{sign}(u_\epsilon(x,t) - k) = 1, \quad \text{sign}(w_\epsilon(x,t) - l) = -1, \quad w_\epsilon(x,t-\epsilon) - w_\epsilon(x,t) > 0,$$

so that, respectively,

$$u_\epsilon(x,t) > k, \quad w_\epsilon(x,t) < l, \quad w_\epsilon(x,t-\epsilon) > w_\epsilon(x,t)$$

or

$$\text{sign}(u_\epsilon(x,t) - k) = -1, \quad \text{sign}(w_\epsilon(x,t) - l) = 1, \quad w_\epsilon(x,t-\epsilon) - w_\epsilon(x,t) < 0,$$

so that, respectively,

$$u_\epsilon(x,t) < k, \quad w_\epsilon(x,t) > l, \quad w_\epsilon(x,t-\epsilon) < w_\epsilon(x,t).$$

It can be easily seen from Figures 2a and 2b that these situations are not possible because of the properties of the hysteresis operator; thus, (15) must be true.

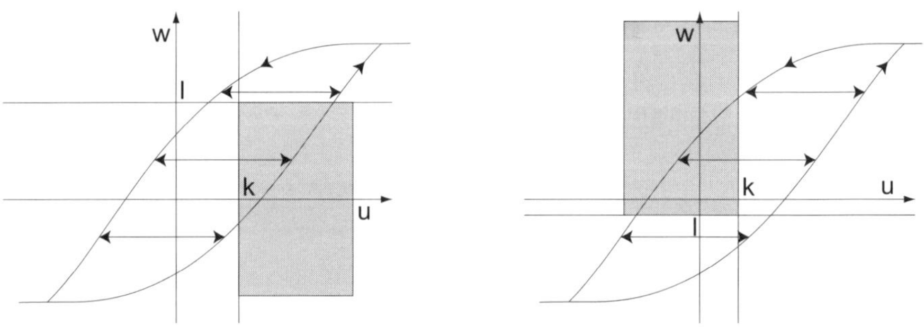

Fig. 8.2. Possible situations in the proof of (15).

If we introduce the notation $g_\epsilon(x,t) := (w_\epsilon(x,t)-l)[\text{sign }(w_\epsilon(x,t)-l)] = |w_\epsilon(x,t) - l|$, then, similarly, we get

$$(w_\epsilon(x,t-\epsilon) - w_\epsilon(x,t))[\text{sign }(u_\epsilon(x,t) - k)] \\ \leq (w_\epsilon(x,t-\epsilon) - w_\epsilon(x,t))[\text{sign }(w_\epsilon(x,t) - l)] \\ = (w_\epsilon(x,t-\epsilon) - l)[\text{sign }(w_\epsilon(x,t) - l)] - (w_\epsilon(x,t) - l)[\text{sign }(w_\epsilon(x,t) - l)] \\ \leq g_\epsilon(x,t-\epsilon) - g_\epsilon(x,t). \quad (16)$$

Therefore, we infer from (13) that

$$\int_\Omega \psi(x,t)\left\{\left[\frac{(h_\epsilon(x,t-\epsilon)-h_\epsilon(x,t))}{\epsilon}+\frac{(g_\epsilon(x,t-\epsilon)-g_\epsilon(x,t))}{\epsilon}\right]\right.$$
$$-[\operatorname{sign}(u_\epsilon(x,t)-k)]\left[\sum_{j=1}^{N}\frac{\partial}{\partial x_j}[b_j(u_\epsilon(x,t)-k)]\right.$$
$$\left.\left.+[c(u_\epsilon(x,t)-k)]+k\left(\sum_{j=1}^{N}\frac{\partial}{\partial x_j}b_j+c\right)-f(x,t)\right]\right\}dx\geq 0.$$

Hence, we can integrate over $[0,T]$ and get the following inequality:

$$0\leq \epsilon^{-1}\int_0^T\int_\Omega[(h_\epsilon(x,t-\epsilon)-h_\epsilon(x,t)]\,\psi(x,t)dxdt$$
$$+\epsilon^{-1}\int_0^T\int_\Omega[g_\epsilon(x,t-\epsilon)-g_\epsilon(x,t)]\,\psi(x,t)dxdt$$
$$-\int_0^T\int_\Omega\left\{\sum_{j=1}^{N}\frac{\partial}{\partial x_j}b_j(|u_\epsilon(x,t)-k|)+c|u_\epsilon(x,t)-k|\right.$$
$$\left.+[\ \operatorname{sign}\ (u_\epsilon(x,t)-k)]\left[k\left(\sum_{j=1}^{N}\frac{\partial}{\partial x_j}b_j+c\right)-f(x,t)\right]\right\}\psi(x,t)dxdt. \quad (17)$$

Now, it turns out that

$$\epsilon^{-1}\int_0^T\int_\Omega\{[h_\epsilon(x,t-\epsilon)-h_\epsilon(x,t)]\,\psi(x,t)\}\,dxdt$$
$$=\epsilon^{-1}\left(\int_0^\epsilon\int_\Omega\{h_\epsilon(x,t-\epsilon)\psi(x,t)\}\,dxdt-\int_{T-\epsilon}^T\int_\Omega\{h_\epsilon(x,t)\psi(x,t)\}dxdt\right)$$
$$+\int_0^{T-\epsilon}\int_\Omega\left\{h_\epsilon(x,t)\epsilon^{-1}(\psi(x,t+\epsilon)-\psi(x,t))\right\}dxdt.$$

The first and the second integrals vanish for ϵ small enough, since ψ is in $C_0^\infty((0,T)\times\Omega)$. The convergence $u_\epsilon(x,t)\to S_1(t)v_1(x)$ in $L^1(\Omega)$, uniformly in t as $\epsilon\to 0$, implies that the third term tends to

$$\int_0^T\int_\Omega|S_1(t)v_1(x)-k|\psi_t(x,t)dxdt \qquad \text{as } \epsilon\downarrow 0.$$

By a similar argument, using the convergence $w_\epsilon(x,t)\to S_2(t)v_2(x)$ in $L^1(\Omega)$, we have that $\int_0^{T-\epsilon}\int_\Omega\left\{g_\epsilon(x,t)\epsilon^{-1}(\psi(x,t+\epsilon)-\psi(x,t))\right\}dxdt$

tends to $\int_0^T \int_\Omega |S_2(t)v_2(x) - l|\psi_t(x,t)dxdt$ as $\epsilon \downarrow 0$. If we now let $\epsilon \downarrow 0$ in (17), we get

$$\int_0^T \int_\Omega |S_1(t)v_1(x) - k|\psi_t(x,t)dxdt + \int_0^T \int_\Omega |S_2(t)v_2(x) - l|\psi_t(x,t)dxdt$$

$$+ \int_0^T \int_\Omega \left\{ \sum_{j=1}^N b_j |S_1(t)v_1(x) - k| \frac{\partial}{\partial x_j}\psi(x,t) - c|S_1(t)v_1(x) - k|\psi(x,t) \right.$$

$$\left. -[\text{sign}(S_1(t)v_1(x) - k)] \left[k\left(\sum_{j=1}^N \frac{\partial}{\partial x_j} b_j + c \right) - f(x,t) \right] \psi(x,t) \right\} dxdt \geq 0,$$

which is the claim of our theorem. □

References

[1] M.G. Crandall, The semigroup approach to first order quasilinear equations in several space variables. *Israel J. Math.* **12**, 108–132 (1972).

[2] J. Kopfová, Entropy condition for a quasilinear hyperbolic equation with hysteresis. *Differential Integral Equations* **18**, 451–467 (2005).

[3] P. Kordulová, An example of discontinous solution for a quasilinear hyperbolic equation with hysteresis. *Submitted*, 2005.

[4] P. Krejčí, *Hysteresis, Convexity and Dissipation in Hyperbolic Equations.* Gakkōtosho, Tokyo 1996.

[5] S.N. Kružkov, Generalized solutions of the Cauchy problem in the large for nonlinear equations of first order. *Soviet. Math. Dokl.* **10**, 785–788 (1969).

[6] S.N. Kružkov, First order quasilinear equations in several independent variables. *Math. U.S.S.R. Sbornik* **10**, 217–243 (1970).

[7] O.A. Olejnik, Uniqueness and stability of the generalized solution of the Cauchy problem for a quasi-linear equation. *Amer. Math. Soc. Transl. Ser.* **33**, 285–290 (1963).

[8] M. Peszyńska and R.E. Showalter, A Transpport Model with Adsorption Hysteresis. *Differential Integral Equations* **11**, 327–340 (1998).

[9] H.-K. Rhee, R. Aris and N.R. Amundson, *First Order Partial Differential Equations. Vol. I.* Prentice-Hall, Englewood Cliffs NJ, 1986.

[10] J. Smoller, *Shock Waves and Reaction-Diffusion Equations.* Springer, New York, 1983.

[11] A. Visintin, *Differential Models of Hysteresis.* Springer-Verlag, Berlin, 1995.

[12] A. Visintin, Quasilinear first order P.D.E.s with hysteresis. *J. Math. Anal. Appl.* (to appear).

APPROXIMATION OF INVERSE PROBLEMS RELATED TO PARABOLIC INTEGRO-DIFFERENTIAL SYSTEMS OF CAGINALP TYPE

Alfredo Lorenzi[1] and Elisabetta Rocca[2]

[1] *Department of Mathematics, University of Milan*
Via Saldini 50, 20133 Milano, Italy
E-mail: lorenzi@mat.unimi.it

[2] *Department of Mathematics, University of Milan*
Via Saldini 50, 20133 Milano, Italy
E-mail: rocca@mat.unimi.it

In this paper, we approximate an identification problem for a partial integro-differential system, related to Caginalp's model for phase transitions, when the data are affected by a known error δ. Since such a problem is weakly ill–posed, we regularize it by a family of *well-posed* identification problems, depending on a small parameter ε. We prove that, as δ and ε accordingly tend to 0, the regularized solution converges to the *exact* one. We also compute the conditioned convergence rate.

1. Introduction

This paper deals with the regularization of an integrodifferential PDE system, related to Caginalp's model for phase transitions, which has recently been studied in [16] from the *inverse-problem* point of view.

In order to introduce the equations we would like to approximate, let us consider a smooth bounded container $\Omega \subset \mathbb{R}^d$, $1 \leq d \leq 3$, occupied by the substance undergoing the phase transition in the time interval $(0, T)$; let moreover $Q := \Omega \times (0, T)$. Name ϑ and χ the basic state variables of the process, corresponding to the *relative* temperature and to the order parameter, respectively. Then, the PDE system governing the phase transition (cf. [2, 9, 10, 11]) is

$$D_t(\vartheta + \ell\chi) - k_0\Delta\vartheta - k * \Delta\vartheta = f, \quad \text{in } \Omega \times (0,T), \tag{1}$$

$$D_t\chi + \Delta^2\chi - \Delta[\beta(\chi) + \sigma'(\chi) - \ell\vartheta] = 0, \quad \text{in } \Omega \times (0,T), \tag{2}$$

$$\vartheta(\cdot,0) = \vartheta_0, \qquad \chi(\cdot,0) = \chi_0, \quad \text{in } \Omega, \tag{3}$$

$$D_n\vartheta = D_n\chi = D_n\Delta\chi = 0, \qquad \text{on } \partial\Omega \times (0,T), \tag{4}$$

where D_t and D_n are, respectively, the time and the normal derivative (in the remainder of the paper D_t will also be indicated by ∂_t), Δ is the Laplace operator, $k*\Delta\vartheta(x,t) = \int_0^t k(t-s)\Delta\vartheta(x,s)\,ds$, where $k : (0,T) \to \mathbb{R}$ is the so-called heat conductivity relaxation kernel, f is a heat source also incorporating an additional term depending on the past history of ϑ up to $t = 0$, which is assumed to be given, k_0 is the heat conductivity coefficient, ℓ is the latent heat coefficient, while β and σ stand here for smooth functions, β being, in addition, increasing.

In (1) we assume, instead of the Fourier law, a Coleman-Gurtin law (cf. [3]) for the heat flux. Let us note here that phase field models of Caginalp type (cf. [2]), possibly accounting for memory effects, have been extensively investigated in the recent years from the viewpoint of *direct problems* (see, e.g., [4, 12, 18]).

Assuming that the *memory kernel k* is itself unknown, in [16] the authors have recently recovered k in (1–4) (under a reasonable a priori condition on χ) in the space $L^2(0,T)$, relying on additional information of the form

$$\Phi[\vartheta(\cdot,t)] := \int_\Omega \varphi(x)\vartheta(x,t)\,dx = g(t), \quad t \in (0,T), \tag{5}$$

involving some measurement of the mean temperature of the system. In this context, the sufficiently regular functions φ and g account, respectively, for a sensor (possibly with a compact support) and for the measured temperature. We point out that the choice of the space $L^2(0,T)$ for k allows to deal with memory kernels having integrable singularities (cf. also [1]). Unfortunately, in order to obtain the global (i.e. on the whole time interval $[0,T]$) existence of k, in [16] the authors need to impose (in the case of a non-Lipschitz nonlinearity β in (2)) an a priori condition on the χ-component of the solution (ϑ,χ) of (1–4). However, let us stress that such a condition is, in particular, satisfied whenever χ fulfils the constraint $\chi \in [0,1]$. The latter is physically meaningful since χ represents (in the physical representation of the model) the volume fraction of the substance subject to the phase transition.

We remark that identifying *memory kernels* in PDE systems arising from phase transition models is a quite new problem. Concerning the *conserved model* considered in this paper (on account of the boundary condition (4), it is straightforward to check that the average of χ is constant in time), let us quote the papers [5] and [13], where the authors (using semigroup techniques) study the *local (in time)* identification of a kernel for system (1–4). Moreover, in [8] the authors prove the local (in time) existence of the kernel k for a PDE system similar to (1–4), but with the Gurtin-Pipkin heat flux law in (1) (i.e., with with $k_0 = 0$ in (1)). This gives rise to a hyperbolic dynamics for the equation analogous to (1). Let us point out once more that all these contributions provide local (in time) existence results for the inverse problem.

In this paper, we assume that the data f, ϑ_0, χ_0, g are affected, in a suitable norm, by a known error δ. So, we deal with a family of approximating problems, obtained by replacing in (1–5) the exact data $(f, \vartheta_0, \chi_0, g)$ with the approximate ones $(f_\delta, \vartheta_{0,\delta}, \chi_{0,\delta}, g_\delta)$, due to the error in our measured data. When dealing with a weakly ill-posed problem with noisy data - such as our identification problem -, it is natural to modify problem (1–5) (with the data f, ϑ_0, χ_0, g replaced by f_δ, $\vartheta_{0,\delta}$, $\chi_{0,\delta}$, g_δ), by a family of well-posed approximating problems, depending on an additional regularizing parameter ε, and to try to determine under which conditions on the parameters of the family the approximating solutions converge to those of the problem with *exact* data.

By the way, we recall that an inverse problem is called *weakly ill-posed* when, after some operator transformations, the problem turns into a *well-posed* one. In our specific case, this can be obtained, as shown in [16], by performing a differentiation with respect to time.

We can now state our approximating and regularizing family of identification problems

$(\mathrm{P})_{\delta,\varepsilon}$: *determine a triplet of functions* $\vartheta_{\delta,\varepsilon}, \chi_{\delta,\varepsilon} : \Omega \times [0,T] \to \mathbf{R}$ *and* $k_{\delta,\varepsilon} : [0,T] \to \mathbf{R}$ *such that*

$$D_t(\vartheta_{\delta,\varepsilon} + \ell\chi_{\delta,\varepsilon}) - k_0\Delta\vartheta_{\delta,\varepsilon} - k_{\delta,\varepsilon} * \Delta\vartheta_{\delta,\varepsilon} = f_\delta, \quad \textit{in } \Omega \times (0,T), \tag{6}$$

$$D_t\chi_{\delta,\varepsilon} + \Delta^2\chi_{\delta,\varepsilon} - \Delta[\beta(\chi_{\delta,\varepsilon}) + \sigma'(\chi_{\delta,\varepsilon}) - \ell\vartheta_{\delta,\varepsilon}] = 0, \quad \textit{in } \Omega \times (0,T), \tag{7}$$

$$\vartheta_{\delta,\varepsilon}(\cdot,0) = \vartheta_{0,\delta}, \qquad \chi_{\delta,\varepsilon}(\cdot,0) = \chi_{0,\delta}, \quad \textit{in } \Omega, \tag{8}$$

$$D_n\vartheta_{\delta,\varepsilon} = D_n\chi_{\delta,\varepsilon} = D_n\Delta\chi_{\delta,\varepsilon} = 0, \quad \textit{on } \partial\Omega \times (0,T), \tag{9}$$

$$\Phi[\vartheta_{\delta,\varepsilon}(\cdot,t)] + \alpha_\delta[P_2(\varepsilon D_t) - I]D_t^{-2}k_{\delta,\varepsilon}(t) = q_{\delta,\varepsilon}(t) + g_\delta, \quad t \in (0,T), \tag{10}$$

where $\delta \in (0, \delta_0]$, $\varepsilon \in (0, \varepsilon_0]$, and

$$P_2(D_t) = \frac{1}{2}(D_t^2 + 3D_t + 2), \quad \alpha_\delta = \Phi[\Delta\vartheta_{0,\delta}] \neq 0, \tag{11}$$

$$q_{\delta,\varepsilon}(t) = \gamma_{0,\delta,\varepsilon} t + \gamma_{1,\delta,\varepsilon}, \tag{12}$$

$$D_t^{-j} k(t) = \frac{1}{(j-1)!} \int_0^t (t-s)^{j-1} k(s)\, ds, \quad j \in \mathbf{N} \backslash \{0\}, \tag{13}$$

$$\|f_\delta - f\|_F + \|\vartheta_{0,\delta} - \vartheta_0\|_\Theta + \|\chi_{0,\delta} - \chi_0\|_X + \|g_\delta - g\|_G \leq \delta. \tag{14}$$

The functional spaces F, Θ, X and G are themselves under investigation, as well as the real numbers $\gamma_{0,\delta,\varepsilon}$ and $\gamma_{1,\delta,\varepsilon}$, while the positive constants δ and ε are given. The characterization of F, Θ, χ, and $\gamma_{0,\delta,\varepsilon}$, $\gamma_{1,\delta,\varepsilon}$ will be given in formulae (38), (40–41).

According to well-known techniques going back to Tikhonov (cf. [20]), the aim of this paper consists in:

1. *showing that there exists a positive exponent σ such that the solution $(\vartheta_{\delta,\delta^\sigma}, \chi_{\delta,\delta^\sigma}, k_{\delta,\delta^\sigma})$ of the approximating problem converges, in a suitable metric, to the solution (ϑ, χ, k) of the exact problem;*
2. *determining the rate of convergence δ^ρ of the approximating solution to the exact one.*

Finally, let us note that this procedure is inspired by the paper [14] of the first author, focusing on heat propagation in materials with memory. Moreover, let us note that the results in the present paper also have some numerical interest, allowing to obtain an explicit reconstruction of the memory kernel.

Here is the plan of the paper. In the next section, we recall a regularity result for the direct problem and a well-posedness result concerning the identification problem for system (1–5) with exact data, both proved in [16]. In Section 3, we show the well-posedness of problem (6–10) (with approximate data) and some uniform estimate (with respect to δ and ε) of $k_{\delta,\varepsilon}$. Finally, making use of some results proved in the Appendix (cf. Section 5), in Section 4 we show our main theorem, establishing the convergence of the solution $(\vartheta_{\delta,\delta^\sigma}, \chi_{\delta,\delta^\sigma}, k_{\delta,\delta^\sigma})$ of the approximating problem to the solution (ϑ, χ, k) of the exact problem, and determining also the rate of convergence δ^ρ of the approximating solution to the exact one.

2. The problem with exact data

In the following two subsections, we recall from [16] some regularity and continuous dependence (with respect to the data) results concerning the *direct* and the *inverse* problems associated with (1–5). Throughout the paper, we will denote, whenever it is more convenient, by H the Hilbert space $L^2(\Omega)$, by $|\cdot|_{0,2}$ or by $|\cdot|_H$ the norm in H, and by $(\cdot,\cdot)_H$ the corresponding scalar product. Moreover, for $\sigma \in (0,1/2)$ and any Banach space $\mathcal{Z}$, we denote by $|\cdot|_{\sigma,2,\mathcal{Z}}$ the norm in $H^\sigma(0,T;\mathcal{Z})$, that is, for $f \in H^\sigma(0,T;\mathcal{Z})$:

$$|f|_{\sigma,2,\mathcal{Z}} := \left(|f|^2_{L^2(0,T;\mathcal{Z})} + \int_0^T dt \int_0^T |t-s|^{-1-2\sigma} |f(t)-f(s)|^2_{\mathcal{Z}}\, ds \right)^{1/2}. \tag{15}$$

2.1. *The direct problem with exact data*

For any arbitrary positive number Ω, we use the notation

$$v_\Omega(t) := \exp(-\Omega t) v(t), \quad \forall t \in [0,T], \tag{16}$$

where v is a generic function defined on $[0,T]$ with values in some Banach space X. Assuming the well-posedness of the *direct* problem (1–4) (cf. Thm 2.1 in [18]), we recall the following regularity result for the direct problem, proved in Thm. 2.11 in [16].

Theorem 1: Let $k \in L^1(0,T)$, $k_0 > 0$, and assume for $s \in \mathbb{N}$ (with the convention $H^0 \equiv H$):

$$f = f_1 + f_2,\ f_1 \in H^1(0,T;H^s(\Omega)) \text{ and } f_2 \in W^{1,1}(0,T;H^{s+1}(\Omega)); \tag{17}$$

$$\beta \in C^{s+3}(\mathbb{R}), \quad \sigma \in C^{s+4}(\mathbb{R}), \quad \beta' \geq 0 \quad \text{in } \mathbb{R}; \tag{18}$$

$$\vartheta_0 \in \{\zeta \in H^{s+4}(\Omega) : D_n\zeta = 0 \quad \text{on } \partial\Omega\}; \tag{19}$$

$$\chi_0 \in \{\zeta \in H^{s+6}(\Omega) : D_n\zeta = 0 \quad \text{on } \partial\Omega\}. \tag{20}$$

Then, the solution (ϑ,χ) to the direct problem (1–4) satisfies

$$\vartheta \in W^{1,\infty}(0,T;H^{s+1}(\Omega)) \cap H^1(0,T;H^{s+2}(\Omega)), \tag{21}$$

$$\chi \in H^2(0,T;H^s(\Omega)) \cap W^{1,\infty}(0,T;H^{s+2}(\Omega)). \tag{22}$$

Moreover, under the assumption

$$\left(1+\frac{2}{k_0}\right)|k_\Omega|^2_{L^1(0,T)} \leq \frac{k_0}{4}, \tag{23}$$

there exists a positive constant $K(\Omega)$, depending on Ω and on the data of the problem (but not on k), such that the pair (ϑ, χ) fulfils

$$|\vartheta_\Omega|_{W^{1,\infty}(0,T;H^{s+1}(\Omega))\cap H^1(0,T;H^{s+2}(\Omega))} + |\chi_\Omega|_{H^2(0,T;H^s(\Omega))\cap W^{1,\infty}(0,T;H^{s+2}(\Omega))} \leq K(\Omega). \tag{24}$$

Remark 2: Note that problem (6–9) differs from (1–4) in the data only. Hence, replacing k with k_∂ in assumption (23), we immediately get an estimate like (24) for the solution $(\vartheta_\partial, \chi_\partial)$ of the *direct* problem (6–9) (cf. also Theorem 8 and its proof below).

A further result concerning the direct problem is the following continuous dependence theorem for the regular solution obtained in Theorem 1, with $s = 2$. We refer to Thm. 2.12 in [16] for its proof.

Theorem 3: Suppose that the assumptions of Theorem 1 hold with $s = 2$ and $f_1 = 0$. Let (ϑ_1, χ_1) and (ϑ_2, χ_2) be two solutions of (1–4), corresponding to data f^1, k_1, ϑ_{01}, χ_{01}, f^2, k_2, ϑ_{02}, χ_{02}, respectively, which enjoy the regularity properties (17–20) stated in Theorem 1, along with

$$k_j \in L^2(0,T), \quad j = 1, 2; \tag{25}$$

$$|(k_j)_\Omega|_{L^1(0,T)} \leq \frac{k_0}{2(2+k_0)^{1/2}}, \quad |k_j|_{L^2(0,T)} \leq m, \qquad j = 1, 2, \tag{26}$$

for some positive constant m.

Then, there exists a positive constant $\overline{K} = \overline{K}(\Omega, T, m)$, depending on Ω, T, m, and on the data of problem (1–4), such that the following estimate holds true for any $\tau \in (0, T]$:

$$\begin{aligned}&|\partial_t\vartheta_2 - \partial_t\vartheta_1|_{L^2(0,\tau;H)} + |\partial_t\chi_2 - \partial_t\chi_1|_{L^2(0,\tau;H)} + |\Delta\vartheta_2 - \Delta\vartheta_1|_{L^2(0,\tau;H)}\\ &+|\Delta^2\chi_2 - \Delta^2\chi_1|_{L^2(0,\tau;H)} \leq \overline{K}\big[|f^2 - f^1|_{L^2(0,\tau;H)} + |k_2 - k_1|_{L^1(0,\tau)}\\ &\qquad +|\vartheta_{02} - \vartheta_{01}|_{H^1(\Omega)} + |\chi_{02} - \chi_{01}|_{H^2(\Omega)}\big].\end{aligned} \tag{27}$$

2.2. *The inverse problem with exact data*

In this subsection, we assume that the memory kernel $k : (0, T) \to \mathbb{R}$ in equation (1) is *unknown* and we provide the *additional information* (5) in order to recover it. From now on, we shall assume that the unknown kernel k is searched for in $L^2(0, T)$.

We make the following assumptions:

$$f \in W^{1,p}(0,T;H), \; g \in W^{2,p}(0,T), \; p \in (2, +\infty); \tag{28}$$

$$\varphi \in W^{4,\infty}(\Omega), \; D_n\Delta^j\varphi = 0 \text{ on } \partial\Omega, \; j = 0, 1. \tag{29}$$

From (3) and (5) we immediately deduce the *consistency* condition

$$\Phi[\vartheta_0] = g(0). \tag{30}$$

Remark 4: From the equality $\Phi[\partial_t \vartheta(0,\cdot)] = g'(0)$ and from equations (1–2), we easily infer the further consistency condition

$$g'(0) = \Phi\big[k_0 \Delta\vartheta_0 + f(0,\cdot) - \ell\{\Delta^2\chi_0 - \Delta[\beta(\chi_0) + \sigma'(\chi_0) - \ell\vartheta_0]\}\big]. \tag{31}$$

We conclude this section by recalling the *conditioned* existence, uniqueness and continuous dependence result stated in Thm. 2.14 in [16].

Theorem 5: Let assumptions listed in Theorem 1 (with $s = 2$ and $f_1 = 0$) be satisfied. Assume also that (28) and (29) hold, along with the consistency conditions (30), (31), and that

$$\Phi[\Delta\vartheta_0] \neq 0, \qquad |\beta' \circ \chi|_{L^2(0,T;L^1(\Omega))} \leq M, \tag{32}$$

for some a priori given constant $M > 0$.

Then, the identification problem (1–5) admits a unique solution $(\vartheta, \chi, k) \in [W^{1,\infty}(0,T;H^3(\Omega)) \cap H^1(0,T;H^4(\Omega))] \times [H^2(0,T;H^2(\Omega)) \cap W^{1,\infty}(0,T;H^4(\Omega))] \times L^2(0,T)$, and there exists $\Omega_0 \in \mathbb{R}^+$ such that

$$\Big(1 + \frac{2}{k_0}\Big)|k_{\Omega_0}|^2_{L^1(0,T)} \leq \frac{k_0}{4}. \tag{33}$$

Further, let $(\vartheta_1, \chi_1, k_1)$ and $(\vartheta_2, \chi_2, k_2)$ be the solutions to (1–5) corresponding, respectively, to the data $(f^1, g_1, \vartheta_{01}, \chi_{01})$ and $(f^2, g_2, \vartheta_{02}, \chi_{02})$. Then, the following continuous dependence estimate holds

$$\begin{aligned}
|\partial_t\vartheta_2 - \partial_t\vartheta_1|_{L^2(0,T;H)} &+ |\partial_t\chi_2 - \partial_t\chi_1|_{L^2(0,T;H)} + |\Delta\vartheta_2 - \Delta\vartheta_1|_{L^2(0,T;H)} \\
&+ |\Delta^2\chi_2 - \Delta^2\chi_1|_{L^2(0,T;H)} + |k_2 - k_1|_{L^2(0,T)} \\
\leq \tilde{K}(\Omega, T, M)\Big\{&|f^2 - f^1|_{W^{1,p}([0,T];H)} + |\vartheta_{02} - \vartheta_{01}|_{H^4(\Omega)} \\
&+ |\chi_{02} - \chi_{01}|_{H^6(\Omega)} + |g_2 - g_1|_{W^{2,p}(0,T)}\Big\},
\end{aligned} \tag{34}$$

where the positive constant $\tilde{K}(\Omega, T, M)$ depends continuously and increasingly on the norms of the data, too.

Remark 6: The latter condition in (32) would be trivially satisfied if we could show that the χ-component of the solution to the *direct* problem, *with the prescribed regularity*, takes its values in $[0, 1]$. Unfortunately this is, at present, an open question, depending on the fact that no maximum principle is known for our fourth-order Caginalp-type system.

3. The inverse problem with approximate data

In this section, we first prove the well-posedness of the problem with the approximate data (6–10), and then we derive some uniform (in δ and ε) estimates on $k_{\delta,\varepsilon}$. Such estimates lead, in the following Section 4, to the proof of the convergence estimates, which is the main goal of this paper.

3.1. *Well-posedness of the problem with approximate data*

We prove here existence and uniqueness of solutions to Problem $(P)_\partial$. First, we note that, from equations (6–10), the following conditions

$$\Phi[\vartheta_{0,\delta}] = -\frac{1}{2}\alpha_\delta \varepsilon^2 k_\partial(0) + \gamma_{1,\partial} + g_\delta(0), \tag{35}$$

$$\ell\Phi[\Delta^2\chi_{0,\delta}] - \ell\Phi[\Delta(\beta(\chi_{0,\delta}) + \sigma(\chi_{0,\delta}))] - \ell^2\Phi[\vartheta_{0,\delta}] + k_0\Phi[\Delta\vartheta_{0,\delta}]$$
$$+\Phi[f_\delta(\cdot,0)] = -\frac{1}{2}\alpha_\delta\varepsilon^2 k'_\partial(0) - \frac{3}{2}\alpha_\delta\varepsilon k_\partial(0) + \gamma_{0,\partial} + g'_\delta(0) \tag{36}$$

can be easily derived. By the way, we note that (35), (36) are *not* actually consistency conditions, since they involve the initial values of the *unknown* function k_∂.

We choose now $\gamma_{0,\delta,\varepsilon}$ and $\gamma_{1,\delta,\varepsilon}$ such as to have

$$k_{\delta,\varepsilon}(0) = k'_\partial(0) = 0. \tag{37}$$

Theorem 7: Let assumptions (12–13), conditions (30–31), (35–37), and the following hypotheses hold true:

$$\chi_{0,\delta} \in H^4(\Omega),\ \vartheta_{0,\delta} \in H^2(\Omega),\ f_\delta \in H^1(0,T;L^2(\Omega)),\ g_\delta \in H^2(0,T), \tag{38}$$

$$\beta'(\chi_{0,\delta}), \beta''(\chi_{0,\delta}) \in L^2(\Omega), \tag{39}$$

$$|\chi_{0,\delta} - \chi_0|_{H^4(\Omega)} \le C\delta, \quad |\vartheta_{0,\delta} - \vartheta_0|_{H^2(\Omega)} \le C\delta,$$
$$|f_\delta - f|_{H^1(0,T;L^2(\Omega))} \le C\delta, \tag{40}$$

$$|g_\delta - g|_{H^2(0,T)} \le C\delta,\ |\beta^{(j)}(\chi_{0,\delta}) - \beta^{(j)}(\chi_0)|_{L^2(\Omega)} \le C\delta,\ j = 1, 2. \tag{41}$$

Then, for any pair $(\delta,\ \varepsilon) \in (0, \delta_0] \times (0, 1]$, problem (6–10) admits a unique solution $(\vartheta_{\delta,\varepsilon}, \chi_{\delta,\varepsilon}, k_{\delta,\varepsilon}) \in [W^{1,\infty}(0,T;H^3(\Omega)) \cap H^1(0,T;H^4(\Omega))] \times [H^2(0,T;H^2(\Omega)) \cap W^{1,\infty}(0,T;H^4(\Omega))] \times L^2(0,T)$.

Proof: Using (30–31) and (35–37), we find

$$\gamma_{1,\partial} = \gamma_{1,\delta} = -\{g_\delta(0) - g(0) - \Phi[\vartheta_{0,\delta} - \vartheta_0]\} \tag{42}$$

$$\begin{aligned}\gamma_{0,\partial} = \gamma_{0,\delta} = \ell\Phi[\Delta^2(\chi_{0,\delta} - \chi_0)] - \ell\Phi[\Delta(\beta(\chi_{0,\delta}) - \beta(\chi_0)\\ +\sigma(\chi_{0,\delta}) - \sigma(\chi_0))] - \ell^2\Phi[\vartheta_{0,\delta} - \vartheta_0] + k_0\Phi[\Delta(\vartheta_{0,\delta} - \vartheta_0)]\\ +\Phi[f_\delta(\cdot,0) - f(\cdot,0)] - [g'_\delta(0) - g'(0)].\end{aligned} \tag{43}$$

According to (38–41) and (42–43), we deduce the bounds

$$|\gamma_{j,\partial}| \le C_j\delta, \qquad j = 0,1,$$

the positive constants C_0 and C_1 being continuous and non-decreasing functions of their arguments. In particular, we have $\gamma_{0,0,0} = \gamma_{1,0,0} = 0$.

We can now explicitly write down equation (10) as a fixed-point equation for $k_{\delta,\varepsilon}$ in $(0,T)$:

$$\begin{aligned}k_{\delta,\varepsilon}(t) = -\frac{3}{\varepsilon}\int_0^t k_{\delta,\varepsilon}(s)\,ds\\ +\frac{2}{\alpha_\delta\varepsilon^2}\big\{ -\Phi[\Theta(k_{\delta,\varepsilon})(\cdot,t)] + q_{\delta,\varepsilon}(t) + g_\delta(t)\big\},\end{aligned} \tag{44}$$

where $\Theta(k_\partial)$ denotes the first component of the solution $(\vartheta_\partial, \chi_\partial)$ of the direct problem (8–12) with an arbitrarily fixed $k_\partial \in L^2(0,T)$.

First, we observe that the solution of the linear integral equation

$$k_{\delta,\varepsilon}(t) = -\frac{3}{\varepsilon}\int_0^t k_{\delta,\varepsilon}(s)\,ds + h_{\delta,\varepsilon}(t) \tag{45}$$

is given by

$$k_{\delta,\varepsilon}(t) = h_{\delta,\varepsilon}(t) - \frac{3}{\varepsilon}\int_0^t e^{-3(t-s)/\varepsilon} h_{\delta,\varepsilon}(s)\,ds, \quad t \in (0,T).$$

Consequently, equation (44) is equivalent to the following one in $(0,T)$:

$$\begin{aligned}k_{\delta,\varepsilon}(t) = -\frac{6}{\alpha_\delta\varepsilon^3}\int_0^t e^{-3(t-s)/\varepsilon}\Big\{ -\Phi[\Theta(k_{\delta,\varepsilon})(\cdot,s)] + q_{\delta,\varepsilon}(s) + g_\delta(s)\Big\}ds\\ +\frac{2}{\alpha_\delta\varepsilon^2}\Big\{ -\Phi[\Theta(k_{\delta,\varepsilon})(\cdot,t)] + q_{\delta,\varepsilon}(t) + g_\delta(t)\Big\} =: N_\partial(k_\partial)(t).\end{aligned} \tag{46}$$

To solve the fixed point equation (46), we need to introduce the closed ball $B(\Omega, k_0)$ in $L^2(0,T)$ defined by

$$B(\Omega,k_0) = \Big\{k \in L^2(0,T) : |k_\Omega|^2_{L^1(0,T)} \le \frac{k_0}{2(k_0+2)^{1/2}}\Big\}, \quad \Omega \in \mathbb{R}^+.$$

To show that N_∂ maps $B(\Omega, k_0)$ into itself for large values of Ω, we need estimate (2.26) in [16], which we rewrite in the following simplified form,

$$\begin{aligned}|\vartheta_\Omega|^2_{L^2(0,T;H)} &\le 2|(\vartheta-\vartheta_0)_\Omega|^2_{L^2(0,T;H)} + 2|(\vartheta_0)_\Omega|^2_{L^2(0,T;H)}\\ &\le \overline{C}_2\Big[|\chi_t(0)|^2_{L^\infty(\Omega)}|(\beta'(\chi))_{2\Omega}|_{L^1(Q)}\\ &+\frac{1}{\Omega}|\chi_t(0)|^2_V + \frac{1}{\Omega}|B\vartheta_0|^2 + |f_\Omega|^2_{L^2(0,T;H)}\Big] + \frac{1}{\Omega}|\vartheta_0|^2, \end{aligned} \tag{47}$$

$\overline{C}_2$ being a positive constant depending on k_0 and the norms of the data of the problem.

From (46) and (47) we deduce that $|(N_\partial(k_\partial))_\Omega| \to 0$ as $\Omega \to +\infty$. Hence N_∂ maps $B(\Omega, k_0)$ into itself if $\Omega \ge \Omega_\partial$, for some suitable positive Ω_∂.

Then, from Young's inequality for convolutions, Thm. 2.5 in [15], and Lemma 2.4 in [15], we deduce the following chain of inequalities, where $k_\partial^{(1)}, k_\partial^{(2)} \in B(\Omega, k_0)$:

$$\begin{aligned}\|N_\partial(k_\partial^{(2)}) - N_\partial(k_\partial^{(1)})\|_{L^2(0,\tau;H)} &\le \frac{4}{|\alpha_\delta|\varepsilon^2}\|\Phi\|\|\Theta(k_{\delta,\varepsilon}^{(2)}) - \Theta(k_{\delta,\varepsilon}^{(1)})\|_{L^2(0,\tau;H)}\\ = \frac{4}{|\alpha_\delta|\varepsilon^2}\|\Phi\|\|1 * [\Theta(k_{\delta,\varepsilon}^{(2)}) - \Theta(k_{\delta,\varepsilon}^{(1)})]\|_{L^2(0,\tau;H)}\\ \le \frac{4\tau^{1/2}}{|\alpha_\delta|\varepsilon^2}\|\Phi\|\Big(\int_0^\tau \|\partial_t\Theta(k_{\delta,\varepsilon}^{(2)}) - \partial_t\Theta(k_{\delta,\varepsilon}^{(1)})\|^2_{L^2(0,s;H)}\,ds\Big)^{1/2}\\ \le \frac{4\tau^{1/2}}{|\alpha_\delta|\varepsilon^2}\|\Phi\|K(\Omega)\Big(\int_0^\tau \|k_\partial^{(2)} - k_\partial^{(1)}\|^2_{L^2(0,s;H)}\,ds\Big)^{1/2}, \quad \tau \in (0,T] \end{aligned} \tag{48}$$

Consequently, the iterates N_∂^m of N_∂ satisfy, for $\tau \in (0,T]$,

$$\begin{aligned}&\|N_\partial^m(k_\partial^{(2)}) - N_\partial^m(k_\partial^{(1)})\|^2_{L^2(0,\tau;H)}\\ &\le \Big(\frac{16\tau}{|\alpha_\delta|^2\varepsilon^4}\|\Phi\|^2K(\Omega)^2\Big)^m \int_0^\tau \frac{(\tau-s)^{m-1}}{(m-1)!}\|k_\partial^{(2)} - k_\partial^{(1)}\|_{L^2(0,\tau;H)}\,ds. \end{aligned} \tag{49}$$

From (49) we conclude that N_∂^m is a contraction map from $B(\Omega, k_0)$ into itself for any m large enough. Consequently, the fixed-point equation $k_\partial = N_\partial^m(k_\partial)$ admits a unique solution $k_\partial \in B(\Omega, k_0)$. This concludes the proof of Theorem 7. □

3.2. *Uniform estimates of the solution to Problem* (P_∂)

In this subsection let us denote, from now on, by c_1, c_2, ... positive constants (depending only on Ω, ℓ, T) playing some special role in our estimates. We prove here the following result.

Theorem 8: Let the assumptions listed in Theorem 7 hold true. Consider the solution $(\vartheta_\partial,\chi_\partial,k_\partial)$ to (6–10) and suppose that

$$|\beta' \circ \chi_\partial|_{L^2(0,T;L^1(\Omega))} \leq M, \tag{50}$$

M being the positive constant (independent, in particular, of δ and ε) appearing in (32). Then, there exists $\Omega_0 \in \mathbb{R}^+$ (independent of δ and ε), such that

$$\Big(1+\frac{2}{k_0}\Big)|(k_\partial)_{\Omega_0}|^2_{L^1(0,T)} \leq \frac{k_0}{4}. \tag{51}$$

Moreover, the pair $(\vartheta_\partial, \chi_\partial)$ fulfils

$$\begin{aligned} &|(\vartheta_\partial)_\Omega|_{W^{1,\infty}(0,T;H^3(\Omega))\cap H^1(0,T;H^4(\Omega))} \\ &+ |(\chi_\partial)_\Omega|_{H^2(0,T;H^2(\Omega))\cap W^{1,\infty}(0,T;H^4(\Omega))} \leq K(\Omega_0), \end{aligned} \tag{52}$$

$K(\Omega_0)$ being the positive constant appearing in Theorem 1.

Proof: First, we observe that the function k_∂ solves the equation

$$\Phi[\partial_t^2\Theta(k_\partial)(\cdot,t)] + \alpha_\delta[P_2(\varepsilon\partial_t) - I]k_\partial(t) = g_\delta''(t), \quad \forall t \in (0,T),$$

i.e.,

$$\begin{aligned} &k_0\Phi[\Delta(\partial_t\Theta(k_\partial))(t)] + k_\partial(t)\Phi[\Delta\vartheta_{0,\delta}] + k * \Phi[\Delta(\partial_t\Theta(k_\partial))(t)] \\ &+ \Phi[\partial_t f_\delta(t)] + \ell\Phi[\Delta^2(\partial_t X(k_\partial)(t))] - \ell\Phi[\Delta((\beta'(X(k_\partial)(t)) \\ &+ \sigma''(X(k_\partial)(t)))\partial_t X(k_\partial)(t))] + \ell^2\Phi[\Delta\partial_t\Theta(k_\partial)(t)] \\ &+ \alpha_\delta[P_2(\varepsilon\partial_t) - I]k_\partial(t) = g_\delta''(t), \quad \forall t \in (0,T), \end{aligned} \tag{53}$$

where $(\Theta(k_\partial), X(k_\partial))$ denotes the solution to the direct problem (6–9) and α_δ is defined in (11).

Solving for $P_2(\varepsilon\partial_t)k_\partial$ and recalling the initial conditions (37), we derive the following operator Cauchy problem for all $t \in (0,T)$

$$\begin{aligned} P_2(\varepsilon\partial_t)k_\partial(t) &= \frac{1}{\alpha_\delta}\Big\{ -k_0\Phi[\Delta(\partial_t\Theta(k_\partial))(t)] - k_\partial * \Phi[\Delta(\partial_t\Theta(k_\partial))(t)] \\ &- \Phi[\partial_t f_\delta(t)] - \ell\Phi[\Delta^2(\partial_t X(k_\partial)(t))] + \ell\Phi[\Delta((\beta'(X(k_\partial)(t)) \\ &+ \sigma''(X(k_\partial)(t))))\partial_t X(k_\partial)(t))] - \ell^2\Phi[\Delta\partial_t\Theta(k_\partial)(t)] + g_\delta''(t)\Big\}, \end{aligned} \tag{54}$$

$$k_\partial(0) = k_{\delta,\varepsilon}'(0) = 0. \tag{55}$$

Setting $\Phi_j[u] = (\Delta^j \varphi, u)_{L^2(\Omega)}$, $j = 0, 1, 2$, and integrating by parts in space, from (29) we get

$$\begin{aligned} &P_2(\varepsilon\partial_t)k_{\delta,\varepsilon}(t) = H_\delta(f_\delta, g_\delta)(t) - \rho_\delta k_{\delta,\varepsilon} * \Phi_1[\partial_t\Theta(k_\partial)(t)] \\ &- \rho_\delta(k_0 + \ell^2)\Phi_1[\partial_t\Theta(k_\partial)(t)] - \rho_\delta\ell\Phi_2[\partial_t X(k_\partial)(t)] \\ &+ \rho_\delta\ell\Phi_1[(\beta'(X(k_\partial)(t)) + \sigma''(X(k_\partial)(t)))\partial_t X(k_\partial)(t)] =: N_\delta(k_\partial)(t), \end{aligned} \tag{56}$$

$$k_\partial(0) = k'_\partial(0) = 0, \tag{57}$$

where

$$\rho_\delta = 1/\alpha_\delta, \quad H_\delta(f_\delta, g_\delta)(t) = \rho_\delta\{-\Phi[\partial_t f_\delta(t)] + g''_\delta(t)\}. \tag{58}$$

We now recall that the *exact* kernel k solves, for all $t \in [0, T]$, the following operator equation (cf. (5.3) in [16])

$$\begin{aligned} &k(t) = H(f, g)(t) - \rho k * \Phi_1[\partial_t\Theta(k)(t)] - \rho(k_0 + \ell^2)\Phi_1[\partial_t\Theta(k)(t)] \\ &- \rho\ell\Phi_2[\partial_t X(k)(t)] + \rho\ell\Phi_1[(\beta'(X(k)(t)) + \sigma''(X(k)(t)))\partial_t X(k)(t)] \\ &=: N_0(k)(t), \end{aligned} \tag{59}$$

where (cf. (32))

$$H(f, g)(t) = \frac{1}{\alpha}\{-\Phi[\partial_t f(t)] + g''(t)\}, \quad \rho = 1/\alpha, \quad \alpha = \Phi(\Delta\vartheta_0). \tag{60}$$

Then, we solve the Cauchy problem (56–57) for all $t \in [0, T]$:

$$k_\partial(t) = \frac{2}{\varepsilon}\int_0^t [e^{-(t-s)/\varepsilon} - e^{-2(t-s)/\varepsilon}]N_\delta(k_\partial)(s)\,ds := K_\varepsilon N_\delta(k_\partial)(t). \tag{61}$$

Now, we derive a uniform bound for k_∂. From (56), recalling notation (16), we get

$$\begin{aligned} &N_\delta(k_\partial)_\Omega(t) = H_\delta(f_\delta, g_\delta)_\Omega(t) - \rho_\delta(k_{\delta,\varepsilon})_\Omega * \Phi_1[(\partial_t\Theta(k_\partial))_\Omega(t)] \\ &-\rho_\delta(k_0 + \ell^2)\Phi_1[(\partial_t\Theta(k_\partial))_\Omega(t)] - \rho_\delta\ell\Phi_2[(\partial_t X(k_\partial))_\Omega(t)] \\ &+\rho_\delta\ell\Phi_1[(\beta'(X(k_\partial)(t)) + \sigma''(X(k_\partial)(t)))(\partial_t X(k_\partial))_\Omega(t)]. \end{aligned} \tag{62}$$

Introduce the auxiliary functions

$$u_\partial = U(k_\partial) = \Theta(k_\partial) - \vartheta_{0,\delta}, \quad v_\partial = V(k_\partial) = X(k_\partial) - \chi_{0,\delta} - t\overline{\chi}_{0,\delta},$$

where

$$\overline{\chi}_{0,\delta} := \partial_t\chi_\partial(0) = -\Delta^2\chi_{0,\delta} + \Delta[\beta(\chi_{0,\delta}) + \sigma'(\chi_{0,\delta}) - \ell\vartheta_{0,\delta}].$$

Observe then that

$$\begin{aligned} N_\delta(k_\partial)_\Omega(t) &= H_\delta(f_\delta, g_\delta)_\Omega(t) - \rho_\delta(k_{\delta,\varepsilon})_\Omega * \Phi_1[(\partial_t U(k_\partial))_\Omega(t)] \\ &- \rho_\delta(k_0 + \ell^2)\Phi_1[(\partial_t U(k_\partial))_\Omega(t)] - \rho_\delta \ell \Phi_2[\overline{\chi}_{0,\delta}]e^{-t\Omega} \\ &- \rho_\delta \ell \Phi_2[(\partial_t V(k_\partial))_\Omega(t)] \\ &+ \rho_\delta \ell e^{-t\Omega}\Phi_1[(\beta' + \sigma'')\,(V(k_\partial) + \chi_{0,\delta} + t\bar{\chi}_{0,\delta})\,\bar{\chi}_{0,\delta}] \\ &+ \rho_\delta \ell \Phi_1[((\beta'(V(k_\partial) + \chi_{0,\delta} + t\bar{\chi}_{0,\delta}))^{1/2} \\ &\times ((\beta'(V(k_\partial) + \chi_{0,\delta} + t\bar{\chi}_{0,\delta}))^{1/2}\,(\partial_t V(k_\partial))_\Omega\,(t)] \\ &+ \rho_\delta \ell \Phi_1[\sigma''(V(k_\partial) + \chi_{0,\delta} + t\bar{\chi}_{0,\delta})\,(\partial_t V(k_\partial))_\Omega\,(t)]. \end{aligned} \tag{63}$$

Moreover, note that, thanks to condition (23) on (k_∂, Ω) and to Thm. 2.4 in [16], we have the following estimate

$$\begin{aligned} &|U(k_\partial)_\Omega|^2_{H^1(0,T;H)\cap C^0([0,T];V)\cap L^2(0,T;W)} + |V(k_\partial)_\Omega|^2_{W^{1,\infty}(0,T;V')\cap H^1(0,T;V)} \\ &\quad + |\beta'(V(k_\partial) + \chi_{0,\delta} + t\bar{\chi}_{0,\delta})^{1/2}(\partial_t V(k_\partial))_\Omega|^2_{L^2(0,T;H)} + \frac{1}{\Omega}|\bar{\chi}_{0,\delta}|^2_V \\ &\le \overline{C}_2\Big[|\bar{\chi}_{0,\delta}|^2_{L^\infty(\Omega)}|[\beta'(V(k_\partial) + \chi_{0,\delta} + t\bar{\chi}_{0,\delta})]_{2\Omega}|_{L^1(Q)} + \frac{1}{\Omega}|\Delta\vartheta_{0,\delta}|^2 \\ &\quad + |(f_\delta)_\Omega|^2_{L^2(0,T;H)}\Big], \end{aligned} \tag{64}$$

$\overline{C}_2$ being a positive constant depending on the data of the problem, but not on k, k_∂, δ, and ε (cf. (2.26) in [16]). In the sequel, the meaning of $\overline{C}_2$ may also differ from line to line. Using now (64), along with assumption $\varphi \in W^{4,\infty}(\Omega)$ and Lemma 11 in Section 5, we derive the inequalities

$$\begin{aligned} &|e^{-t\Omega}K_\varepsilon N_\delta(k_\partial)|_{L^2(0,T)} \le |e^{-t\Omega}N_\delta(k_\partial)|_{L^2(0,T)} \\ &\le c_1|(k_\partial)_\Omega|_{L^1(0,T)}|(\partial_t U(k_\partial)_\Omega|_{L^2(0,T;H)} \\ &+ \Big\{|H_\delta(f_\delta, g_\delta)_\Omega|_{L^2(0,T)} \\ &+ c_2\big(|(\partial_t U(k_\partial))_\Omega|_{L^2(0,T;H)} + |(\partial_t V(k_\partial))_\Omega|_{L^2(0,T;H)}\big) \\ &+ \frac{c_3}{\Omega^{1/2}}|\bar{\chi}_{0,\delta}| + c_4(1+M)\Omega^{-1/2}|\bar{\chi}_{0,\delta}|_{L^\infty(\Omega)} \\ &+ c_5 M^{1/2}|(\beta'(\chi_\partial))^{1/2}(\partial_t V(k_\partial))_\Omega|_{L^2(0,T;H)}\Big\} \\ &\le c_1 T^{1/2} I_2(f_\delta, g_\delta, \vartheta_{0,\delta}, \chi_{0,\delta}, \Omega)|(k_\partial)_\Omega|_{L^2(0,T)} \\ &+ I_3(f_\delta, g_\delta, \vartheta_{0,\delta}, \chi_{0,\delta}, M, \Omega) \end{aligned} \tag{65}$$

from (61) and (63), and using the fact that $\sigma''(\chi)$ is bounded due to (18) and (24). Here, we have set

$$\begin{aligned} &I_2(f_\delta, g_\delta, \vartheta_{0,\delta}, \chi_{0,\delta}, \Omega) \\ &:= \overline{C}_2 \Big[\frac{1}{\Omega} |\overline{\chi}_{0,\delta}|_{L^\infty(\Omega)} + \frac{1}{\Omega} |\overline{\chi}_{0,\delta}|_H + \frac{1}{\Omega} |\Delta \vartheta_{0,\delta}|_H \\ &+ \Omega^{-(p-2)/(2p)} |f_\delta|_{L^p(0,T;H)} \Big], \end{aligned} \tag{66}$$

$$\begin{aligned} &I_3(f_\delta, g_\delta, \vartheta_{0,\delta}, \chi_{0,\delta}, M, \Omega) := \Omega^{-(p-2)/(2p)} |H_\delta(f_\delta, g_\delta)|_{L^p(0,T)} \\ &+ \frac{c_3}{\Omega^{1/2}} |\overline{\chi}_{0,\delta}| + (c_2 + c_5) I_2(f_\delta, g_\delta, \vartheta_{0,\delta}, \chi_{0,\delta}, \Omega) \\ &+ c_4 (1+M) \Omega^{-1/2} |\overline{\chi}_{0,\delta}|_{L^\infty(\Omega)}. \end{aligned} \tag{67}$$

Since the right-hand side in (66) tends to 0 uniformly with respect to $\delta \in (0, \delta_0]$ as $\Omega \to +\infty$, we can choose $\Omega \in \mathbb{R}_+$ independent of δ and so large as to satisfy

$$I_2(f_\delta, g_\delta, \vartheta_{0,\delta}, \chi_{0,\delta}, \Omega) \le \frac{1}{2 c_1 T^{1/2}}, \qquad \forall \delta \in [0, \delta_0]. \tag{68}$$

As a consequence, from (61) and (65), (66–68), under conditions (23) for (k_∂, Ω), we deduce the estimate

$$\begin{aligned} |(k_\partial)_\Omega|_{L^2(0,T)} =& |e^{-t\Omega} K_\varepsilon N_\delta(k_\partial)|_{L^2(0,T)} \le \frac{1}{2} |(k_\partial)_\Omega|_{L^2(0,T)} \\ &+ I_3(f_\delta, g_\delta, \vartheta_{0,\delta}, \chi_{0,\delta}, M, \Omega). \end{aligned} \tag{69}$$

Therefore, under the same conditions, all the solutions to the fixed-point equation

$$k_\partial = K_\varepsilon N_\delta(k_\partial) \tag{70}$$

satisfy the estimate

$$\begin{aligned} |e^{-t\Omega} k_\partial|_{L^2(0,T)} &= |(k_\partial)_\Omega|_{L^2(0,T)} \le 2 I_3(f_\delta, g_\delta, \vartheta_{0,\delta}, \chi_{0,\delta}, M, \Omega) \\ &=: 2 I_3(f_\delta, g_\delta, \vartheta_{0,\delta}, \chi_{0,\delta}, M, \Omega), \end{aligned} \tag{71}$$

where $I_3(f_\delta, g_\delta, \vartheta_{0,\delta}, \chi_{0,\delta}, M, \Omega) \to 0$ uniformly with respect to $\delta \in [0, \delta_0]$ as $\Omega \to +\infty$. Note that such a limit is uniform also with respect to the data varying in every (fixed) closed ball.

Summing up, we have proved that, under conditions (23) for (k_∂, Ω) and (68), the operator $K_\varepsilon(N_\delta)$ satisfies the estimate (69), while each solution to the fixed-point equation (70) satisfies (71).

Conversely, choose any Ω_0 independent of $\delta \in [0, \delta_0]$ and satisfying the system of inequalities

$$\begin{cases} T^{1/2} I_3(f_\delta, g_\delta, \vartheta_{0,\delta}, \chi_{0,\delta}, M, \Omega_0) \le \frac{k_0}{2(2+k_0)^{1/2}}, \\ I_2(f_\delta, g_\delta, \vartheta_{0,\delta}, \chi_{0,\delta}, M, \Omega_0) \le 1/(2c_1 T^{1/2}). \end{cases} \tag{72}$$

Of course, Ω_0 is *independent* of k, k_∂, δ, and ε, since I_2 and I_3 are. Moreover, let us choose

$$X = \{k \in L^2(0,T) : |k_{\Omega_0}|_{L^2(0,T)} \le 2I_3(f_\delta, g_\delta, \vartheta_{0,\delta}, \chi_{0,\delta}, M, \Omega_0)\}.$$

Then, for any $k \in X$, we get

$$\begin{aligned} |k_{\Omega_0}|_{L^1(0,T)} &\le T^{1/2}|k_{\Omega_0}|_{L^2(0,T)} \le T^{1/2} I_3(f_\delta, g_\delta, \vartheta_{0,\delta}, \chi_{0,\delta}, M, \Omega_0) \\ &\le \frac{k_0}{2(2+k_0)^{1/2}}. \end{aligned}$$

Consequently, according to Theorem 1 with $s = 2$ and Remark 2, for any kernel $k_\partial \in L^2(0,T)$ and any $\Omega = \Omega_0$ satisfying (72), the solution $(\vartheta_\partial, \chi_\partial) = (\Theta(k_\partial), X(k_\partial))$ to the direct problem (6–9) satisfies the following estimates (cf. (24))

$$\begin{aligned} &|\Theta(k_\partial)|_{W^{1,\infty}(0,T;H^3(\Omega))\cap H^1(0,T;H^4(\Omega))} \\ &+ |X(k_\partial)|_{W^{1,\infty}(0,T;H^4(\Omega))\cap H^2(0,T;H^2(\Omega))} \le K(\Omega_0), \end{aligned} \tag{73}$$

K being the positive constant appearing in estimate (24), and depending only on the norms of the data and on Ω_0. Finally, we also note that for any $\Omega_0 \in \mathbb{R}^+$ (satisfying (72)) the estimate

$$\left(1 + \frac{2}{k_0}\right) |(k_\partial)_{\Omega_0}|^2_{L^1(0,T)} \le \frac{k_0}{4},$$

holds, k_∂ being the solution of the fixed-point equation (70). □

4. Convergence estimates

We can now prove our main theorem, namely a convergence result including an estimate of the convergence rate of the approximating solution to the exact one.

Theorem 9: Let assumptions listed in Theorems 5 and 8 be satisfied. Suppose, moreover, that there exists $\overline{\sigma} \in (0, 1/2)$ and $M_{f,g} > 0$ such that

$$|H(f,g)|_{\overline{\sigma},2,\mathbb{R}} \le M_{f,g}, \tag{74}$$

$H(f,g)$ being defined by (60) and $|\cdot|_{\overline{\sigma},2,\mathbb{R}}$ defined as in (15). Then, the solution $(\vartheta_\partial, \chi_\partial, k_\partial)$ to Problem (P_∂) converges (as δ and ε tend to 0) to the solution (ϑ, χ, k) of (1–5) in the following sense:

$$\begin{aligned}\vartheta_\partial \to \vartheta \quad &\text{weakly star in } W^{1,\infty}(0,T;H^3(\Omega)) \cap H^1(0,T;H^4(\Omega)),\\ &\text{and strongly in } C^0([0,T];H^2(\Omega)) \cap L^2(0,T;H^3(\Omega)),\end{aligned} \tag{75}$$

$$\begin{aligned}\chi_\partial \to \chi \quad &\text{weakly star in } W^{1,\infty}(0,T;H^4(\Omega)) \cap H^2(0,T;H^2(\Omega)),\\ &\text{and strongly in } C^1([0,T];H^3(\Omega)) \cap L^2(0,T;H^4(\Omega)),\end{aligned} \tag{76}$$

$$k_\partial \to k \quad \text{strongly in } L^2(0,T). \tag{77}$$

Moreover, taking $\Omega_0 \in \mathbb{R}^+$ as in (51), there exists a positive constant $\overline{C}(\Omega_0, \overline{\sigma}, T)$, also depending on the data of the problem, but not on δ and ε, such that

$$|k_\partial - k|_{0,2}^2 \leq \overline{C}(\Omega_0, \overline{\sigma}, T)\left(\varepsilon^{2\overline{\sigma}} + \delta^2\right). \tag{78}$$

Remark 10: Choosing $\varepsilon = \delta^{1/\overline{\sigma}}$, we immediately deduce that the rate of convergence of the approximating solution to the exact one is δ, i.e. $\rho = 1$ (cf. the end of Section 1).

Proof: First of all, let us note that the weak convergence relations (75–76) immediately follow from estimate (73). To derive (77) and the convergence estimate (78), we first recall that the *exact* kernel k solves the following operator equation (cf. (5.24) in [16])

$$k = N_{0,2}(k). \tag{79}$$

As far as k_∂ is concerned, observe that it solves the fixed-point equation

$$k_\partial = K_\varepsilon N_{\delta,2}(k_\partial). \tag{80}$$

By comparison with (62), we deduce that the nonlinear operator $N_{\delta,2}$ is defined by

$$\begin{aligned}
&N_{\delta,2}(k_\partial) = H_\delta(f_\delta, g_\delta) - \rho_\delta k_\partial * \Phi_1[\partial_t \vartheta_\partial] \\
&- \rho_\delta(k_0 + \ell^2)\{\ell\Phi_2[-\Delta\chi_\partial + \ell\beta(\chi_\partial) + \ell\sigma'(\chi_\partial) - (\ell^2 + k_0)\vartheta_\partial \\
&- k_\partial * \vartheta_\partial] + \Phi_1[f_\delta]\} - \rho_\delta\Phi_3[-\Delta\chi_\partial \\
&+ \beta(\chi_\partial) + \sigma'(\chi_\partial) - \ell\vartheta_\partial] + \rho_\delta((\beta(\chi_\partial) + \sigma'(\chi_\partial))\Delta\varphi_1, -\Delta\chi_\partial \\
&+ (\beta(\chi_\partial) + \sigma'(\chi_\partial)) - \ell\vartheta_\partial)_H + 2\rho_\delta((\beta'(\chi_\partial) + \sigma''(\chi_\partial))\nabla\chi_\partial \cdot \nabla\varphi_1, \\
&- \Delta\chi_\partial + (\beta(\chi_\partial) + \sigma'(\chi_\partial)) - \ell\vartheta_\partial)_H + \rho_\delta(\varphi_1(\beta'(\chi_\partial) \\
&+ \sigma''(\chi_\partial))\Delta\chi_\partial, -\Delta\chi_\partial + (\beta(\chi_\partial) + \sigma'(\chi_\partial)) - \ell\vartheta_\partial)_H \\
&+ \rho_\delta(\varphi_1(\beta''(\chi_\partial) + \sigma'''(\chi_\partial))|\nabla\chi_\partial|^2, -\Delta\chi_\partial \\
&+ (\beta(\chi_\partial) + \sigma'(\chi_\partial)) - \ell\vartheta_\partial)_H,
\end{aligned} \tag{81}$$

and $H_\delta(f_\delta, g_\delta)$ is defined by (58). Further, observe that the following relation holds

$$k_\partial - k = K_\varepsilon N_{\delta,2}(k_\partial) - N_{0,2}(k). \tag{82}$$

From Lemma 12 in the appendix and equations (79–80) we deduce that the following estimate holds for all $\tau \in (0, T]$ and all $\overline{\sigma} \in (0, 1/2)$ satisfying (74),

$$\begin{aligned}
|K_\varepsilon N_{\delta,2}(k_\partial) - N_{0,2}(k)|^2_{L^2(0,\tau)} \le C(\overline{\sigma}, T)^2 \big(&|N_{\delta,2}(k_\partial) - N_{0,2}(k)|^2_{L^2(0,\tau)} \\
&+ \varepsilon^{2\overline{\sigma}} |k|^2_{\overline{\sigma},2,\mathbb{R}}\big) \quad \forall \tau \in (0, T],
\end{aligned} \tag{83}$$

where $C(\overline{\sigma})$ is a positive constant depending only on $\overline{\sigma}$.

Setting then $\gamma := \beta + \sigma'$, $\alpha_j := \alpha_j(V(k), V(k_\partial))$ (cf. (43) and formula (4.9)–(4.11) in [16]), and

$$\widetilde{\Theta}(k,\, k_\partial) := \Theta(k_\partial) - \Theta(k), \qquad \widetilde{X}(k, k_\partial) := X(k_\partial) - X(k), \tag{84}$$

from (81) and by formula (5.24) in [23], we get

$$
\begin{aligned}
N_{\delta,2}(k_\partial) - N_{0,2}(k) &= H_\delta(f_\delta, g_\delta) - H(f,g) - \rho_\delta k_\partial * \Phi_1[\partial_t \widetilde{\Theta}(k,k_\partial)] \\
&- \rho_\delta \widetilde{k} * \Phi_1[\partial_t \Theta(k)] - \rho_\delta (k_0 + \ell^2)\ell \Phi_2[-\Delta \widetilde{X}(k,k_\partial) + \ell \alpha_2 \widetilde{X}(k,k_\partial)] \\
&+ \rho_\delta (k_0 - l) l \Phi_2[(k_0 + \ell^2)\widetilde{\Theta}(k,k_\partial) + k_\partial * \widetilde{\Theta}(k,k_\partial) + (k_\partial - k) * \Theta(k)] \\
&+ \Phi_1[f_2 - f_1] - \rho_\delta \Phi_3[-\Delta \widetilde{X}(k,k_\partial) + \alpha_2 \widetilde{X}(k,k_\partial) - \ell \widetilde{\Theta}(k,k_\partial)] \\
&+ \rho_\delta(\alpha_2 \widetilde{X}(k,k_\partial)\Delta\varphi_1, -\Delta X(k_\partial) + \gamma(X(k_\partial)) - \ell\Theta(k_\partial))_H \\
&+ \rho_\delta(\gamma(X(k))\Delta\varphi_1, -\Delta \widetilde{X}(k,k_\partial) + \alpha_2 \widetilde{X}(k,k_\partial) - \ell \widetilde{\Theta}(k,k_\partial))_H \\
&+ 2\rho_\delta(\alpha_3 \widetilde{X}(k,k_\partial)\nabla X(k_\partial)\cdot\nabla\varphi_1 + \gamma'(X(k_\partial))\nabla \widetilde{X}(k,k_\partial)\cdot\nabla\varphi_1, \\
&\qquad - \Delta X(k_\partial) + \gamma(X(k_\partial)) - \ell\Theta(k_\partial))_H \\
&+ 2\rho_\delta(\gamma'(X(k)\nabla X(k)\cdot\nabla\varphi_1, -\Delta\widetilde{X}(k,k_\partial) + \alpha_2\widetilde{X}(k,k_\partial) - \ell\widetilde{\Theta}(k,k_\partial))_H \\
&+ \rho_\delta(\varphi_1(\alpha_3\widetilde{X}(k,k_\partial) - \Delta X(k_\partial) + \gamma'(X(k_\partial))\Delta\widetilde{X}(k,k_\partial)), \\
&\qquad - \Delta X(k_\partial) + \gamma(X(k_\partial)) - \ell\Theta(k_\partial))_H \\
&+ \rho_\delta(\varphi_1\gamma'(X(k)) - \Delta X(k), -\Delta\widetilde{X}(k,k_\partial) + \alpha_2\widetilde{X}(k,k_\partial) - \ell\widetilde{\Theta}(k,k_\partial))_H \\
&+ \rho_\delta(\varphi_1(\alpha_4\widetilde{X}(k,k_\partial)|\nabla X(k_\partial)|^2 + \gamma''(X(k_\partial))\nabla(X(k) \\
&\qquad + X(k_\partial))\cdot\nabla\widetilde{X}(k,k_\partial)), -\Delta X(k_\partial) + \gamma(X(k_\partial)) - \ell\Theta(k_\partial))_H \\
&+ (\rho-\rho_\delta)(\varphi_1\gamma''(X(k))|\nabla X(k)|^2, -\Delta\widetilde{X}(k,k_\partial) + \alpha_2\widetilde{X}(k,k_\partial) \\
&\qquad - \ell\widetilde{\Theta}(k,k_\partial))_H - (\rho-\rho_\delta)k * \Phi_1[\partial_t\vartheta] \\
&- (\rho-\rho_\delta)(k_0+\ell^2)\{\ell\Phi_2[-\Delta\chi + \ell\beta(\chi) + \ell\sigma'(\chi) - (\ell^2+k_0)\vartheta - k*\vartheta] \\
&+ \Phi_1[f]\} - (\rho-\rho_\delta)\Phi_3[-\Delta\chi + \beta(\chi) + \sigma'(\chi) - \ell\vartheta] \\
&+ (\rho-\rho_\delta)((\beta(\chi)+\sigma'(\chi))\Delta\varphi_1, -\Delta\chi + (\beta(\chi)+\sigma'(\chi)) - \ell\vartheta)_H \\
&+ 2(\rho-\rho_\delta)((\beta'(\chi)+\sigma''(\chi))\nabla\chi\cdot\nabla\varphi_1, -\Delta\chi + (\beta(\chi)+\sigma'(\chi)) - \ell\vartheta)_H \\
&+ (\rho-\rho_\delta)(\varphi_1(\beta'(\chi)+\sigma''(\chi))\Delta\chi, -\Delta\chi + (\beta(\chi)+\sigma'(\chi)) - \ell\vartheta)_H \\
&+ (\rho-\rho_\delta)(\varphi_1(\beta''(\chi)+\sigma'''(\chi))|\nabla\chi|^2, -\Delta\chi + (\beta(\chi)+\sigma'(\chi)) - \ell\vartheta)_H,
\end{aligned}
\tag{85}
$$

where $H(f,g)$ is defined by (60) and $\rho = \alpha^{-1} = (\Phi[\Delta\vartheta_0])^{-1}$. Next, before estimating the left-hand side in (85), we recall the following estimates, holding for any $\tau \in (0,T]$:

$$|\widetilde{X}(k,k_\partial)(\tau)|_V \le C|\Delta\widetilde{X}(k,k_\partial)(\tau)|,$$

$$|k_\partial * \partial_t\widetilde{\Theta}(k,k_\partial)|^2_{L^2(0,\tau,H)} \le |k_\partial|^2_{L^1(0,\tau)} \int_0^\tau |\partial_t\widetilde{\Theta}(k,k_\partial)(s)|^2_H\, ds,$$

$$\begin{aligned}|(k_\partial - k) * \partial_t\Theta(k)|^2_{L^2(0,\tau,H)} &\le |\partial_t\Theta(k)|^2_{L^\infty(0,\tau;H)}|1 * (k_\partial - k)|^2_{L^2(0,\tau)}\\ &\le |\partial_t\Theta(k)|^2_{L^\infty(0,\tau;H)} T \int_0^\tau |k_\partial - k|^2_{L^2(0,s)}\, ds. \qquad (86)\end{aligned}$$

Then, we observe that the coefficients α_j, $j = 2,3,4$, defined in formula (4.9)–(4.11) in [16], belong to $L^\infty((0,T)\times\Omega)$, according to Theorem 1 with $s = 2$ and $\Omega = \Omega_0$. Moreover, the α_j's can be estimated *independently* of k due to (29) and (24), with $\Omega = \Omega_0$. Consequently, from (84–86) and the following estimates (cf. (4.44–4.46) in [16]):

$$\begin{aligned}&|\partial_t\widetilde{\Theta}(k,k_\partial)|_{L^2(0,\tau;H)} + |\partial_t\widetilde{X}(k,k_\partial)|_{L^2(0,\tau;H)} + |\Delta\widetilde{\Theta}(k,k_\partial)|_{L^2(0,\tau;H)}\\ &\quad + |\Delta^2\widetilde{X}(k,k_\partial)|_{L^2(0,\tau;H)} \le \kappa(\Omega_0,m,T)\Big\{|\vartheta_{0,\delta} - \vartheta_0|_{H^1(\Omega)}\\ &\quad + |\chi_{0,\delta} - \chi_0|_{H^2(\Omega)} + |f_\delta - f|_{L^2(0,\tau;H)} + |k_\partial - k|_{L^1(0,\tau)}\Big\}, \qquad (87)\end{aligned}$$

$$\begin{aligned}|\widetilde{\Theta}(k,k_\partial)|_{L^2(0,\tau;H)} &\le C[|\vartheta_{0,\delta} - \vartheta_0|_H + |\chi_{0,\delta} - \chi_0|_H]\\ &\quad + \kappa(\Omega_0,m,T)\int_0^\tau [|f_\delta - f|_{L^2(0,s;H)} + |k_\partial - k|_{L^1(0,s)}]\, ds, \qquad (88)\end{aligned}$$

$$\begin{aligned}&|\Delta\widetilde{X}(k,k_\partial)|_{L^2(0,\tau;H)} \le C|\Delta(\chi_{0,\delta} - \chi_0)|_H\\ &\quad + \kappa(\Omega_0,m,T)\int_0^\tau (\tau - s)^{-1/2}[|f_\delta - f|_{L^2(0,s;H)} + |k_\partial - k|_{L^1(0,s)}]\, ds, \qquad (89)\end{aligned}$$

we easily deduce the following inequality for any $\tau \in (0,T]$

$$|N_{\delta,2}(k_\partial) - N_{0,2}(k)|^2_{L^2(0,\tau)} \le C^2\delta^2 + \int_0^\tau h(\tau - s)|k_\partial - k|^2_{L^2(0,s)}\, ds, \qquad (90)$$

where

$$\begin{aligned}h(\tau) &= \kappa(\Omega_0,m,T)(\tau^{-1/2} + 1)\\ &\le \max(1,T^{1/2})\kappa(\Omega_0,m,T)\tau^{-1/2}. \qquad \forall\tau \in (0,T]. \qquad (91)\end{aligned}$$

Then, from (83), (90) and Lemma 13 in the last section we immediately deduce the following chain of inequalities for all $\tau \in (0,T]$:

$$\begin{aligned}
&|k_\partial - k|^2_{L^2(0,T)} = |K_\varepsilon N_{\delta,2}(k_\partial) - N_{0,2}(k)|^2_{L^2(0,\tau)} \\
&\le C(\overline{\sigma},T)^2\big[|N_{\delta,2}(k_\partial) - N_{0,2}(k)|^2_{L^2(0,\tau)}\big] + C(\overline{\sigma})^2\varepsilon^{2\overline{\sigma}}|k|^2_{\overline{\sigma},2,\mathbb{R}} \\
&\le C(\overline{\sigma},T)^2\Big\{\varepsilon^{2\overline{\sigma}}|k|^2_{\overline{\sigma},2,\mathbb{R}} + C^2\delta^2 + \int_0^\tau h(\tau-s)|k_\partial - k|^2_{L^2(0,s)}\,ds\Big\} \\
&\le C(\overline{\sigma},\tau)^2\big\{\kappa(\Omega_0)\varepsilon^{2\overline{\sigma}} + C^2\delta^2 + \int_0^\tau h(\tau-s)|k_\partial - k|^2_{L^2(0,s)}\,ds\big\}, \qquad (92)
\end{aligned}$$

$C(\overline{\sigma},T)$ being the same positive constant depending only on $\overline{\sigma}$ and T as in formula (83). Exploiting the above Gronwall inequality and using the uniform estimate of k_∂, obtained in Theorem 8, we get

$$\begin{aligned}
|k_\partial - k|^2_{L^2(0,\tau)} &\le C(\overline{\sigma},T)^2[\kappa(\Omega_0)\varepsilon^{2\overline{\sigma}} + C^2\delta^2]\Big[1 + \sum_{j=1}^{\infty}(h*)^{j-1}h(\tau)\Big] \\
&\le \kappa(\Omega_0,m,M,T,\overline{\sigma})(\varepsilon^{2\overline{\sigma}} + \delta^2), \qquad \forall\tau \in (0,T], \qquad (93)
\end{aligned}$$

where the symbol $(h*)^{j-1}h = h*h*\cdots*h$ denotes the convolution of h with itself $j-1$ times.

Setting $\tau = T$, we finally obtain the desired convergence estimate

$$|k_\partial - k|^2_{0,2} \le \kappa(\Omega_0,m,M,T,\overline{\sigma})(\varepsilon^{2\overline{\sigma}} + \delta^2). \qquad (94)$$

Then, the convergence (77) immediately follows and we can pass to the limit as δ and ε tend to 0 in (6–12), obtaining the limit problem (1–5).

Finally, we deduce the weak-star convergence (75–76) from Theorem 1 with $s = 2$, while the strong convergence of ϑ_∂ in (75) and the strong convergence of χ_∂ in $C^1([0,T];H^3(\Omega))$ can be recovered using the standard compactness result proved in [19] (cf. Cor. 8, p. 89). The last strong convergence of χ_∂ in (76) in $L^2(0,T;H^4(\Omega))$ is a consequence of Theorem 3. Notice that in all these passages to the limit we obtain the same limit pair (ϑ,χ), due to the fact that each limit pair turns out to solve problem (1–4) which admits only one solution due to Thm. 2.1 in [18]. This concludes the proof of Theorem 9. □

5. Apppendix

In this appendix, we shall prove some results that have been used in the previous sections.

First of all, concerning the operator K_ε defined in (61), we can show the following property.

Lemma 11: *For any $\varepsilon, \Omega \in \mathbb{R}_+$ the following estimate holds*

$$|(K_\varepsilon h)_\Omega|_{0,2} \leq |h_\Omega|_{0,2}, \qquad \forall h \in L^2(0,T). \tag{95}$$

Proof: The assertion is implied by Young's inequality for convolutions and the following relations

$$\begin{aligned}&\frac{2}{\varepsilon}\int_0^{+\infty}\{\exp[-t(\Omega+\varepsilon^{-1})] - \exp[-t(\Omega+2\varepsilon^{-1})]\}\,dt \\ &= \frac{2}{\varepsilon}\Big[\frac{1}{(\Omega+\varepsilon^{-1})} - \frac{1}{(\Omega+2\varepsilon^{-1})}\Big] = \frac{2}{(\varepsilon\Omega+1)(\varepsilon\Omega+2)} \leq 1,\end{aligned} \tag{96}$$

holding for all $\varepsilon, \Omega \in \mathbb{R}^+$. □

We now need another fundamental property of operator K_ε.

Lemma 12: *Let $h_\partial \in L^2(0,T)$, $h_{0,0} \in H^{\overline{\sigma}}(0,T)$ with $\overline{\sigma} \in (0,1/2)$. Then, the operator K_ε defined in (61) satisfies the following estimate*

$$|K_\varepsilon h_\partial - h_{0,0}|_{0,2} \leq C(\overline{\sigma},T)\big(|h_\partial - h_{0,0}|_{0,2} + \varepsilon^{\overline{\sigma}}|h_{0,0}|_{\overline{\sigma},2,\mathbb{R}}\big), \tag{97}$$

where $|f|_{\overline{\sigma},2,\mathbb{R}}$ denotes the $H^{\overline{\sigma}}(0,T)$-norm of f defined in (15) *and $C(\overline{\sigma},T)$ is a positive constant depending only on $\overline{\sigma}$ and T.*

Proof: From (61) we easily deduce the identity for all $t \in [0,T]$

$$\begin{aligned}K_\varepsilon h_\partial(t) - h_{0,0}(t) &= -h_{0,0}(t)\big(2e^{-t/\varepsilon} - e^{-2t/\varepsilon}\big) \\ &+\frac{2}{\varepsilon}\int_0^t (e^{-s/\varepsilon} - e^{-2s/\varepsilon})[h_\partial(t-s) - h_{0,0}(t-s)]\,ds \\ &+\frac{2}{\varepsilon}\int_0^t (e^{-s/\varepsilon} - e^{-2s/\varepsilon})[h_{0,0}(t-s) - h_{0,0}(t)]\,ds.\end{aligned} \tag{98}$$

Recall now that $H^{\overline{\sigma}}(0,T)$ is continuously embedded in $L^2(0,T;t^{-2\overline{\sigma}}\,dt)$ if $\overline{\sigma} \in (0,1/2)$. Then, from (96) and (98), via Young and Hölder's inequalities, we get

$$|K_\varepsilon h_\partial - h_{0,0}|_{0,2} \le \Big(\int_0^T t^{-2\overline{\sigma}}|h_{0,0}(t)|^2 t^{2\overline{\sigma}}\big(2e^{-t/\varepsilon}-e^{-2t/\varepsilon}\big)^2\,dt\Big)^{1/2}$$

$$+|h_\partial - h_{0,0}|_{0,2} + \sqrt{2}\varepsilon^{-1/2}\Big(\int_0^T dt\int_0^t s^{1+2\overline{\sigma}}\big(2e^{-s/\varepsilon}-e^{-2s/\varepsilon}\big)s^{-1-2\overline{\sigma}}\times$$

$$\times|h_{0,0}(t-s)-h_{0,0}(t)|^2\,ds\Big)^{1/2} \le C(\overline{\sigma})\varepsilon^{\overline{\sigma}}\Big(\int_0^T t^{-2\overline{\sigma}}|h_{0,0}(t)|^2\,dt\Big)^{1/2}$$

$$+|h_\partial - h_{0,0}|_{0,2} + C(\overline{\sigma})\varepsilon^{\overline{\sigma}}\Big(\int_0^T dt\int_0^t s^{-1-2\overline{\sigma}}|h_{0,0}(t-s)-h_{0,0}(t)|^2\,ds\Big)^{1/2}$$

$$\le C(\overline{\sigma},T)\big[|h_\partial - h_{0,2}|_{0,2} + \varepsilon^{\overline{\sigma}}|h_{0,0}|_{\overline{\sigma},2,\mathbb{R}}\big]. \tag{99}$$

From (99) we easily deduce the desired estimate (97). □

Hence, we can prove the following result.

Lemma 13: *Let the assumptions in Theorem 1 (with $s=2$) and Theorem 5 be fulfilled. Moreover, let (ϑ,χ,k) be a solution of (1–5) and let $\Omega_0 \in \mathbb{R}^+$ be as in* (33)*. Suppose that there exist $\overline{\sigma}\in(0,1/2)$ and $M_{f,g}>0$ such that*

$$|H(f,g)|_{\overline{\sigma},2,\mathbb{R}} \le M_{f,g}, \tag{100}$$

$H(f,g)$ being defined by (60)*. Then, there exists a positive constant $\overline{\kappa}(\Omega_0)$, depending on Ω_0 and the data, but not on k_∂, δ, ε, such that*

$$|k|_{\overline{\sigma},2,\mathbb{R}} \le M_{f,g} + \overline{\kappa}(\Omega_0). \tag{101}$$

Proof: According to Thm. 2.11 in [16] with $s=2$, for any kernel $k\in L^2(0,T)$ and any $\Omega=\Omega_0$ satisfying (72), the solution $(\vartheta,\chi)=(\Theta(k),X(k))$ to the direct problem satisfies the estimate (cf. (24))

$$|\Theta(k)|_{W^{1,\infty}(0,T;H^3(\Omega))\cap H^1(0,T;H^4(\Omega))} + |X(k)|_{W^{1,\infty}(0,T;H^4(\Omega))\cap H^2(0,T;H^2(\Omega))} \le \kappa(\Omega_0). \tag{102}$$

Therefore, from (59) and (102) we easily deduce, for $t\in(0,T)$, the integral inequality

$$|k(t)| \le |H(f,g)(t)| + \kappa(\Omega_0)\big[1+|\beta'(X(k)(t))|_{0,2}\big] + \kappa(\Omega_0)\int_0^t k(s)\,ds. \tag{103}$$

We can now show that k belongs to $H^{\overline{\sigma}}(0,T)$. To this aim, we first observe that equations (1), (4) can be rewritten in the form

$$\partial_t\vartheta - k_0\Delta\vartheta - k*\Delta\vartheta = \widetilde{f}, \quad \text{in } \Omega\times(0,T), \tag{104}$$

$$D_n\vartheta = 0, \quad \text{on } \partial\Omega\times(0,T), \tag{105}$$

where

$$\widetilde{f} = f - \ell\partial_t X(k). \tag{106}$$

Thanks to (100), since

$$|\partial_t X(k)|_{L^\infty(0,T;H^4(\Omega))\cap H^1(0,T;H^2(\Omega))} \le \kappa(\Omega_0), \tag{107}$$

we conclude that $\widetilde{f}\in H^{\overline{\sigma}}(0,T;L^2(\Omega))$ and satisfies

$$|\widetilde{f}|_{\overline{\sigma},2,H} \le \kappa(\Omega_0) + |f|_{\overline{\sigma},2,H}. \tag{108}$$

From well-known properties concerning parabolic equations, we deduce that the solution ϑ to (104–105), with $k=0$, belongs to $H^{1+\overline{\sigma}}(0,T;L^2(\Omega))\cap H^{\overline{\sigma}}(0,T;H^2(\Omega))$ (cf., e.g., Thm. 30 in [6]). Moreover, ϑ satisfies the estimate

$$|\partial_t\vartheta|_{\overline{\sigma},2,H} \le C\left(\kappa(\Omega_0) + |f|_{\overline{\sigma},2,H}\right). \tag{109}$$

Then, it is easy to check that problem (104–105) is equivalent to the following integrodifferential operator equation for ϑ:

$$\vartheta(\cdot,t) = e^{tk_0\Delta}\vartheta_0 + \int_0^t e^{(t-s)k_0\Delta}\widetilde{f}(s)\,ds + \int_0^t e^{(t-s)k_0\Delta}k*\Delta\vartheta(\cdot,s)\,ds. \tag{110}$$

Setting $z = k_0\Delta\vartheta$, we immediately deduce that z solves the following operator equation in $H^{\overline{\sigma}}(0,T;L^2(\Omega))$:

$$\begin{aligned} z(\cdot,t) &= k_0 e^{tk_0\Delta}\Delta\vartheta_0 + k_0\Delta\int_0^t e^{(t-s)k_0\Delta}\widetilde{f}(s)\,ds \\ &\quad + (k_0e^{tk_0\Delta} - I)k*w(\cdot,t) \\ &\quad + k_0\int_0^t \Delta e^{(t-s)k_0\Delta}[k*z(\cdot,s) - k*z(\cdot,t)]\,ds \\ &= k_0e^{tk_0\Delta}\Delta\vartheta_0 + [e^{(t-s)k_0\Delta} - I]\widetilde{f}(t) - k*z(\cdot,t) \\ &\quad + \partial_t\int_0^t e^{(t-s)k_0\Delta}k*z(\cdot,s)\,ds =: z_0(\cdot,t) + L(k*z)(t). \end{aligned} \tag{111}$$

We now need the following estimate, holding for any $k\in L^2(0,T)$ and $z\in H^{\sigma}(0,T;H)$:

$$|k*z|_{\overline{\sigma},2,H} \le T^{(1-2\overline{\sigma})/2}(4T^{3/2}+1)^{\overline{\sigma}}|k|_{L^2(0,T)}|z|_{\overline{\sigma},2,H}.$$

Such an estimate can be easily proved by interpolation, taking advantage of the formula $\partial_t(k * z) = k * z' + kz(0)$ and of the estimates

$$\begin{aligned}|k * z|_{L^2(0,T;H)} &\leq |k|_{L^1(0,T)}|z|_{L^2(0,T;H)} \leq T^{1/2}|k|_{L^2(0,T)}|z'|_{L^2(0,T;H)},\\ |\partial_t(k * z)|_{L^2(0,T;H)} &\leq |k|_{L^1(0,T)}|z|_{L^2(0,T;H)} + |k|_{L^2(0,T)}|z(0)|\\ &\leq |k|_{L^2(0,T)}\{2T^{1/2}|z'|_{L^2(0,T;H)} + T^{-1/2}|z|_{L^2(0,T;H)}\}\\ &\leq |k|_{L^2(0,T)}(4T + T^{-1/2})|z|_{H^1(0,T;H)}.\end{aligned}$$

Then, since (cf. Lemma 3.1 in [7])

$$|\{\partial_t[e^{tk_0\Delta} * k * z]\}_\Omega|_{H^{\overline{\sigma}}(0,T;L^2(\Omega))} \leq C|(k * z)_\Omega|_{H^{\overline{\sigma}}(0,T;L^2(\Omega))}, \tag{112}$$

C being a positive constant *independent* of (Ω, k, z), we easily deduce the estimates

$$\begin{aligned}|[L(k * z)]_\Omega|_{\overline{\sigma},2,H} &\leq (C+1)|(k * z)_\Omega|_{\overline{\sigma},2,H}\\ &\leq (C+1)T^{(1-2\sigma)/2}(4T^{3/2}+1)^{\overline{\sigma}/2}\\ &\qquad \times|k_\Omega|_{L^2(0,T)}|z_\Omega|_{\overline{\sigma},2,H}.\end{aligned} \tag{113}$$

Consequently, L turns out to be a contracting operator whenever

$$|k_\Omega|_{0,2} \leq \frac{1}{2(C+1)T^{(1-2\sigma)/2}(4T^{3/2}+1)^{\overline{\sigma}/2}}. \tag{114}$$

Of course, we can add this condition in the definition of Ω_0 in (70) with $\delta = 0$ without any trouble - at most this new condition increases Ω_0 -. Therefore, we can conclude that equation (111) admits a unique solution $z \in H^{\overline{\sigma}}(0,T;L^2(\Omega))$, satisfying the estimate

$$|z_\Omega|_{\overline{\sigma},2,H} \leq \kappa(\Omega_0). \tag{115}$$

Since (104) implies $(\partial_t\vartheta)_\Omega = z_\Omega + k_0^{-1}k_\Omega * z_\Omega + \widetilde{f}_\Omega$, (115), in turn, yields

$$|(\partial_t\vartheta)_\Omega|_{\overline{\sigma},2,H} \leq \kappa(\Omega_0). \tag{116}$$

Finally, from (59), (107), and (116) we deduce the inequalities

$$\begin{aligned}|k|_{\overline{\sigma},2,\mathbb{R}} \leq\, &|H(f,g)|_{\overline{\sigma},2,\mathbb{R}} + \kappa(\Omega_0)(|k|_{0,2,\mathbb{R}}+1)\\ &+ |\beta'(X(k))\partial_t X(k)|_{\overline{\sigma},2,H}.\end{aligned} \tag{117}$$

Recall now that the estimate

$$|ah|_{\overline{\sigma},2,H} \leq C(\overline{\sigma},T)|a|_{C^1(\overline{Q})}|h|_{\overline{\sigma},2,H}, \tag{118}$$

holds true for all $h \in H^{\overline{\sigma}}(0,T;L^2(\Omega))$ and $a \in C^1(\overline{Q})$, $Q = \Omega \times (0,T)$. Then, from (102) and (118) with $a = \beta'(X(k))$ and $h = \partial_t X(k)$, we obtain

$$|\beta'(X(k))\partial_t\chi|_{\overline{\sigma},2,H} \leq \kappa(\Omega_0). \tag{119}$$

Consequently, from (117), we deduce

$$|k|_{\overline{\sigma},2,\mathbb{R}} \leq |H(f,g)|_{\overline{\sigma},2,\mathbb{R}} + \kappa(\Omega_0). \tag{120}$$

This, along with (100), leads exactly to (101) with $\overline{\kappa}(\Omega_0) := \kappa(\Omega_0)$ as in (120). □

References

[1] G. Bonfanti, P. Colli, M. Grasselli and F. Luterotti, Nonsmooth kernels in a phase relaxation problem with memory. *Nonlinear Anal.* **32**, 455–465 (1998).

[2] G. Caginalp, An analysis of a phase field model of a free boundary. *Arch. Rational Mech. Anal.* **92**, 205–245 (1986).

[3] B.D. Coleman and M.E. Gurtin, Thermodynamics and wave propagation. *Quart. Appl. Math.* **24**, 257–262 (1966).

[4] P. Colli, G. Gilardi and M. Grasselli, Global smooth solution to the standard phase-field model with memory. *Adv. Differential Equations* **2**, 453–486 (1997).

[5] F. Colombo and D. Guidetti, An inverse problem for a phase-field model. *Preprint* (2004).

[6] G. Di Blasio, Linear parabolic evolution equations in L^p-spaces. *Ann. Mat. Pura Appl. (4)* **138**, 55–104 (1984).

[7] A. Favini and A. Lorenzi, Singular integrodifferential equations of parabolic type and inverse problems. *Math. Models Methods Appl. Sci.* **13**, 1745–1766 (2003).

[8] D. Guidetti and A. Lorenzi, A mixed type identification problem related to a phase field model with memory. *Preprint* (2005).

[9] M.E. Gurtin and A.C. Pipkin, A general theory of the heat conduction with finite wave speeds. *Arch. Rational Mech. Anal.* **31**, 113–126 (1968).

[10] D.D. Joseph and L. Preziosi, Heat waves. *Rev. Modern Phys.* **61**, 41–73 (1989).

[11] D.D. Joseph and L. Preziosi, Addendum to the paper: *Heat waves* [Rev. Modern Phys. **61**, 41–73 (1989)]. *Rev. Modern Phys.* **62**, 375 (1990).

[12] N. Kenmochi, Systems of non linear PDEs arising from dynamical phase transitions. In: *Phase transitions and hysteresis.* Ed. A. Visintin, (Springer-Verlag, Berlin, 1994), p. 9.

[13] A. Lorenzi, An identification problem for a conserved phase-field model with memory. *Math. Methods Appl. Sci.* **28**, 1315–1339 (2005).

[14] A. Lorenzi, Regularization and approximation of weakly ill-posed integrodifferential identification problem. *Preprint* (2005).

[15] A. Lorenzi and E. Paparoni, Direct and inverse problems in the theory of materials with memory. *Rend. Sem. Mat. Univ. Padova* **87**, 105–138 (1992).

[16] A. Lorenzi, E. Rocca and G Schimperna, Direct and inverse problems for a parabolic integro-differential system of Caginalp type. *Adv. Math. Sci Appl.* **15**, 227–263 (2005).

[17] A. Lunardi, *Analytic semigroups and optimal regularity in parabolic problems.* Birkhäuser, Basel, 1995.

[18] E. Rocca, Existence and uniqueness for the parabolic conserved phase-field model with memory. *Comm. Appl. Anal.* **8**, 27–46 (2004).

[19] J. Simon, Compact sets in the space $L^p(0,T;B)$. *Ann. Mat. Pura Appl. (4)* **146**, 65–96 (1987).

[20] A. Tikhonov and V. Arsénine, *Méthodes de resolution de problèmes mal posés.* Mir, Moscow, 1976.

GRADIENT FLOW REACTION/DIFFUSION MODELS IN PHASE TRANSITIONS

John Norbury[1] and Christophe Girardet[2]

[1] *Institute of Mathematics, Oxford University*
24-29 St Giles', OX1 4LB Oxford, UK
E-mail: john.norbury@lincoln.ox.ac.uk

[2] *Institute of Mathematics, Oxford University*
24-29 St Giles', OX1 4LB Oxford, UK
E-mail: girardet@maths.ox.ac.uk

In this article we consider a nonlinear large reaction small diffusion problem which has two (or more) stable states, and we analyse it using two different methods.

- First method: an approach based on (asymptotic) expansions;
- and second method: an approach based on the notion of Γ-convergence.

The analysis of such a problem shows that the two methods are complementary. It is well-known that, for such a problem, time-dependent solutions are characterised by (moving) layers or vortices. Here, we are specially interested in the existence, the shape and the motion of such layers or vortices, with respect to the inhomogeneous coefficients appearing in the problem as well as the domain Ω. We generalise, in the limit of small diffusion, the usual motion by mean curvature laws found for homogeneous problems.

1. Introduction

The general form of the problem we are interested in is

$$(\mathcal{P}) \quad \begin{cases} \epsilon^2 \frac{\partial \mathbf{w}}{\partial t} = \epsilon^2 \mathrm{div}(k\nabla \mathbf{w}) + W_{\mathbf{w}}(x, \mathbf{w}) & \text{for all } x \in \Omega \text{ and } t > 0, \\ \frac{\partial \mathbf{w}}{\partial n} = 0 & \text{for all } x \in \partial\Omega \text{ and } t > 0, \\ \mathbf{w}(x, 0) = \psi(x) & \text{for all } x \in \Omega, \end{cases}$$

where the function W is a multi-well potential, $\mathbf{w} = \mathbf{w}(x,t)$ is a function of $x \in \Omega \subset \mathbb{R}^n, n \in \{1,2\}$, and $t \in \mathbb{R}^+$.

1.1. *Structure and results*

Let us first consider the simple scalar problem

$$(\mathcal{P}_{scal}) \begin{cases} \epsilon^2 w_t = \epsilon^2 (k w_x)_x + f^2 w(g^2 - w^2) & \text{for all } x \in \Omega = (a,b) \subset \mathbb{R} \\ & \text{and } t > 0, \\ w_x(a,t) = w_x(b,t) = 0 & \text{for all } t > 0, \\ w(x,0) = \psi(x) & \text{for all } x \in (a,b), \end{cases}$$

where the functions $f = f(x)$, $g = g(x)$, $k = k(x)$ and $\psi(x)$ are smooth. We take f, g, k strictly positive in $[a,b]$, and are interested in the case when ψ takes both signs in (a,b). The potential $W(x,w)$ is equal to $\frac{f^2}{2}(g^2 - w^2)^2$ up to a term independent of w, and has two wells of equal depth located at $w = \pm g$.

Problem $(\mathcal{P}_{scal})$ has been considered by many mathematicians, see [2,7,8,10] for instance. Traditionally, their analyses are based on one of the two following approaches.

Approach based on the energy functional

The usual achievement of this method is the following result. (Note that for the sake of simplicity we provide in this introduction only a very restricted version. A more general version with weaker hypotheses is given at the beginning of Section 5.)

Assuming that

- the strictly positive functions $f, g, k \in C^\infty([a,b])$;
- g satisfies the homogeneous Neumann boundary conditions $g_x(a) = g_x(b) = 0$;
- ϵ is a small positive constant;
- w_0, of the form $w_0 = \pm g\chi_{(a,c_1)} \pm g\chi_{(c_1,c_2)} \pm \cdots \pm g\chi_{(c_d,b)}$ for a fixed $d \in \mathbb{N}$ and $a < c_1 < \cdots < c_d < b$, is an isolated local minimiser of

$$E_0(\tilde{c}_1, \ldots, \tilde{c}_d) = \frac{2}{3}\sum_{i=1}^{d} \sqrt{k(\tilde{c}_i)} f(\tilde{c}_i) g(\tilde{c}_i)^3$$

with respect to $\tilde{c}_i$, $i \in \{1, \ldots, d\}$;

then there exist $\epsilon_0 > 0$ and a family $(w_\epsilon)_{0<\epsilon<\epsilon_0}$ of steady solutions of Problem $(\mathcal{P}_{scal})$ such that for each $\epsilon \in (0, \epsilon_0)$

- $w_\epsilon \in C^2(a, b)$;
- $\|w_\epsilon - w_0\|_{L^1(a,b)} \to 0$ as $\epsilon \to 0$;
- w_ϵ is an asymptotically stable solution of Problem $(\mathcal{P}_{scal})$, i.e. for any $\rho > 0$ small enough, there exists $\delta > 0$ such that for any $\psi \in C^2(a, b)$ with $\|\psi - w_\epsilon\|_{H^1(a,b)} < \delta$ we have

$$\|T(t)\psi - w_\epsilon\|_{H^1(a,b)} < \rho, \quad 0 < t < \infty.$$

Here $T(t)\psi$ denotes the solution of Problem $(\mathcal{P}_{scal})$ with $T(0)\psi = \psi$ as initial condition (satisfying the boundary conditions).
Moreover there exists a value $\tilde{\delta} > 0$ such that

$$\|\psi - w_\epsilon\|_{H^1(a,b)} < \tilde{\delta} \text{ implies that } \|T(t)\psi - w_\epsilon\|_{H^1(a,b)} \to 0 \text{ as } t \to \infty.$$

The steady solution w_ϵ of Problem $(\mathcal{P}_{scal})$ is an isolated local minimiser (in the Sobolev space $H^1(a, b)$) of the energy functional

$$E_\epsilon(w) = \int_a^b \frac{\epsilon}{2} k w_x^2 + \frac{1}{2\epsilon} f^2 (w^2 - g^2)^2 \mathrm{d}x,$$

which Γ-converges to E_0 as $\epsilon \to 0$.

The energy functional decreases (as time increases) on solutions $w(x, t) = T(t)\psi$ of the time-dependent problem. This is a gradient flow, and solutions decrease energy at the maximum rate by means of the relation

$$-\frac{\epsilon}{2} \|w_t\|^2_{L^2(\Omega)} = DE_\epsilon(w)(w_t),$$

where DE_ϵ stands for the usual (variational) derivative of the energy functional E_ϵ. In this sense, the energy acts as a Lyapounov function for this dynamical system; the solution is quickly attracted to a slow manifold, and we wish to describe this slow manifold in the limit $\epsilon \to 0$.

Approach based on asymptotics

A serious drawback in the previous approach consists in the lack of information concerning the solution of Problem $(\mathcal{P}_{scal})$ that we denoted $T(t)\psi$, where ψ is the initial condition, and especially concerning the slow manifold mentioned above. Obviously, the solution $T(t)\psi$ depends on ψ. Nevertheless, for $\epsilon > 0$ small enough, it has been observed that the behaviour of the time-dependent solution is roughly made of two distinct phases.

- The first phase is the evolution on a fast timescale of the solution towards a solution with "layer behaviour", see Remark 2 later on.
- The second phase is the motion of this (these) layer(s) towards a steady stable location (this is the slow manifold behaviour).

To illustrate this, let us consider Problem $(\mathcal{P}_{scal})$ with $(a,b) = (-4,4)$, $\epsilon = \frac{1}{\sqrt{10}}$, $g(x) = \frac{9}{10} - \frac{1}{10}\cos(\frac{\pi}{4}x)$, $\psi(x) = g(x)\tanh(\frac{x+1}{2})$, and $f \equiv k \equiv 1$. The solution is plotted in Figure 10.1. We clearly observe the formation of a layer in the time interval $(0,\bar{t})$ with $\bar{t} \sim \frac{1}{5}$, and then the motion of this layer towards its stable location (here at $x = 0$) in the time interval $(\bar{t}, \infty)$. We also observe that even if the first locations where layers occur depend

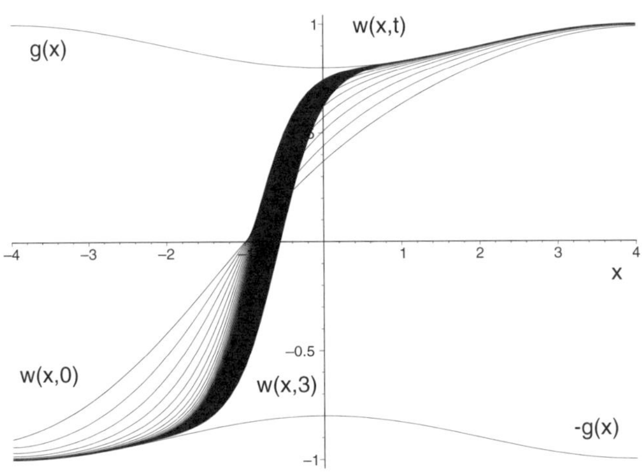

Fig. 10.1. Solution $w(x,t)$ of Problem $(\mathcal{P}_{scal})$ with $(a,b) = (-4,4)$, $\epsilon = \frac{1}{\sqrt{10}}$, $g(x) = \frac{9}{10} - \frac{1}{10}\cos(\frac{\pi}{4}x)$, $\psi(x) = g(x)\tanh(\frac{x+1}{2})$, and $f \equiv k \equiv 1$. The curves in this figure are $w(x,0), w(x,\frac{1}{25}), w(x,\frac{2}{25}), \ldots, w(x,3)$.

strongly on the initial condition, the initial condition does (almost) not influence the motion of the layers from the first location where they occur to their stable steady limit (if they have one). In Figure 10.1, the particular initial condition $\psi(x) = g(x)\tanh(\frac{x+1}{2})$ creates a layer located at $x = -1$ which moves towards its stable steady limit located at $x = 0$.

As a consequence, for an isolated single layer, for $\epsilon > 0$ small enough we can model the second phase of the solution by an expansion of the form

$$w(x,t) = g(x)\tanh(\xi) + o(\epsilon^0),$$

where

$$\xi = \xi(x,t,\epsilon) = \frac{1}{\sqrt{2}\epsilon}\int_{S(t)}^{x} \frac{f(s)g(s)}{\sqrt{k(s)}} \mathrm{d}s$$

is the fast spacescale, and where $S(t)$ (the zero of w, which corresponds to the location of the layer), satisfies

$$\frac{dS(t)}{dt} = -k(S(t))\frac{d}{dx}\ln(\sqrt{k}fg^3)\big|_{x=S(t)} + o(\epsilon^0)$$

(here $o(\epsilon^0)$ stands for terms vanishing in the limit $\epsilon \to 0$).

Remark 1: When considering the method of (asymptotic) expansions for small values of ϵ, we commonly refer to problems stated without any initial conditions. This is to stress the fact that we consider the evolution of a layer once it has occurred (even if, from a purely mathematical point of view, such problems would be completely and correctly defined only with suitable initial conditions stated). The fact that a layer exists, and its location, imply that initial data suitable to form such a layer were involved.

Remark 2: From a purely abstract mathematical point of view, it is rather technical to give a precise definition of a layer. In the sequel, we consider that a layer is a fast transition between the neighbourhood of two stable states (also called phases). (For instance, in the example given in Figure 10.1, the layer is the fast transition between the neighbourhood of $-g(x)$ and that of $g(x)$.) As $\epsilon \to 0$, a solution (divided by g) with a layer converges in $L^1(\Omega)$ to a combination of indicator functions of sets with reasonable boundaries.

Let us now consider Problem $(\mathcal{P})$, that is a generalised version of $(\mathcal{P}_{scal})$. For the sake of simplicity, we assume that $\Omega \subset \mathbb{R}^2$ (the case $\Omega \subset \mathbb{R}$ can be deduced easily).

Depending on the following characteristics of Problem ($\mathcal{P}$), solutions $\mathbf{w}$ *will be markedly different.*

- The dimension of $\mathbf{w}(x,t)$ (either scalar or vector);
- the inhomogeneities of W (i.e. the dependence of W on x);
- the number of wells (possibly not finite);
- the depth of the wells.

In the next sections we cover, by developing both approaches (energy functional and expansions) as far as possible, some of the possible different combinations through one example of wells of the same depth. The case of wells of different depth does not create any difficulties, but is different. Indeed, if the difference of the depth of the wells does not vanish as $\epsilon \to 0$, then the motion takes place on another timescale and the use of the Γ-limit is no longer relevant.

We consider a vector problem (i.e. $\mathbf{w}(x,t)$ is a vector), of dimension 2: $\mathbf{w}(x,t) = (u(x,t), v(x,t))$

$$(\mathcal{P}_{vect}) \quad \begin{cases} \frac{\epsilon^2}{2} u_t = \frac{\epsilon^2}{2} \operatorname{div}(k\nabla u) + f^2 u(g^2 - u^2 - \alpha v^2), & \forall x \in \Omega,\ t > 0, \\ \frac{\epsilon^2}{2} v_t = \frac{\epsilon^2}{2} \operatorname{div}(k\nabla v) + f^2 v(g^2 - v^2 - \alpha u^2), & \\ \frac{\partial u}{\partial n} = \frac{\partial v}{\partial n} = 0 & \forall x \in \partial\Omega,\ t > 0. \end{cases}$$

The factor $\frac{1}{2}$ has been added to simplify the notation later. Another reason is that the "shape" of the layer (i.e. tanh in the expansion of a solution of Problem ($\mathcal{P}_{scal}$) given above), satisfies $\frac{1}{2}\tanh'' + \tanh(1 - \tanh^2) = 0$, which is the canonical equation for the interface or layer behaviour for this problem.

For $\alpha \in (0,1)$ constant, the potential corresponding to Problem ($\mathcal{P}_{vect}$) has four wells of equal depth $\left(\pm\frac{g}{\sqrt{1+\alpha}}, \pm\frac{g}{\sqrt{1+\alpha}}\right)$. First, we have the simple case where one component (either u or v) has a layer but the other does not. For interesting f, g, k, we see that either the layer disappears in finite time, or that the layer is attracted to a stable steady limit.

The most interesting (and complicated) situation occurs when both components u and v have a layer and that both layers are forced by the x-

dependent coefficients k, f and g to move towards the same location. The difficulty comes from the fact that the two layers (locally) repel each other. (Here and later "locally" means sufficiently close given a fixed $\epsilon > 0$.) Indeed, it can be shown that the configuration which consists of two layers located exactly at the same place, is unstable. Therefore, for the limiting steady solution, we have a trade-off between the fact that the coefficients k, f and g force the layers to move towards a common place and the fact that they locally repel each other. See the following example for an illustration.

Example 3: In the case $\Omega = (0,1) \times (0,1)$, $\alpha = 0.9$, $\epsilon = \frac{1}{200}$, $f \equiv k \equiv 1$, and

$$g(x,y) = \begin{cases} 1 - 3\cosh(\pi r)^2 \exp(-\frac{1}{1-r}) \exp(-\frac{1}{r}) & \text{if } r := \sqrt{x^2+y^2} \leq 1; \\ 1 & \text{otherwise,} \end{cases}$$

we find four steady solutions for Problem $(\mathcal{P}_{vect})$.

First and second (stable) solutions, the latter by interchanging u and v: (note that v has a "ridge" where u has a layer)

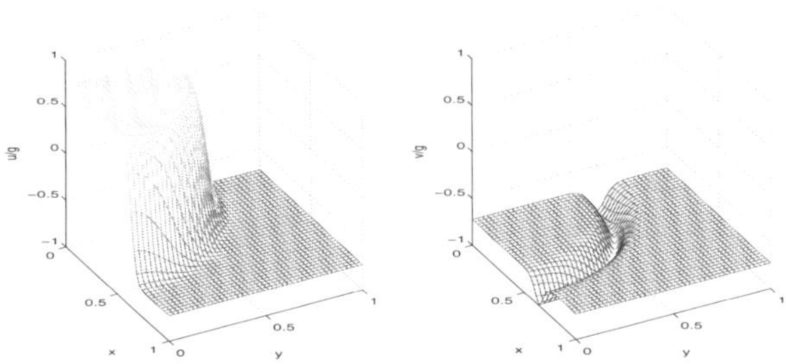

third (unstable) solution:
(here $u = -v$)

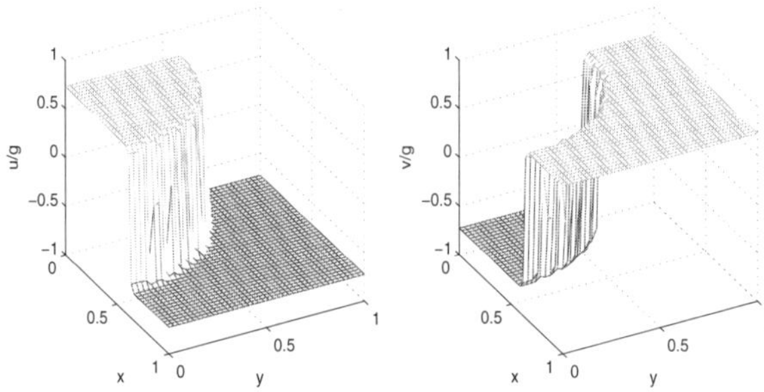

fourth (stable) solution:
(here u and v both have layers and nearby ridges)

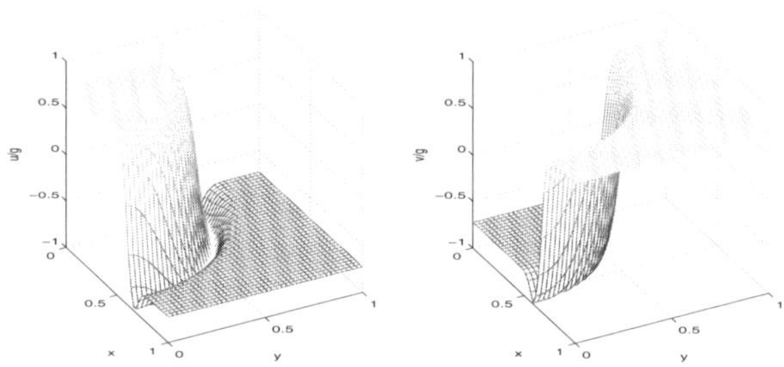

In the approach based on expansions for small values of $\epsilon > 0$, this difficulty (of the competition between larger scale attraction and local repulsion of pairs of layers) appears in the following fact: what is a stable local minimiser under the constraint $u = \pm v$ turns out to be a saddle point without any constraint, with additional local minimisers created. More technically, there is no solution to the canonical equations that arise for the double interface

$$\frac{1}{2}U_{0\xi\xi} + U_0(1 - U_0^2 - \alpha V_0^2) = 0 \quad \text{and} \quad \frac{1}{2}V_{0\xi\xi} + V_0(1 - V_0^2 - \alpha U_0^2) = 0,$$

for $-\infty < \xi < \infty$, such that $V_0(\xi) = U_0(-\xi)$ for all $\xi \in \mathbb{R}$, that

$$\lim_{\xi\to\infty} U_0(\xi) = -\lim_{\xi\to-\infty} U_0(\xi) = \frac{1}{\sqrt{1+\alpha}},$$

and such that the unique zero of U_0 is different from 0. Thus, to model a situation as the one illustrated in Figure 10.2 (right panel), we can only use for the "shape" of a layer, instead of tanh, the solution $(U_0, V_0) = (\mathcal{U}, \mathcal{V})$

illustrated in Figure 10.2 (left panel) of the canonical equations. But the boundary conditions are

$$\lim_{\xi\to\infty}\mathcal{U}(\xi) = -\lim_{\xi\to-\infty}\mathcal{U}(\xi) = \lim_{\xi\to\infty}\mathcal{V}(\xi) = \lim_{\xi\to-\infty}\mathcal{V}(\xi) = \frac{1}{\sqrt{1+\alpha}},$$

and when matching two expansions (one in the neighbourhood of $S^u(t,\epsilon)$ and one in the neighbourhood of $S^v(t,\epsilon)$), we realise that the distance between the two limiting steady curves $S^u_\infty(\epsilon) = \lim_{t\to\infty} S^u(t,\epsilon)$ and $S^v_\infty(\epsilon) = \lim_{t\to\infty} S^v(t,\epsilon)$ must be of larger order in ϵ than the order of the width of a layer (i.e. $\mathcal{O}(\epsilon)$). This indicates that very accurate expansions for the time-independent and time-dependent solutions will be needed. Especially, we note that the accurate expansion for the time-independent problem can be used to get a more explicit expression for the Γ-limit of the associated energy functional than what we can obtain directly from a result on the relaxation of functionals given by Fonseca and Müller [4].

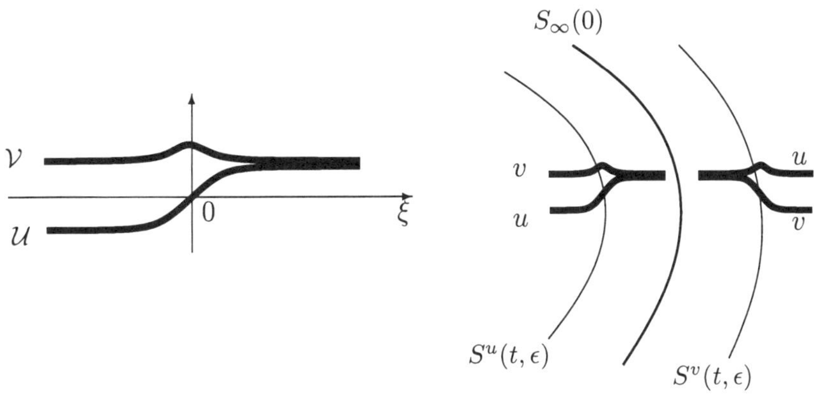

Fig. 10.2. Scheme of the two matched expansions in the case of two colliding layers. $S^u(t,\epsilon)$ (resp. $S^v(t,\epsilon)$) represents the nodal curve of $u(x,t)$: $\{x\in\Omega : u(x,t)=0\}$ (resp. $v(x,t)$: $\{x\in\Omega : v(x,t)=0\}$). $S_\infty(0)$ represents the common location of both nodal curves in the limit $t\to\infty$ and $\epsilon\to 0$.

- In Section 2 we consider the canonical equations (i.e. the steady one space dimensional homogeneous version of Problem $(\mathcal{P}_{vect})$). The existence and the properties of new (families of) solutions are discussed. The lack of a better solution than $(\mathcal{U},\mathcal{V})$ to model two repelling layers, as well as the existence and properties of $(\mathcal{U},\mathcal{V})$, are also commented upon.

- In Section 3 we provide an extension of Norbury and Yeh's expansion, see [8,9,11], together with another expansion based on the particular coordinate $z(x,t)$ defined below in (1). These two different expansions illustrate the trade-off we face between the lack of smoothness (of the first two leading terms), the lack of information as $t \to \infty$, and the limitation of the domain of validity in Ω of such (asymptotic) expansions as $\epsilon \to 0$. More precisely, for Problem $(\mathcal{P}_{scal})$ for instance, the extended version of Norbury and Yeh's expansion for the time-independent solutions is such that

• either the expansion is valid only in a neighbourhood of the nodal curve $S(t)$ of bandwidth $\mathcal{O}(\epsilon|\ln\epsilon|)$;

• or the expansion is valid in the entire domain Ω (without matching), but has a lack of smoothness. The second term in the expansion $w(x) = g(x)\tanh(\xi) + \epsilon U_1(\xi, x) + o(\epsilon)$, where $\xi = \xi(x,\epsilon)$ is the fast length-scale, is such that

$$\frac{\epsilon^2}{2}\Big[\mathrm{div}(k\nabla\epsilon U_1(\xi(x,\epsilon),x))\Big]_{S_\infty^-(\epsilon)}^{S_\infty^+(\epsilon)} = \mathcal{O}(\epsilon^2).$$

The only smooth expansion such that the leading order term is $\mathcal{O}(\epsilon)$-accurate uniformly in the entire domain Ω is the following second expansion, corresponding only to time-dependent solutions:

$$w(x,t) = g(x)\tanh\left(\frac{f(x)g(x)}{\sqrt{k(x)}|\nabla z|}\xi\right) + \mathcal{O}(\epsilon),$$

where $\xi = \xi(x,t,\epsilon) = \frac{z(x,t)}{\epsilon}$, and where z satisfies

$$\frac{z_t}{|\nabla z|} = \frac{\sqrt{k}}{fg^3}\mathrm{div}\left(\sqrt{k}fg^3\frac{\nabla z}{|\nabla z|}\right) \tag{1}$$

(as $\mathrm{div}\left(\frac{\nabla z}{|\nabla z|}\right)$ is the mean curvature of the level curve of z and as $\frac{z_t}{|\nabla z|}$ is the normal velocity of such a level curve with respect to the normal $-\frac{\nabla z}{|\nabla z|}$, this corresponds to an extended motion by mean curvature law, where the motion is influenced by the inhomogeneous terms $\sqrt{k}fg^3$ and $\sqrt{k}$). Comparisons and numerical computations show the necessity of considering both expansions.

- In Section 4 we develop the approach based on the energy functional.

We first give the main theoretical results we need, and then we give the Γ-limit of the energy functional associated with the steady version of Problem $(\mathcal{P}_{vect})$. This gives us a necessary condition for a curve to be the limit, as $\epsilon \to 0$, of a steady nodal curve.

The structure of this Γ-convergence theory enables us to avoid the analysis of the differences between the two possibilities in the first expansion, but then does not allow us to analyse properly all the details of the solutions as $\epsilon \to 0$. The approach based on the (asymptotic) expansion of the solutions gives a complete understanding of the problem for solutions quickly forming one or several layers, their motion being perfectly well characterised, whereas the theory based on the Γ-convergence idea justifies formally the results obtained with the expansions. The main results attained with the Γ-convergence theory for steady problems are formal proofs of existence of solutions under some basic assumptions, as well as the convergence (in $L^1(\Omega)$) of these solutions to a limit in $BV(\Omega)$ (the Banach space of bounded variation functions on Ω) as $\epsilon \to 0$.

- Finally, in Section 5 we extend the results of Section 4 to time-dependent problems, to get laws for the motion of layers. It is unusual to use the notion of Γ-limit for time-dependent problems. Nevertheless this gives interesting results.

Remark 4: In the case $\alpha = 1$ (α being as in the definition of $(\mathcal{P}_{vect})$), the potential W admits a circle $\{u^2 + v^2 = g^2\}$ as the set of local minimisers. Layers do not occur anymore, but vortices occur, say at $a_1, \ldots, a_d$ (a vortex is a point singularity and we say that a solution (u_ϵ, v_ϵ) admits a vortex at $a_i \in \Omega$ if, in the limit $\epsilon \to 0$, the solution (u_0, v_0) is twice continuously differentiable in a neighbourhood of a_i excluding a_i, that $u_0^2 + v_0^2 = g^2$ holds for all x in this neighbourhood (except at $x = a_i$), and that the winding number of (u_0, v_0) around a_i is different from zero). Similarly to what has been done for the layers, by giving an expansion for small values of $\epsilon > 0$ of a solution with vortices, and using this expansion with the energy functional approach, the motion of distinct vortices can be computed with respect to the coefficients f, g and k. We get, for $a_i = a_i(t)$,

$$\frac{da_i}{dt} = -k\nabla(kg^2)\big|_{x=a_i}, \quad i = 1, \ldots, d.$$

Note that we also have a trade-off here, as two vortices a_i, a_j, $i \neq j$, can be forced to move towards a common location by the coefficients k, f, g, but repel each other locally (the repelling effect for vortices when f, g, k are constant has been characterised by Bethuel, Brezis and Hélein [1]).

2. Canonical equations

Let us consider the canonical problem:

$$(\mathcal{P}_\alpha) \begin{cases} \frac{1}{2}u_{xx} + u[1 - u^2 - \alpha v^2] = 0, & \text{for all } x \in (-\infty, \infty), \\ \frac{1}{2}v_{xx} + v[1 - v^2 - \alpha u^2] = 0, & \\ \lim_{x\to\pm\infty} u_x(x), v_x(x) = 0. & \end{cases}$$

This is a perturbed Ginzburg-Landau system (or two coupled Allen-Cahn equations), in which all the coefficients are constant and which is defined over the entire real line $\Omega = \mathbb{R}$.

Any solution (u, v) of Problem $(\mathcal{P}_\alpha)$ is in fact $C^\infty(-\infty, \infty) \times C^\infty(-\infty, \infty)$. Moreover, if (u, v) is a solution of Problem $(\mathcal{P}_\alpha)$, then

$$(u, v), \quad (u, -v), \quad (-u, v), \quad (-u, -v)$$

are solutions, as well as

$$(u(-x), v(-x)) \text{ and } (u(x + c), v(x + c)),$$

where $c \in \mathbb{R}$ is a constant. Thus we introduce the notion of family of solutions.

Definition 5: By a family of solutions we mean a set of solutions $\{(u_j, v_j),\ j \in J\}$ such that any solution (u_{j_0}, v_{j_0}) can be mapped into any other solution (u_{j_1}, v_{j_1}) of the same family by using one or more of the mappings: reflection with respect to the x-axis or the y-axis, exchange of the names u and v, and shift along the x-axis.

In the sequel, we shall identify a family with one of its representatives.

The analysis of this problem provides necessary results for the understanding of the more general case and shows different numerical solutions such as those illustrated in Figures 10.3 and 10.4.

Remark 6: When we refer to the first, second and third families of solutions, we point out that we don't have any list of all existing solutions. The question of a classification of all solutions with respect to a particular characteristic (such as the total number of zeros of u and v for instance), remains an open issue.

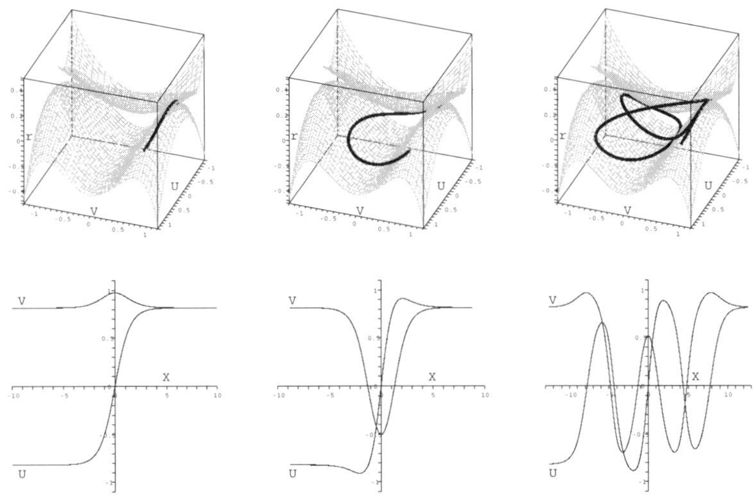

Fig. 10.3. First, second and third families of solutions, for $\alpha = \frac{1}{2}$, in the (u, v, r)-space (upper panels) and in the usual plane (lower panels). The surfaces represent forbidden zones, and the dark curves orbits of solutions that begin and end as $x \to \pm\infty$. (The third variable r is defined up to a constant by $r_x = \frac{1}{2}(u^2 v v_x - v^2 u u_x)$.)

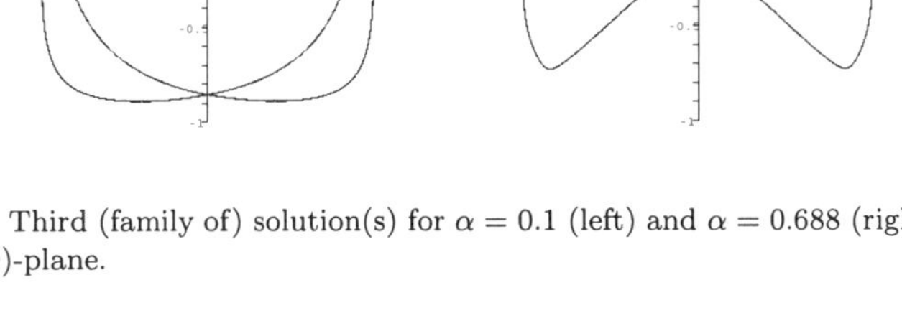

Fig. 10.4. Third (family of) solution(s) for $\alpha = 0.1$ (left) and $\alpha = 0.688$ (right), plotted in the (u, v)-plane.

Theorem 7: Problem $(\mathcal{P}_\alpha)$, $\alpha \in (0, 1)$, admits a family of solutions for which one of its representatives, say $(\mathcal{U}, \mathcal{V})$, satisfies

$$\mathcal{U}(x) \leq 0 \text{ for } x \leq 0 \text{ and } \mathcal{U}(x) \geq 0 \text{ for } x \geq 0,$$

$$\mathcal{V}(x) \geq 0 \text{ for all } x \in \mathbb{R},$$

$$\lim_{x \to \pm\infty} \mathcal{U}(x) = \pm \frac{1}{\sqrt{1+\alpha}}, \quad \lim_{x \to \pm\infty} \mathcal{V}(x) = \frac{1}{\sqrt{1+\alpha}}.$$

$(\mathcal{U}, \mathcal{V})$ also satisfies the equality

$$\frac{1}{4}(\mathcal{U}_x^2 + \mathcal{V}_x^2) = \frac{1}{4}(\mathcal{U}^4 + \mathcal{V}^4) - \frac{1}{2}(\mathcal{U}^2 + \mathcal{V}^2) + \frac{\alpha}{2}\mathcal{U}^2\mathcal{V}^2 + \frac{1}{2(1+\alpha)}. \quad (2)$$

See Figure 10.5 for an illustration of $\mathcal{U}$ and $\mathcal{V}$ when $\alpha = 0.1, 0.5$ and 0.9.

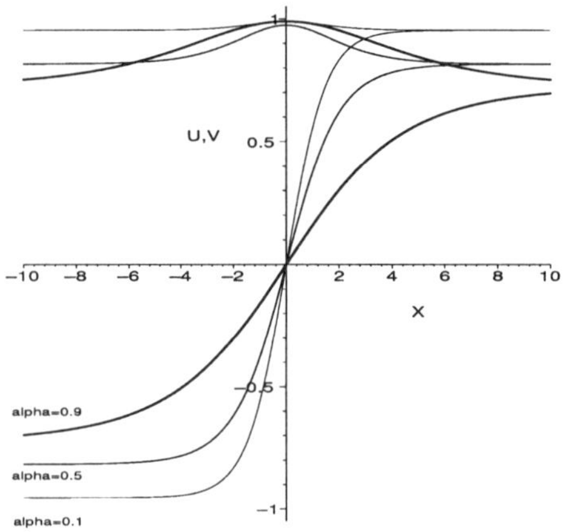

Fig. 10.5. First family of solutions for $\alpha = 0.1, 0.5$ and 0.9.

Remark 8: There is no solution to Problem $(\mathcal{P}_\alpha)$, $\alpha \in (0,1)$, such that

$$\frac{1}{\sqrt{1+\alpha}} = -\lim_{x\to-\infty} u(x) = \lim_{x\to\infty} u(x),$$

$v(x) = u(-x)$, and such that the unique zero of u is different from 0. (The observation from Norbury and Yeh's result [9], tells us that if such a (stable) solution did exist, then it should be such that $v_x(S^u) = 0$, but this would mean that u (resp. v) is odd (resp. even) about $x = S^u$, which is in contradiction with the fact that v has only one zero distinct from S^u.)

Then, we get for the time-dependent problem the following result.

Theorem 9: The solutions of the time-dependent problem

$$\frac{1}{2}u_t = \frac{1}{2}u_{xx} + u(1 - u^2 - \alpha v^2),$$
$$\frac{1}{2}v_t = \frac{1}{2}v_{xx} + v(1 - v^2 - \alpha u^2),$$

for all $x \in \mathbb{R}$, $\alpha \in (0,1)$, with boundary conditions

$$\lim_{x\to\pm\infty} u(x)^2, v(x)^2 = \frac{1}{1+\alpha},$$

and starting with the initial conditions

$$\begin{pmatrix} u(x,0) \\ v(x,0) \end{pmatrix} = \begin{pmatrix} \mathcal{U}(x - S^u(0)) \\ \mathcal{V}(x - S^u(0)) \end{pmatrix} + \begin{pmatrix} \mathcal{V}(S^v(0) - x) \\ \mathcal{U}(S^v(0) - x) \end{pmatrix} - \begin{pmatrix} \frac{1}{\sqrt{1+\alpha}} \\ \frac{1}{\sqrt{1+\alpha}} \end{pmatrix},$$

where $0 < -S^u(0) = S^v(0)$ is large, behave like travelling waves for large t, and

$$\lim_{t\to\infty} -S^u(t) = \lim_{t\to\infty} S^v(t) = \infty.$$

This is what we call hereafter the "repelling effect".

3. Expansions

Notation 10: Let $\gamma_{x,p}$ stand for a curve with endpoints x and p. We use the metric $\frac{fg}{\sqrt{k}}$ in the sense that we define the signed distance between two points x and p, computed along the (oriented) curve $\gamma_{x,p}$, by$d_{\gamma_{x,p}}(x,p)$

$$d_{\gamma_{x,p}}(x,p) = \int_{\gamma_{x,p}} \frac{fg}{\sqrt{k}} \mathrm{d}s,$$

where s stands for the arc length. We define the minimal positive distance $d(x,p)$ between two points x and p as the minimum of the positive distance $|d_{\gamma_{x,p}}(x,p)|$ over all possible curves $\gamma_{x,p}$ starting at x and ending at p. For any point x and smooth curve S, we denote by $p(x,S)$ the projection assumed unique of x on S, i.e. the point of S whose positive distance to x is the least. (The results don't change if the uniqueness does not hold and the precise formulation then only adds technical details.) We denote the signed distance between x and S by $d(x,S)$ (such that $|d(x,S)| = d(x,p(x,S))$). We also define the two curves $\hat{c}$ and $\check{c}$ as follows. For S and a particular point $p \in S$, $\hat{c}(p,S)$ stands for the set of points x whose projection on S is p, i.e.

$$\hat{c}(p,S) := \{x \in \Omega : p(x,S) = p\},$$

and $\check{c}$ stands for the parallel curve to S passing through x, i.e.

$$\check{c}(x,S) := \{y \in \Omega : d(y,S) = d(x,S)\}.$$

In order to simplify our integrations later, we introduce, for $p \in S$ and $x \in \hat{c}(p,S)$, the notation $\hat{c}_S(p,x)$ denoting the part of the curve $\hat{c}(p,S)$

located between the two points p and x. Finally, $S(t,\epsilon)$ (with $S_\infty(\epsilon) := \lim_{t\to\infty} S(t,\epsilon)$), stands for the nodal curve of a solution (i.e. the set of zeros of w for Problem $(\mathcal{P}_{scal})$), $n = n(x,t,\epsilon)$ stands for the unit normal to the curve $\check{c}(x, S(t,\epsilon))$, $\nu = \nu(x,t,\epsilon)$ for the normal velocity of $\check{c}(x, S(t,\epsilon))$, and $\kappa = \kappa(x,t,\epsilon)$ for the mean curvature of the curve $\check{c}(x, S(t,\epsilon))$. (For the orientation, n is the outward unit normal if $\check{c}(x, S(t,\epsilon))$ is closed, or either of the two possible orientations otherwise. The normal velocity ν refers to n, as well as the mean curvature $\kappa = \mathrm{div}(n)$.)

See Figure 10.6 for an illustration.

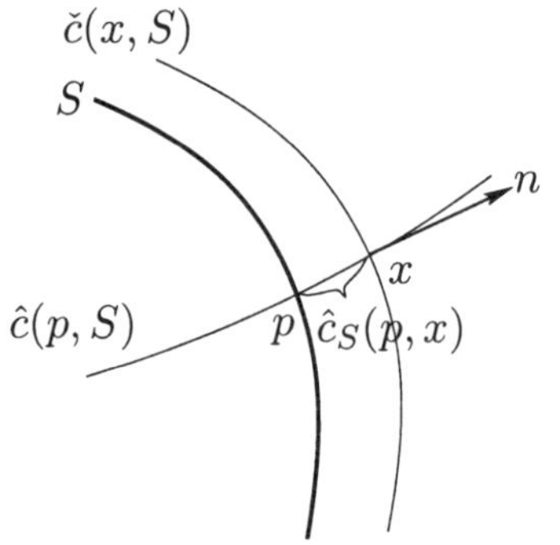

Fig. 10.6. Illustration of the notation. If $S = S(t,\epsilon)$, then $n(x,t,\epsilon)$ corresponds to n in this figure.

Then, the following two expansions hold.

Theorem 11: (First expansion) A solution $(u(x,t), v(x,t))$ of Problem $(\mathcal{P}_{vect})$, with a single layer for u, admits an expansion of the form

$$u(x,t) = \pm g(x)\mathcal{U}(\xi) + \underbrace{\epsilon U_1(\xi,x,t) + o(\epsilon)}_{R_u(\xi,x,t,\epsilon)},$$

$$v(x,t) = \pm g(x)\mathcal{V}(\xi) + \underbrace{\epsilon V_1(\xi,x,t) + o(\epsilon)}_{R_v(\xi,x,t,\epsilon)},$$

where

$$\xi = \xi(x,t,\epsilon) = \frac{1}{\epsilon}\int_{\hat{c}_{S^u(t,\epsilon)}(p(x,S^u(t,\epsilon)),x)} \frac{fg}{\sqrt{k}}\mathrm{d}s,$$

and

$$\nu_u = -k\left(\frac{\partial}{\partial n}\ln(\sqrt{k}fg^3) + \kappa_u\right)\bigg|_{x\in S^u(t,\epsilon)} + o(1). \qquad (3)$$

Here

- either the functions $U_1(.,x,t), V_1(.,x,t)$ are unbounded, but under the assumption that $d(x, S^u(t,\epsilon)) = \mathcal{O}(\epsilon|\ln\epsilon|)$ we have $R_{u,v} = \mathcal{O}(\epsilon)$ with the accuracy in (3) being $\mathcal{O}(\epsilon|\ln\epsilon|)$ instead of $o(1)$;
- or the functions $U_1(.,x,t), V_1(.,x,t)$ are bounded, $R_{u,v}(\xi,x,t,\epsilon) = \mathcal{O}(\epsilon)$, the accuracy in (3) is $\mathcal{O}(\epsilon)$ instead of $o(1)$ but $U_1(\xi(x,t,\epsilon),x,t), V_1(\xi(x,t,\epsilon),x,t) \in C^2(\Omega\setminus S^u(t,\epsilon))\cap C^1(\Omega)$ for all $t>0$, and the jump of the second derivative is of order ϵ^2:

$$\frac{\epsilon^2}{2}\Big[\mathrm{div}(k\nabla\ \epsilon U_1(\xi(x,t,\epsilon),x,t))\Big]_{S^{u-}(t,\epsilon)}^{S^{u+}(t,\epsilon)} = \mathcal{O}(\epsilon^2),$$

$$\frac{\epsilon^2}{2}\Big[\mathrm{div}(k\nabla\ \epsilon V_1(\xi(x,t,\epsilon),x,t))\Big]_{S^{u-}(t,\epsilon)}^{S^{u+}(t,\epsilon)} = \mathcal{O}(\epsilon^2).$$

This first expansion provides in general a good approximation of the solutions, especially for the steady solutions. Nevertheless, it is natural to look for a smooth and accurate expansion. The next expansion provides an answer to this issue.

Theorem 12: (Second expansion) Solutions of Problem $(\mathcal{P}_{vect})$, with a single layer for u, admit an expansion of the form

$$u(x,t,\epsilon) = g(x)U_0(\xi,x,t) + \epsilon U_1(\xi,x,t) + \mathcal{O}(\epsilon^2),$$
$$v(x,t,\epsilon) = g(x)V_0(\xi,x,t) + \epsilon V_1(\xi,x,t) + \mathcal{O}(\epsilon^2),$$

where

$$U_0(\xi,x,t) = \pm\mathcal{U}\Big(\frac{f(x)g(x)}{\sqrt{k(x)}|\nabla z|}\xi\Big),$$
$$V_0(\xi,x,t) = \pm\mathcal{V}\Big(\frac{f(x)g(x)}{\sqrt{k(x)}|\nabla z|}\xi\Big),$$
$$\xi = \xi(x,t,\epsilon) = \frac{z(x,t)}{\epsilon},$$

and where $z(x,t)$ is defined (implicitly) by

$$\frac{z_t}{|\nabla z|} = \frac{\sqrt{k}}{fg^3}\mathrm{div}\left(\sqrt{k}fg^3\frac{\nabla z}{|\nabla z|}\right), \qquad (4)$$

together with some initial conditions $z(x,0) = z_{init}(x)$ for all $x \in \Omega$ and some boundary condition $z(x,t) = z_{bc}(x,t)$ for all x belonging to a specified subset of $\partial\Omega$ and all $t > 0$. The functions $U_1(\xi, x, t), V_1(\xi, x, t)$ are uniformly bounded and smooth with respect to $\xi \in \mathbb{R}$, but this expansion is valid only under the assumption that the solution z of (4) is $\mathcal{O}(1)$.

This expansion is uniformly convergent on the space $\times$ time domain $\Omega \times (0, \infty)$. It consists in deriving a new system of coordinates, which enables us to compute the motion of layers and the evolution of the solutions, but does not allow us to compute details of the time-independent solutions (i.e. to get anything better than a correctly located step function in the limit $t \to \infty$).

Remark 13: Recalling that the term

$$\frac{z_t}{|\nabla z|}$$

corresponds to the normal velocity of the curves $\{x \in \Omega : z(x,t) = c\}$ for any fixed $t > 0$ and any constant $c \in \mathbb{R}$, with respect to the unit normal $-\frac{\nabla z}{|\nabla z|}$, we observe that

$$\nu(x,t) = -\frac{z_t}{|\nabla z|}$$

is the normal velocity of such a level set with respect to the unit vector $\frac{\nabla z}{|\nabla z|}$. Recalling also that (for each fixed $t > 0$), the term

$$\operatorname{div}\left(\frac{\nabla z}{|\nabla z|}\right)$$

is the mean curvature $\kappa(x,t)$ of the curves $\{x \in \Omega : z(x,t) = c\}$, we observe that (4) gives (3) when evaluated on $S^u(t, \epsilon)$. Moreover, in one dimension $\Omega \subset \mathbb{R}$, (4) can be solved by computing its characteristics as a first order hyperbolic PDE for $z(x,t)$.

4. Γ-convergence – Steady problem

For the reader's convenience, we recall the definition of Γ-convergence of the functionals E_ϵ to E_0 as $\epsilon \to 0$ (see e.g. [3] and references therein for more on Γ-convergence).

Definition 14: We say that the functionals E_ϵ defined on $L^1(\Omega) \times L^1(\Omega)$ Γ-converge to the functional E_0 as ϵ tends to 0 if the next two conditions hold.

- **The lower bound**: for any fixed $(u, v) \in L^1(\Omega) \times L^1(\Omega)$ and for any sequences $(p_j, q_j)_{j \geq 1} \in L^1(\Omega) \times L^1(\Omega)$ and $(\epsilon_j)_{j \geq 1} \subset \mathbb{R}_+$ such that, as $j \to \infty$: $\epsilon_j \to 0$ and

$$(p_j, q_j) \to (u, v) \in L^1(\Omega) \times L^1(\Omega),$$

  we have the inequality

$$\liminf_{j \to \infty} E_{\epsilon_j}(p_j, q_j) \geq E_0(u, v);$$

- **Attaining the lower bound**: for any fixed $(u, v) \in L^1(\Omega) \times L^1(\Omega)$ and for any sequence $(\epsilon_j)_{j \geq 1} \in \mathbb{R}^+$ such that $\epsilon_j \to 0$ as $j \to \infty$, there exists at least one sequence $(P_j, Q_j)_{j \geq 1} \in L^1(\Omega) \times L^1(\Omega)$ satisfying

$$(P_j, Q_j) \to (u, v) \in L^1(\Omega) \times L^1(\Omega) \text{ as } j \to \infty,$$

  and

$$\lim_{j \to \infty} E_{\epsilon_j}(P_j, Q_j) = E_0(u, v).$$

The link between the minimisers of a family of functionals E_ϵ and its Γ-limit has been established by Kohn and Sternberg [7] in the case of functionals defined on $L^1(\Omega)$. The extension of their result to the case of our functionals defined on $L^1(\Omega) \times L^1(\Omega)$ below is direct.

Theorem 15: Suppose that the functionals E_ϵ Γ-converge to the functional E_0 and that the two following hypotheses are satisfied:

- every sequence $(p_j, q_j)_{j \geq 1}$ and positive $\epsilon_j \to 0$ as $j \to \infty$ for which $E_{\epsilon_j}(p_j, q_j) \leq C < \infty$ for all j has a $L^1(\Omega) \times L^1(\Omega)$-convergent subsequence;
- there exists an isolated local (or global) minimiser $(u_0, v_0) \in L^1(\Omega) \times L^1(\Omega)$ of E_0.

Then, there exists $\epsilon_0 > 0$ and a family $(u_\epsilon, v_\epsilon)_{0 < \epsilon \leq \epsilon_0}$ such that:

- (u_ϵ, v_ϵ) is a local minimiser of E_ϵ;
- $\|(u_\epsilon, v_\epsilon) - (u_0, v_0)\|_{L^1(\Omega) \times L^1(\Omega)} \to 0$ as $\epsilon \to 0$.

Here is the energy functional we are looking for. (Note that for this energy functional, we can prove that the first hypothesis in the properties of the Γ-limit given above is satisfied.)

Theorem 16: Suppose that Ω is an open bounded domain whose boundary is C^{2+m}, and that $k, f^2, g^2 \in C^m(\bar{\Omega})$, where $m \in \mathbb{N}$ with $m \geq 2$. Then, an isolated local minimiser $(u_\epsilon, v_\epsilon) \in H^1(\Omega) \times H^1(\Omega)$ of the functional E_ϵ, defined on $L^1(\Omega) \times L^1(\Omega)$ for $\epsilon > 0$ by

$$E_\epsilon(u,v) = \begin{cases} \int_\Omega \frac{\epsilon}{4} k \left(|\nabla u|^2 + |\nabla v|^2\right) + \frac{1}{\epsilon} W(x,u,v) \mathrm{d}x & \textit{if } u, v \in H^1(\Omega), \\ \infty & \textit{otherwise,} \end{cases}$$

is a steady solution of Problem $(\mathcal{P}_{vect})$. Moreover, the minimiser $(u_\epsilon, v_\epsilon) \in C^m(\bar{\Omega}) \times C^m(\bar{\Omega})$.

In order to simplify our notation, we define

$$\bar{g}(x) := \frac{g(x)}{\sqrt{1+\alpha}}, \quad \mathcal{A}_u := \{x : u(x) = \bar{g}(x)\}, \quad \mathcal{A}_v := \{x : v(x) = \bar{g}(x)\}.$$

With such a notation, we can give explicitly the Γ-limit E_0 of the functionals E_ϵ.

Theorem 17: The Γ-limit E_0 of the functionals E_ϵ is given by

$$E_0(u,v) = \begin{cases} K(\alpha) \left(\int_\Omega \sqrt{k} f g^3 |D\chi_{\mathcal{A}_u}| + \int_\Omega \sqrt{k} f g^3 |D\chi_{\mathcal{A}_v}| \right) & \text{for } u, v \in BV(\Omega) \text{ s.t. } (u,v) \in \{(\pm\bar{g}, \pm\bar{g})\} \ \mathcal{L}^2\text{-a.e.,} \\ \infty & \text{otherwise,} \end{cases}$$

where by definition

$$K(\alpha) = \Psi\left(-\frac{1}{\sqrt{1+\alpha}}, \frac{1}{\sqrt{1+\alpha}}\right) = \Psi\left(\frac{1}{\sqrt{1+\alpha}}, -\frac{1}{\sqrt{1+\alpha}}\right)$$
$$\left(= 2\int_{-\infty}^{\infty} \mathcal{W}(\mathcal{U}(\xi), \mathcal{V}(\xi)) \mathrm{d}\xi\right),$$

$$\Psi(u,v) = \inf\left\{ \int_{-1}^{1} \sqrt{\mathcal{W}(l(s))} |l'(s)| \mathrm{d}s : l \textit{ is piecewise } C^1, \right.$$
$$\left. l(-1) = \left(-\frac{1}{\sqrt{1+\alpha}}, -\frac{1}{\sqrt{1+\alpha}}\right),\ l(1) = (u,v) \right\},$$

and

$$\mathcal{W}(u,v) = \frac{1}{4}(u^4 + v^4) - \frac{1}{2}(u^2 + v^2) + \frac{\alpha}{2} u^2 v^2 + \frac{1}{2(1+\alpha)}.$$

5. Γ-convergence – Time-dependent problem

Theorem 18: Let us assume for Problem $(\mathcal{P}_{vect})$ that

- Ω is a smooth (C^4) bounded connected domain of $\mathbb{R}^2$;
- $f^2, g^2 \in C^{0,\beta}(\bar{\Omega})$, $\sqrt{k} \in C^{1,\beta}(\bar{\Omega})$, $\beta \in (0,1)$, $g \in BV(\Omega)$;
- g satisfies the homogeneous Neumann boundary conditions;
- ϵ and α are constant, with ϵ small and $\alpha \in (0,1)$;
- (u_0, v_0) is a local isolated minimiser of E_0 (defined above).

Then, there exist $\epsilon_0 > 0$ and a family $(u_\epsilon, v_\epsilon)_{0<\epsilon<\epsilon_0}$ of steady solutions of Problem $(\mathcal{P}_{vect})$ such that, for each $\epsilon \in (0, \epsilon_0)$

- $(u_\epsilon, v_\epsilon) \in C^{2,\beta}(\bar{\Omega}) \times C^{2,\beta}(\bar{\Omega})$;
- $\|u_\epsilon - u_0\|_{L^1(\Omega)} + \|v_\epsilon - v_0\|_{L^1(\Omega)} \to 0$ as $\epsilon \to 0$;
- for each $\lambda > 0$ small enough, we have that $\mathcal{L}^2(\Omega_\epsilon^\lambda) \to 0$ as $\epsilon \to 0$, where

$$\Omega_\epsilon^\lambda = \{x \in \Omega : -\frac{g(x)}{\sqrt{1+\alpha}} + \lambda < u_\epsilon(x) < \frac{g(x)}{\sqrt{1+\alpha}} - \lambda\}$$
$$\cup\{x \in \Omega : -\frac{g(x)}{\sqrt{1+\alpha}} + \lambda < v_\epsilon(x) < \frac{g(x)}{\sqrt{1+\alpha}} - \lambda\};$$

- (u_ϵ, v_ϵ) is an asymptotically stable solution of Problem $(\mathcal{P}_{vect})$.

Looking for the motion of the limits of the nodal curves $S^u(t,0) = \lim_{\epsilon\to 0} S^u(t,\epsilon)$, $S^v(t,0) = \lim_{\epsilon\to 0} S^v(t,\epsilon)$ of the solutions $(u,v) = (T_1(t)(\psi_1,\psi_2), T_2(t)(\psi_1,\psi_2))$ of Problem $(\mathcal{P}_{vect})$, we find the following result.

Theorem 19: The nodal curves $S^u(t,0), S^v(t,0)$ move according to

$$\nu_u = -k\frac{\partial}{\partial n_u}\ln(\sqrt{k}fg^3) - k\kappa_u,$$
$$\nu_v = -k\frac{\partial}{\partial n_v}\ln(\sqrt{k}fg^3) - k\kappa_v,$$

where by definition ν_u (resp. ν_v) is the normal velocity of the curve $S^u(t,0)$ (resp. $S^v(t,0)$) and n_u (resp. n_v) is the unit normal to the curve $S^u(t,0)$ (resp. $S^v(t,0)$), pointing in the same direction as ν_u (resp. ν_v). κ_u (resp. κ_v) is the signed curvature of the nodal curve $S^u(t,0)$ (resp. $S^v(t,0)$) at the point x.

The result above proves that the coupled problem ($\alpha \neq 0$) behaves like the uncoupled one ($\alpha = 0$) as $\epsilon \to 0$, from the point of view of the nodal curves. Moreover, we can find a necessary condition for a curve to be the stable limit as $t \to \infty$ of a nodal curve.

Theorem 20: A necessary condition for a nodal curve $S_\infty^{u,v}(0)$ to be the steady stable limit as $t \to \infty$ of the (limit) nodal curves $S^{u,v}(t,0)$ is

$$\kappa_{u,v} = -\frac{1}{\sqrt{k} f g^3} \frac{\partial}{\partial n}(\sqrt{k} f g^3) = -\frac{\partial}{\partial n} \ln(\sqrt{k} f g^3),$$

which is one of Norbury and Yeh's results [8] for the uncoupled problem, obtained by asymptotic expansion methods. It expresses the possible steady limits of nodal curves as (extremal) weighted geodesics of Ω using the metric $\sqrt{k} f g^3$ (strictly positive in $\bar{\Omega}$). To be stable we require this geodesic to be minimal in the sense that any variation of the extremal geodesic leads to an increase of the energy functional.

6. Conclusion

As we have seen, there are essentially two ways to analyse the type of multi-well large reaction small diffusion problems we have considered.

One way is based on the notion of Γ-convergence, which appears to be very efficient for nonlocal analyses of the solutions, providing proofs of results on the existence and the convergence of steady solutions for $\epsilon > 0$. It helps in analysing the time-dependent problems in some particular cases, as for example the motion of distinct vortices located at $\{a_i(t)\}_{i=1}^d \in \Omega$ and given by

$$\frac{da_i(t)}{dt} = -k \nabla \ln(kg^2) \bigg|_{x=a_i(t)} \qquad \text{for } i = 1, \ldots, d,$$

but fails when the inner layer behaviour (characterised by the fast length-scale ξ) is not known from any of the possible optimal sequences in the definition of the Γ-limit.

The other way, which consists in transforming the large reaction small diffusion problem into an optimal coordinate $z(x,t)$ problem of the form

$$\frac{z_t}{|\nabla z|} = \frac{\sqrt{k}}{f g^3} \text{div} \left(\sqrt{k} f g^3 \frac{\nabla z}{|\nabla z|} \right),$$

is a powerful method for computing and/or producing examples of interfaces and models with desired properties for the interfaces between the states. The above formula generalises the motion by mean curvature results that have been established for homogeneous vectorial problems by Rubinstein, Sternberg and Keller [10].

At this stage, one "challenging" problem remains open for the theory presented here. It is the number of families of solutions of Problem $(\mathcal{P}_\alpha)$, with respect to $\alpha \in (-1, \infty)$. If it were possible to find the number of such families, together with a characterisation of some subspaces in which they appear as stable, then these solutions would be applicable to the transmission of data through optical fibres, and lead to a massive improvement of the speed of transmission. (See [5,6] for instance.) Such a result is difficult, but should attract future attention.

This work has been partially submitted in three separate articles (which include proofs of results stated here):

- New solutions for perturbed steady Ginzburg-Landau equations in $\mathbb{R}$, *Journal of Differential Equations;*
- New steady solutions (and their stability) for a large reaction small diffusion problem, *Nonlinearity, Institute of Physics;*
- New coordinates for time-dependent nonlinear reaction diffusion problems, *Nonlinearity, Institute of Physics.*

References

[1] F. Bethuel, H. Brezis and F. Hélein, *Ginzburg-Landau vortices.* Birkhäuser, Boston, 1994.

[2] G. Bouchitté, Singular perturbations of variational problems arising from a two-phase transition model. *Appl. Math. Optim.* **21**, 289–314 (1990).

[3] G. Dal Maso, *An introduction to Γ-convergence.* Birkhäuser, Boston, 1993.

[4] I. Fonseca and S. Müller, Relaxation of quasiconvex functionals in $BV(\Omega, \mathbb{R}^p)$ for integrands $f(x, u, \nabla u)$. *Arch. Rational Mech. Anal.* **123**, 1–49 (1993).

[5] M. Haelterman and A.H. Sheppard, Vector soliton associated with polarization modulational instability in the normal-dispersion regime. *Phys. Rev. E* **49**, 3389–3399 (1994).

[6] P. Kockaert, *Dynamique non linéaire vectorielle de la propagation lumineuse en fibres optiques et caractérisation des phénomènes ultracourts associés.* Thèse de Doctorat, Université Libre de Bruxelles, 2001.

[7] R.V. Kohn and P. Sternberg, Local minimisers and singular perturbations. *Proc. Roy. Soc. Edinburgh Sect. A* **111**, 69–84 (1989).

[8] J. Norbury and L-C. Yeh, Inhomogeneous fast reaction, slow diffusion and weighted curve shortening. *Nonlinearity* **14**, 849–862 (2001).

[9] J. Norbury and L-C. Yeh, The location and stability of interface solutions of

an inhomogeneous parabolic problem. *SIAM J. Appl. Math.* **61**, 1418–1430 (2001).

[10] J. Rubinstein, P. Sternberg and J.B. Keller, Fast reaction, slow diffusion, and curve shortening. *SIAM J. Appl. Math.* **49**, 116–133 (1989).

[11] L-C. Yeh, *Patterned solutions in a reaction diffusion problem with convex nonlinearity.* D.Phil. thesis, University of Oxford, 1999.

NEW EXISTENCE RESULT FOR A 3-D SHAPE MEMORY MODEL

Irena Pawłow[1,3] and Wojciech M. Zajączkowski[2,3]

[1] *Systems Research Institute, Polish Academy of Sciences, Newelska 6, 01-447 Warsaw, Poland*
E-mail: pawlow@ibspan.waw.pl

[2] *Institute of Mathematics, Polish Academy of Sciences, Śniadeckich 8, 00-956 Warsaw, Poland*
E-mail: wz@impan.gov.pl

[3] *and Institute of Mathematics and Cryptology, Cybernetics Faculty, Military University of Technology, S. Kaliskiego 2, 00-908 Warsaw, Poland*

This paper presents a new existence result for a three-dimensional (3-D) shape memory model which has the form of a nonlinear thermoelasticity system, with viscosity $\nu > 0$ and capillarity $\varkappa > 0$. In contrast to the authors' previous results, proved under the assumption $0 < 2\sqrt{\varkappa} < \nu$, here we admit $\varkappa > 0$ and $\nu > 0$ possibly arbitrarily small. Thus, the existence result obtained becomes more adequate for shape memory problems where viscosity effects are negligible small. Moreover, we consider a broader class of boundary conditions. The main new part of the present paper is given by the solvability analysis of the initial-boundary-value problem for the viscoelasticity system with capillarity.

1. Introduction

The goal of this paper is to present a new existence result for a three-dimensional (3-D) shape memory model, which has been previously studied by the authors under more restrictive assumptions in [11, 10]. The model, which was first introduced and studied in [13, 14, 9], is given by the following nonlinear thermoelasticity system, with a viscosity coefficient $\nu > 0$ and a strain-gradient coefficient (called capillarity) $\varkappa > 0$:

$$\begin{array}{ll} \boldsymbol{u}_{tt} - \nu \boldsymbol{Q}\boldsymbol{u}_t + \varkappa \boldsymbol{Q}^2\boldsymbol{u} = \nabla \cdot F_{\boldsymbol{\varepsilon}}(\boldsymbol{\varepsilon},\theta) + \boldsymbol{b} & \text{in } \Omega^T = \Omega \times (0,T), \\ \boldsymbol{u}|_{t=0} = \boldsymbol{u}_0, \quad \boldsymbol{u}_t|_{t=0} = \boldsymbol{u}_1 & \text{in } \Omega, \\ \boldsymbol{B}(\partial_x)\boldsymbol{u} = \boldsymbol{0} & \text{on } S^T = S \times (0,T), \end{array} \tag{1}$$

$$\begin{array}{ll} c_0(\boldsymbol{\varepsilon},\theta)\theta_t - k_0\Delta\theta = \theta F_{,\theta\boldsymbol{\varepsilon}}(\boldsymbol{\varepsilon},\theta) + \nu(\boldsymbol{A}\boldsymbol{\varepsilon}_t)\cdot\boldsymbol{\varepsilon}_t + g & \text{in } \Omega^T, \\ \theta|_{t=0} = \theta_0 & \text{in } \Omega, \\ \boldsymbol{n}\cdot\nabla\theta = 0 & \text{on } S^T, \end{array} \tag{2}$$

where

$$c_0(\boldsymbol{\varepsilon},\theta) = c_v - \theta F_{,\theta\theta}(\boldsymbol{\varepsilon},\theta), \tag{3}$$

and $\boldsymbol{B}(\partial_x)\boldsymbol{u}$ stands for one of the following two types of boundary conditions:

$$\boldsymbol{u} = \boldsymbol{0}, \quad \boldsymbol{Q}\boldsymbol{u} = \boldsymbol{0} \qquad \text{on } S^T,$$

or

$$\boldsymbol{u} = \boldsymbol{0}, \quad (\boldsymbol{A}\boldsymbol{\varepsilon}(\boldsymbol{u}))\boldsymbol{n} = \boldsymbol{0} \qquad \text{on } S^T. \tag{4}$$

Here, $\Omega \subset \mathbb{R}^3$ is a bounded domain with a smooth boundary S, occupied by a solid body in a reference configuration, with constant mass density ($\rho = 1$); $\boldsymbol{n}$ is the unit outward normal vector to S; $T > 0$ is an arbitrary fixed time; $\boldsymbol{u} : \Omega^T \to \mathbb{R}^3$ is the displacement and $\theta : \Omega^T \to \mathbb{R}_+$ the absolute temperature. The second order tensors

$$\boldsymbol{\varepsilon} = \boldsymbol{\varepsilon}(\boldsymbol{u}) = \frac{1}{2}(\nabla\boldsymbol{u} + (\nabla\boldsymbol{u})^T) \quad \text{and} \quad \boldsymbol{\varepsilon}_t = \boldsymbol{\varepsilon}(\boldsymbol{u}_t) = \frac{1}{2}(\nabla\boldsymbol{u}_t + (\nabla\boldsymbol{u}_t)^T)$$

respectively denote the linearized strain and the strain rate. The operator $\boldsymbol{Q}$ stands for the linearized elasticity operator, defined by

$$\boldsymbol{Q}\boldsymbol{u} = \nabla\cdot(\boldsymbol{A}\boldsymbol{\varepsilon}(\boldsymbol{u})) = \mu\Delta\boldsymbol{u} + (\lambda+\mu)\nabla(\nabla\cdot\boldsymbol{u}), \tag{5}$$

where $\boldsymbol{A} = (A_{ijkl})$ is the fourth order elasticity tensor representing the linear isotropic Hooke's law

$$\boldsymbol{A}\boldsymbol{\varepsilon}(\boldsymbol{u}) := \lambda \mathrm{tr}\boldsymbol{\varepsilon}(\boldsymbol{u})\boldsymbol{I} + 2\mu\boldsymbol{\varepsilon}(\boldsymbol{u}), \tag{6}$$

$\boldsymbol{I}$ being the identity tensor, and λ, μ the Lamé constants, such that $\mu > 0$ and $3\lambda + 2\mu > 0$.

Correspondingly, the fourth order operator $\boldsymbol{Q}^2 = \boldsymbol{Q}\boldsymbol{Q}$ is given by

$$\boldsymbol{Q}^2\boldsymbol{u} = \nabla\cdot(\boldsymbol{A}\boldsymbol{\varepsilon}(\boldsymbol{Q}\boldsymbol{u})) = \mu^2\Delta^2\boldsymbol{u} + (\lambda+\mu)(\lambda+3\mu)\nabla\nabla\cdot(\Delta\boldsymbol{u}). \tag{7}$$

Moreover, $F(\boldsymbol{\varepsilon},\theta)$ denotes the elastic energy, which is a nonconvex (multi-well) function of $\boldsymbol{\varepsilon}$, with the shape strongly depending on θ. The remaining quantities in (1), (2) have the following meaning: $c_0(\varepsilon, \theta)$ is the specific heat coefficient, c_v, k_0, ν and $\varkappa$ are positive numbers denoting respectively the thermal specific heat, the heat conductivity, the viscosity and the capillarity.

System (1), (2) is made up by balance laws for the linear momentum and the internal energy. The underlying free energy density has the Landau-Ginzburg form

$$f(\varepsilon(\boldsymbol{u}), \nabla\varepsilon(\boldsymbol{u}), \theta) = -c_v\theta\log\theta + F(\varepsilon(\boldsymbol{u}), \theta) + \frac{\varkappa}{2}|\boldsymbol{Q}\boldsymbol{u}|^2, \tag{8}$$

with the three terms representing respectively thermal, elastic and strain-gradient (capillarity) energy. The corresponding stress tensor is given by

$$\begin{aligned} \boldsymbol{S} &= \frac{\delta f}{\delta\varepsilon}(\varepsilon(\boldsymbol{u}), \nabla\varepsilon(\boldsymbol{u}), \theta) + \boldsymbol{S}^v \\ &= F_{,\varepsilon}(\varepsilon(\boldsymbol{u}), \theta) - \varkappa\boldsymbol{A}\varepsilon(\boldsymbol{Q}\boldsymbol{u}) + \nu\boldsymbol{A}\varepsilon(\boldsymbol{u}_t), \end{aligned} \tag{9}$$

where $\delta f/\delta\varepsilon = f_{,\varepsilon} - \nabla\cdot f_{,\nabla\varepsilon}$ denotes the first variation of f with respect to ε, and $\boldsymbol{S}^v = \nu\boldsymbol{A}\varepsilon(\boldsymbol{u}_t)$ is the viscous stress according to Hooke's-like law. For the thermodynamical background of the model, we refer to [9, 14].

Let us add few remarks on model (1), (2) and its solvability. Firstly, we point out that the dynamics $(1)_1$ is in accordance with the so-called viscosity-capillarity criterion, justified by several authors, among whom Slemrod [15] and Abeyaratne-Knowles [1], as a proper model for the dynamics of phase transitions in van der Waals fluids and for propagating phase boundaries in solids. By this criterion, originally formulated in the case of one space dimension, a proper constitutive relation for the stress has the form (see e.g. [1, eq. (2.8)])

$$s = F_{,u_x}(u_x) - \varkappa u_{xxx} + \nu u_{xt}, \tag{10}$$

where u_x is the strain, $F(u_x)$ is a nonconvex double-well elastic energy, and $\nu \geq 0$ and $\varkappa \geq 0$ are the viscosity and the strain-gradient coefficient, respectively. We can see that equation (9) generalizes the stress-strain relation (10) to the case of three space dimensions.

Secondly, we remark that in the case of vanishing viscosity $\nu = 0$, problem (1), (2) represents a 3-D analog of the well-known Falk model for one-dimensional martensitic phase transitions of shear type (see [6, 4]). Unfortunately, neither our previous theory [11, 10] nor the present one cover the case $\nu = 0$. The existence proofs in [11] and [10], as well as the earlier one in [14], were based on the following condition, involving the viscosity and capillarity coefficients

$$0 < 2\sqrt{\varkappa} < \nu. \tag{11}$$

Such a condition allows for the decomposition of the elasticity system $(1)_1$ into two second order parabolic problems

$$\begin{aligned} &\boldsymbol{w}_t - \beta \boldsymbol{Q}\boldsymbol{w} = \nabla \cdot F_{,\boldsymbol{\varepsilon}}(\boldsymbol{\varepsilon},\theta) + \boldsymbol{b} && \text{in } \Omega^T, \\ &\boldsymbol{w}|_{t=0} = \boldsymbol{u}_1 - \alpha \boldsymbol{Q}\boldsymbol{u}_0 && \text{in } \Omega, \\ &\boldsymbol{w} = \boldsymbol{0} && \text{on } S^T, \end{aligned} \tag{12}$$

$$\begin{aligned} &\boldsymbol{u}_t - \alpha \boldsymbol{Q}\boldsymbol{u} = \boldsymbol{w} && \text{in } \Omega^T, \\ &\boldsymbol{u}|_{t=0} = \boldsymbol{u}_0 && \text{in } \Omega, \\ &\boldsymbol{u} = \boldsymbol{0} && \text{on } S^T, \end{aligned} \tag{13}$$

where α, β are numbers satisfying

$$\alpha + \beta = \nu, \quad \alpha\beta = \varkappa.$$

Due to condition (11), these numbers are real and positive, $\alpha, \beta \in \mathbb{R}_+$. The decomposition (12), (13) was the main idea underlying the existence proofs in the above mentioned papers. It is known, however, that in structural phase transitions in shape memory alloys the strain-gradient effect is observable but not the viscous one (see e.g. [4]). For that reason, condition (11) is not appropriate for shape memory models.

In view of this, it important to construct an existence theory allowing for a relaxation of condition (11). In the present paper, we replace (11) by

$$\varkappa > 0 \quad \text{and} \quad \nu > 0, \tag{14}$$

allowing the viscosity to be arbitrarily small but positive. In such a case, system (1) is parabolic and the theory of parabolic equations can be applied (see [16, 5]).

We mention a similar study due to Yoshikawa [19], which is also concerned with the existence of solutions to problem (1), (2) under assumption (14). Unlike the present paper, however, the result in [19] concerns model (1), (2), in which the energy equation $(2)_1$ is simplified. This simplification consists in neglecting the nonlinear term $-\theta F_{,\theta\theta}(\varepsilon,\theta)$ in the specific heat coefficient $c_0(\varepsilon,\theta)$ by assuming that $c_0(\varepsilon,\theta) = c_v = \text{const} > 0$. Obviously, such a simplification destroys the thermodynamic structure of the model but makes the mathematical analysis much simpler. We add that the same simplification was used in the first result on the global in time unique solvability of system (1), (2), in the 2-D and 3-D cases, obtained in [14].

The existence theorem due to Yoshikawa [19] generalizes the result in [14] by admitting weaker assumptions on the data, in particular (14) instead of (11), and a more general solutions class. Note that the technique used in [19] is different from the classical methods for parabolic systems exploited in [14, 11, 10]. Indeed, it is based on the so-called maximal regularity theory for abstract parabolic equations.

The authors' papers [11] and [10] generalize the result of [14], respectively in the 2-D and the 3-D case, by removing the aforementioned simplification of the energy equation. We stress that the presence of a non-linearity in the leading coefficient of the heat conduction equation introduces essential difficulties in the existence proof.

As it has been already observed, in the case of one-space dimension problem (1), (2) with $\varkappa > 0$ and $\nu = 0$ is identical with the Falk model. In such a case, in contrast with the three-dimensional one, there are several results on the existence and uniqueness of solutions, in particular due to Sprekels and Zheng [17], Aiki [2] and Yoshikawa [18]. The latter paper includes an up-to-date list of references related to the 1-D Falk's model. For a survey of diffused-interface models of shape alloys and the related mathematical results we refer to [12]. Let us also point out that problem (1), (2) without capillarity but with viscosity, i.e. $\varkappa = 0$, $\nu > 0$, and with the simplified energy equation discussed above, has been studied by Zimmer [20].

Finally, we add a remark concerning the boundary conditions in $(1)_2$. In [14] and later in [11, 10], the no-displacement boundary condition $\boldsymbol{u} = \mathbf{0}$ on S^T was chosen in order to apply the result due to Nečas [7] on the ellipticity property of the operator $\boldsymbol{Q}$ whereas the condition $\boldsymbol{Q}\boldsymbol{u} = \mathbf{0}$ on S^T resulted in a compatibility with the parabolic decomposition (12), (13).

In the present paper, besides $\boldsymbol{u} = \mathbf{0}$, $\boldsymbol{Q}\boldsymbol{u} = \mathbf{0}$ on S^T, we admit the other type of boundary conditions $(4)_2$.

The plan of the paper is as follows. In Section 2 we formulate the assumptions and state the existence and uniqueness theorems. These theorems generalize Theorems 2.1 and 2.2 in [10] by allowing for assumption (14) and for the broader class of boundary conditions (4). In Section 3 we examine the solvability of the initial-boundary-value problem defined by the differential operator on the left-hand side of $(1)_1$, with the initial conditions $(1)_2$ and the boundary conditions (4). We show that this differential operator is parabolic in the sense of Solonnikov, and that the initial and boundary conditions satisfy the Shapiro-Lopatinskij conditions (complementarity condition). In Section 4 we present some auxiliary results on the solvability of parabolic problems of fourth and second order. These results play a key role in the new existence proof. Section 5 is devoted to the outline of the existence proof.

1.1. *Notation*

We shall use the following notation:

$$f_{,i}=\frac{\partial f}{\partial x_i},\quad i=1,2,3,\quad f_t=\frac{df}{dt},\quad \boldsymbol{\varepsilon}=(\varepsilon_{ij})_{i,j=1,2,3},$$

$$F_{,\boldsymbol{\varepsilon}}(\boldsymbol{\varepsilon},\theta)=\left(\frac{\partial F(\boldsymbol{\varepsilon},\theta)}{\partial\varepsilon_{ij}}\right)_{i,j=1,2,3},\quad F_{,\theta}(\boldsymbol{\varepsilon},\theta)=\frac{\partial F(\boldsymbol{\varepsilon},\theta)}{\partial\theta},$$

where the space and time derivatives are material.

Vectors (tensors of the first order), tensors of the second order (referred simply to as tensors), and tensors of higher order are denoted by bold letters. Tensors of the second order represent linear transformations of vectors into vectors; $\boldsymbol{S}^T$, tr$\boldsymbol{S}$, $\boldsymbol{S}^{-1}$ and det$\boldsymbol{S}$, respectively, denote the transpose, trace, inverse, and determinant of a tensor $\boldsymbol{S}$.

A dot designates the inner product, irrespective of the space in question: for instance, $\boldsymbol{u}\cdot\boldsymbol{v}$ is the inner product of the vectors $\boldsymbol{u}=(u_i)$ and $\boldsymbol{v}=(v_i)$, $\boldsymbol{S}\cdot\boldsymbol{R}=\mathrm{tr}(\boldsymbol{S}^T\boldsymbol{R})$ is the inner product of the tensors $\boldsymbol{S}=(S_{ij})$ and $\boldsymbol{R}=(R_{ij})$, $\boldsymbol{A}^m\cdot\boldsymbol{B}^m$ is the inner product of the m-th order tensors $\boldsymbol{A}^m=(A^m_{i_1\dots i_m})$ and $\boldsymbol{B}^m=(B^m_{i_1\dots i_m})$. In Cartesian components,

$$\begin{aligned}&(\boldsymbol{S}\boldsymbol{u})_i=S_{ij}u_j,\quad (\boldsymbol{S}^T)_{ij}=S_{ji},\quad \mathrm{tr}\boldsymbol{S}=S_{ii},\\ &\boldsymbol{u}\cdot\boldsymbol{v}=u_iv_i,\quad \boldsymbol{S}\cdot\boldsymbol{R}=S_{ij}R_{ij},\\ &\boldsymbol{A}^m\cdot\boldsymbol{B}^m=A^m_{i_1\dots i_m}B^m_{i_1\dots i_m}.\end{aligned}$$

Here and in the sequel, the summation convention over repeated indices is used. By $\boldsymbol{A}=(A_{ijkl})$ we denote the fourth order elasticity tensor, which represents a symmetric linear transformation of symmetric tensors into symmetric tensors. We write $(\boldsymbol{A}\boldsymbol{\varepsilon})_{ij}=A_{ijkl}\varepsilon_{kl}$.

The symbols ∇ and $\nabla\cdot$ denote the material gradient and the divergence. For the divergence, we use the convention of the contraction over the last index, e.g. $(\nabla\cdot\boldsymbol{S})_i=\partial S_{ij}/\partial x_j$.

We use the Sobolev spaces notation of [8]. Throughout the paper, c and $c(T)$ denote generic constants, different in various occurrences, depending on the data of the problem and on the domain Ω. The symbol T indicates time horizon dependence, which is always of the form T^a, $a\in\mathbb{R}_+$.

2. Assumptions and main results

Problem (1), (2) is studied under the following assumptions (A1)–(A5) (the same as in [10]):

(A1) $\Omega \subset \mathbb{R}^3$ is a domain with the boundary of class C^4. The C^4 - regularity is needed to apply the classical regularity result for parabolic systems.
(A2) The coefficients of the operator $\boldsymbol{Q}$ satisfy

$$\mu > 0, \qquad 3\lambda + 2\mu > 0.$$

These conditions ensure the following properties:

(i) Coercivity and boundedness of the operator $\boldsymbol{A}$, i.e.

$$\underline{c}|\boldsymbol{\varepsilon}|^2 \leq (\boldsymbol{A}\boldsymbol{\varepsilon}) \cdot \boldsymbol{\varepsilon} \leq \bar{c}|\boldsymbol{\varepsilon}|^2, \tag{15}$$

where $\underline{c} = \min\{3\lambda + 2\mu, 2\mu\}, \bar{c} = \max\{3\lambda + 2\mu, 2\mu\}$;

(ii) Strong ellipticity of the operator $\boldsymbol{Q}$ (see [14, Sec. 7]). Thanks to this property, the following estimate, due to Nečas [7], holds true:

$$c\|\boldsymbol{u}\|_{\boldsymbol{W}_2^2(\Omega)} \leq \|\boldsymbol{Q}\boldsymbol{u}\|_{\boldsymbol{L}_2(\Omega)} \quad \text{for} \ \left\{\boldsymbol{u} \in \boldsymbol{W}_2^2(\Omega) |\ \boldsymbol{u}|_s = \boldsymbol{0}\right\}; \tag{16}$$

(iii) Parabolicity in the general (Solonnikov) sense of the system defined by the differential operator on the left-hand side of $(1)_1$ (see Lemma 3).

Our next assumption concerns the structure of the elastic energy.
(A3) The function $F(\boldsymbol{\varepsilon}, \theta) : \boldsymbol{S}^2 \times [0, \infty) \to \mathbb{R}$ is of class C^3, $\boldsymbol{S}^2$ denoting the set of the symmetric second order tensors in $\mathbb{R}^3$. We assume the splitting

$$F(\boldsymbol{\varepsilon}, \theta) = F_1(\boldsymbol{\varepsilon}, \theta) + F_2(\boldsymbol{\varepsilon}),$$

where F_1 and F_2 are subject to the following conditions:
(A3-1) Conditions on $F_1(\boldsymbol{\varepsilon}, \theta)$

(i) concavity with respect to θ

$$-F_{1,\theta\theta}(\boldsymbol{\varepsilon}, \theta) \geq 0 \quad \text{for} \quad (\boldsymbol{\varepsilon}, \theta) \in \boldsymbol{S}^2 \times [0, \infty). \tag{17}$$

(ii) Nonnegativity

$$F_1(\boldsymbol{\varepsilon}, \theta) \geq 0 \quad \text{for} \quad (\boldsymbol{\varepsilon}, \theta) \in \boldsymbol{S}^2 \times [0, \infty).$$

(iii) Boundedness of the norm

$$\|F_1\|_{C^3(\boldsymbol{S}^2 \times [0,\infty))} < \infty.$$

(iv) Growth conditions. There exist a positive constant c and numbers $s, K_1 \in (0, \infty)$ such that

$$|\partial_{\boldsymbol{\varepsilon}}^j \partial_\theta^i F_1| \leq c(1 + \theta^{s-i}|\boldsymbol{\varepsilon}|^{K_1 - j})$$

for $0 \le i+j \le 2$, $i,j \in \mathbb{N}$, and $i=2$, $j=1$, as well as for large values of θ and ε_{ij}, where admissible ranges of s and K_1 are given by

$$0 < s < \frac{2}{3}, \qquad 0 < K_1 < \frac{15}{4}.$$

Moreover, in the case $K_1 > 1$ the numbers s and K_1 are linked by the equality

$$15s + 4K_1 = 15.$$

(A3-2) Conditions on $F_2(\boldsymbol{\varepsilon})$

(i) Nonnegativity

$$F_2(\boldsymbol{\varepsilon}) \ge 0 \quad \text{for} \quad \boldsymbol{\varepsilon} \in \boldsymbol{S}^2.$$

(ii) Boundedness of the norm

$$\|F_2\|_{C^2(\boldsymbol{S}^2)} < \infty.$$

(iii) Growth conditions

$$|\partial_{\boldsymbol{\varepsilon}}^i F_2| \le c(1+|\boldsymbol{\varepsilon}|^{K_2-i}), \quad 0 \le i \le 2, \quad i \in \mathbb{N},$$

for large values of ε_{ij}, where

$$0 < K_2 \le \frac{9}{2}.$$

Before formulating our assumptions on the data, let us note some consequences of assumption (A3-1), which are of relevance for the existence proof. In view of (A3-1) (i), by definition of $c_0(\boldsymbol{\varepsilon},\theta)$ we have

$$0 < c_v \le c_0(\boldsymbol{\varepsilon},\theta) \quad \text{for} \quad (\boldsymbol{\varepsilon},\theta) \in \boldsymbol{S}^2 \times [0,\infty). \tag{18}$$

Moreover, (A3-1) (iii) and (iv) imply the bounds

$$\begin{aligned}
&|c_0(\boldsymbol{\varepsilon},\theta)|, \quad |c_{0,\theta}(\boldsymbol{\varepsilon},\theta)| \le c(1+|\boldsymbol{\varepsilon}|^{K_1}),\\
&|c_{0,\boldsymbol{\varepsilon}}(\boldsymbol{\varepsilon},\theta)| \le c(1+|\boldsymbol{\varepsilon}|^{\max\{0,K_1-1\}}) \;\; \text{for} \;\; (\boldsymbol{\varepsilon},\theta) \in \boldsymbol{S}^2 \times [0,\infty).
\end{aligned} \tag{19}$$

From (A3-1)(i) and (ii) it follows that

$$F_1(\boldsymbol{\varepsilon},\theta) - \theta F_{1,\theta}(\boldsymbol{\varepsilon},\theta) \ge 0 \quad \text{for} \quad (\boldsymbol{\varepsilon},\theta) \in \boldsymbol{S}^2 \times [0,\infty), \tag{20}$$

and, owing to (A3-2) (i),

$$(F_1(\boldsymbol{\varepsilon},\theta) - \theta F_{1,\theta}(\boldsymbol{\varepsilon},\theta)) + F_2(\boldsymbol{\varepsilon}) \ge 0 \quad \text{for} \quad (\boldsymbol{\varepsilon},\theta) \in \boldsymbol{S}^2 \times [0,\infty), \tag{21}$$

which means that the elastic part of the internal energy is nonnegative. The latter bound is used in the derivation of energy estimates.

(A4) The data satisfy

$$\begin{aligned}
&\boldsymbol{b} \in \boldsymbol{L}_p(\Omega^T), \quad 5 < p < \infty,\\
&g \in L_q(\Omega^T), \quad 5 < q < \infty, \quad \text{and} \quad g \geq 0 \ \text{ a.e. in } \ \Omega^T,\\
&\boldsymbol{u}_0 \in \boldsymbol{W}_p^{4-2/p}(\Omega), \quad \boldsymbol{u}_1 \in \boldsymbol{W}_p^{2-2/p}(\Omega), \quad 5 < p < \infty,\\
&\theta_0 \in W_q^{2-2/q}(\Omega), \quad 5 < q < \infty, \quad \text{and} \quad \theta_* = \min_\Omega \theta_0 > 0.
\end{aligned}$$

Moreover, the initial data are supposed to satisfy the compatibility conditions for the classical solvability of parabolic problems. We note that, by Sobolev's imbeddings,

$$\theta_0 \in C^{1,\alpha_0}(\Omega), \quad \boldsymbol{\varepsilon}_0 \in \boldsymbol{C}^{2,\alpha_0'} \quad \text{with} \quad 0 < \alpha_0, \ \ \alpha_0' < 1. \tag{22}$$

Like in [10], we introduce an additional technical assumption which requires a special separable form of $F_1(\boldsymbol{\varepsilon}, \theta)$:
(A5) The function $F_1(\boldsymbol{\varepsilon}, \theta)$ has the form

$$F_1(\boldsymbol{\varepsilon}, \theta) = \sum_{i=1}^{N} \tilde{F}_{1i}(\theta)\tilde{F}_{2i}(\boldsymbol{\varepsilon}),$$

where $N \in \boldsymbol{N}$ is a finite number, and in accordance with (A3-1) (i)–(iii), (A5-1)

$$\begin{aligned}
&\tilde{F}_{1i} \in C^2([0,\infty)), && \tilde{F}_{2i} \in C^2(\boldsymbol{S}^2), && i = 1, \ldots, N,\\
&\tilde{F}_{1i}(\theta) \geq 0, && \tilde{F}_{2i}(\boldsymbol{\varepsilon}) \geq 0, && i = 1, \ldots, N,\\
&-\tilde{F}_{1i,\theta\theta}(\theta) \geq 0, && && i = 1, \ldots, N.
\end{aligned}$$

Moreover, the functions $\tilde{F}_i(\theta), \ \ i = 1, \ldots, N$, are given by
(A5-2)

$$\tilde{F}_{1i}(\theta) = \begin{cases} 0 & \text{for } 0 \leq \theta \leq \theta_1,\\ \varphi_i(\theta) & \text{for } \theta_1 \leq \theta \leq \theta_2,\\ \theta^{s_i} & \text{for } \theta \geq \theta_2, \end{cases}$$

where the numbers $s_i, \ \ i = 1, \ldots, N$, and θ_1, θ_2 satisfy the following conditions:

$$0 < s_i \leq s < 1, \qquad 1 < \theta_1 < \theta_2 \leq \theta_1^{1/s_i} \qquad \text{for } i = 1, \ldots, N.$$

The requirements in (A5-1) imply that

$$\begin{aligned}
&\varphi_i(\theta_1) = \theta_1, && \varphi_i'(\theta_1) = 1, && \varphi_i''(\theta_1) = 0,\\
&\varphi_i(\theta_2) = \theta_2^{s_i}, && \varphi_i'(\theta_2) = s_i\theta_2^{s_i-1}, && \varphi_i''(\theta_2) = s_i(s_i-1)\theta_2^{s_i-2},\\
&\theta^{s_i} \leq \varphi_i(\theta) \leq \theta, && -\varphi_i''(\theta) \geq 0, && \text{for } \theta \in (\theta_1, \theta_2),
\end{aligned}$$

where $i = 1, \dots, N$. We note that the functions $\tilde{F}_{1i}(\theta)$ in (A5-2) satisfy the growth conditions (A3-1) (iv), which now read as follows

$$|\partial_\theta^j \tilde{F}_{1i}| \le c(1+\theta^{s-j}),$$
$$|\partial_{\boldsymbol{\varepsilon}}^j \tilde{F}_{2i}| \le c(1+|\boldsymbol{\varepsilon}|^{K_1-j}),$$

where $i = 1, \dots, N$ and $j = 0, 1, 2$.

We point out that, in [10], the above separable form of $F_1(\boldsymbol{\varepsilon}, \theta)$ has been used to prove the key L_∞-norm estimate for θ. In our argument, this part of the proof will remain unchanged. We also add that such separable form of $F_1(\boldsymbol{\varepsilon}, \theta)$ is in accordance with the known Falk-Konopka elastic energy model (for more detailed account see [10]).

The main result of the present paper is the following existence theorem.

Theorem 1: Let assumptions (A1)–(A5) be satisfied and the coefficients $\varkappa$, ν fulfil condition (14). Then, for any $T > 0$ there exists a solution $(\boldsymbol{u}, \theta)$ to problem (1), (2), with the boundary conditions (4), in the space

$$\boldsymbol{V}(p,q) \equiv \{(\boldsymbol{u},\theta) | \boldsymbol{u} \in \boldsymbol{W}_p^{4,2}(\Omega^T), \quad \theta \in W_q^{2,1}(\Omega^T), \quad 5 < p \le q < \infty\}, \tag{23}$$

such that

$$\|\boldsymbol{u}\|_{\boldsymbol{W}_p^{4,2}(\Omega^T)} \le c(T), \quad \|\theta\|_{W_q^{2,1}(\Omega^T)} \le c(T), \tag{24}$$

with a positive constant $C(T)$ depending on the data of the problem and T^a, $a \in \mathbb{R}_+$. Moreover, there exists a positive finite number ω satisfying

$$[g + \nu(\boldsymbol{A}\boldsymbol{\varepsilon}_t) \cdot \boldsymbol{\varepsilon}_t] \exp(\omega t) + [\omega c_0(\boldsymbol{\varepsilon}, \theta) + F_{,\theta\boldsymbol{\varepsilon}}(\boldsymbol{\varepsilon}, \theta) \cdot \boldsymbol{\varepsilon}_t]\theta_* \ge 0 \quad \text{in} \quad \Omega^T,$$

such that

$$\theta \ge \theta_* \exp(-\omega t) \quad \text{in} \quad \Omega^T. \tag{25}$$

We point out that this theorem generalizes the result in [10, Theorem 2.1], by admitting a weaker assumption on the coefficients $\varkappa$, ν and a broader class of boundary conditions. We also remark that the solutions specified in Theorem 1 enjoy, by virtue of Sobolev's imbeddings, the following properties: the functions

$$\boldsymbol{u}, \boldsymbol{u}_t, \boldsymbol{\varepsilon}, \nabla\boldsymbol{\varepsilon}, \nabla^2\boldsymbol{\varepsilon}, \boldsymbol{\varepsilon}_t, \theta, \nabla\theta \tag{26}$$

are Hölder continuous in Ω^T and satisfy the corresponding a priori bounds with constant $c(T)$.

For completeness, we also recall the uniqueness result which follows by repeating the arguments used in [11] in the study of problem (1), (2), $(4)_1$ in 2-D case. The proof is based on a direct comparison of two solutions and on the use of energy estimates, together with the regularity properties (26). The parabolic decomposition of the elasticity system (1) is not applied in the uniqueness proof.

Theorem 2: Let the assumptions of Theorem 1 be satisfied and, in addition, suppose that

(A6) $F(\varepsilon,\theta) : \boldsymbol{S}^2 \times [0,\infty) \to \mathbb{R}$ is of class C^4, and $g \in L_\infty(\Omega^T)$.

Then, the solution $(\boldsymbol{u},\theta) \in \boldsymbol{V}(p,q)$ to problem (1), (2) is unique.

3. Parabolicity of the elasticity system with viscosity and capillarity

We consider the following problem

$$\begin{aligned} &\boldsymbol{u}_{tt} - \nu \boldsymbol{Q}\boldsymbol{u}_t + \varkappa \boldsymbol{Q}^2\boldsymbol{u} = \boldsymbol{f} && \text{in } \Omega^T,\\ &\boldsymbol{u}|_{t=0} = \boldsymbol{u}_0, \quad \boldsymbol{u}_t|_{t=0} = \boldsymbol{u}_1, && \text{in } \Omega,\\ &\boldsymbol{B}(\partial_x)\boldsymbol{u} = \boldsymbol{0} && \text{on } S^T, \end{aligned} \tag{27}$$

where $\boldsymbol{Q}$ is the linear elasticity operator defined by (5) and $\boldsymbol{B}(\partial_x)\boldsymbol{u}$ stands for one of the following two types of boundary conditions

$$\begin{aligned} &\boldsymbol{u} = \boldsymbol{0}, \quad \boldsymbol{Q}\boldsymbol{u} = \boldsymbol{0} && \text{on } S^T,\\ \text{or} &\\ &\boldsymbol{u} = \boldsymbol{0}, \quad (\boldsymbol{A}\varepsilon(\boldsymbol{u}))\boldsymbol{n} = \boldsymbol{0} && \text{on } S^T. \end{aligned} \tag{28}$$

In view of (5), (7) system $(27)_1$ can be expressed in the explicit form

$$\begin{aligned} &\boldsymbol{u}_{tt} + \Delta(-\nu\mu\boldsymbol{u}_t + \varkappa\mu^2\Delta\boldsymbol{u})\\ &\quad + \nabla\nabla\cdot[-\nu(\lambda+\mu)\boldsymbol{u}_t + \varkappa(\lambda+\mu)(\lambda+3\mu)\Delta\boldsymbol{u}] = \boldsymbol{f} \end{aligned} \tag{29}$$

or, equivalently, in the matrix form

$$\sum_{j=1}^{3} l_{kj}(\partial_t,\partial_x)u_j = f_k, \quad k = 1,2,3, \tag{30}$$

where

$$\begin{aligned} l_{kj}(\partial_t,\partial_x) &= \delta_{kj}[\partial_t^2 + \Delta(-\nu\mu\partial_t + \varkappa\mu^2\Delta)]\\ &\quad + \partial_{x_k}\partial_{x_j}[-\nu(\lambda+\mu)\partial_t + \varkappa(\lambda+\mu)(\lambda+3\mu)\Delta]. \end{aligned} \tag{31}$$

By assumption (A2), $\mu > 0$, $\lambda + \mu > 0$.
We write (30) in the short form

$$\mathcal{L}(\partial_t,\partial_x)\boldsymbol{u} = \boldsymbol{f}, \tag{32}$$

where $\mathcal{L}(\partial_t, \partial_x)$ is the matrix operator with elements $\{l_{kj}(\partial_t, \partial_x)\}_{k,j=1,2,3}$. Moreover, let $L = \det \mathcal{L}$.

Lemma 3: *System $(27)_1$ is parabolic.*

Proof: By the Fourier-Laplace transform

$$\hat{u}(\boldsymbol{\xi}, p) = \int_0^\infty e^{pt} dt \int_{\mathbb{R}^3} e^{i\boldsymbol{\xi}\cdot\boldsymbol{x}} u(\boldsymbol{x}, t) dx,$$

system (32) takes the form

$$\sum_{j=1}^{3} l_{kj}(p, i\boldsymbol{\xi})\hat{u}_j = \hat{f}_k, \quad k = 1, 2, 3.$$

Then,

$$L = d^2[d + (\lambda + \mu)b|\boldsymbol{\xi}|^2]$$

where

$$d = p^2 + \mu a|\boldsymbol{\xi}|^2, \quad a = \nu p + \varkappa\mu|\boldsymbol{\xi}|^2, \quad b = \nu p + \varkappa(\lambda + 3\mu)|\boldsymbol{\xi}|^2.$$

The roots of the equation $L = 0$ are

(i) the double-root $d = 0$;
(ii) $d + (\lambda + \mu)b|\boldsymbol{\xi}|^2 = 0$.

Solving (i), we get

$$p^2 + \nu\mu|\boldsymbol{\xi}|^2 p + \varkappa\mu^2|\boldsymbol{\xi}|^4 = 0,$$

so that

$$p = \frac{-\nu \pm \sqrt{\nu^2 - 4\varkappa}}{2}\mu|\boldsymbol{\xi}|^2.$$

In the case (ii), we have

$$p^2 + (\lambda + 2\mu)\nu p|\boldsymbol{\xi}|^2 + \varkappa(\lambda + 2\mu)^2|\boldsymbol{\xi}|^4 = 0.$$

Hence,

$$p = \frac{-\nu \pm \sqrt{\nu^2 - 4\varkappa}}{2}(\lambda + 2\mu)|\boldsymbol{\xi}|^2.$$

In both cases, there exists a real positive number δ such that

$$\mathrm{Re}\, p \le -\delta|\boldsymbol{\xi}|^2.$$

This ends the proof. □

Now, we examine the boundary conditions.

Lemma 4: *The boundary conditions $(28)_1$ and $(28)_2$ satisfy the Shapiro-Lopatinskij conditions for the system (27) (complementarity condition).*

Proof: We examine the system $(27)_1$ with vanishing right-hand side $\boldsymbol{f} = \boldsymbol{0}$ and with initial conditions $\boldsymbol{u}_0 = \boldsymbol{u}_1 = \boldsymbol{0}$. Let us denote this problem by (P).

First, we examine problem (P) locally in the half-space $x_3 > 0$. Looking for solutions vanishing at $x_3 \to \infty$, the Shapiro-Lopatinskij condition means that we have only a solution identically equal to zero (see [16, Chap. 2, § 8]). Hence, we can replace this condition by a coercivity argument. For this purpose, we derive an estimate for the weak solutions of problem (P). Multiplying $(27)_1$ by $\boldsymbol{u}_t$ and integrating over Ω, we get

$$\frac{1}{2}\frac{d}{dt}\int\limits_{\Omega} |\boldsymbol{u}_t|^2 dx - \nu \int\limits_{\Omega} \boldsymbol{Q}\boldsymbol{u}_t \cdot \boldsymbol{u}_t dx + \varkappa \int\limits_{\Omega} \boldsymbol{Q}^2 \boldsymbol{u} \cdot \boldsymbol{u}_t dx = 0.$$

Integrating by parts and then integrating with respect to time yields

$$\begin{aligned}
&\frac{1}{2}\int\limits_{\Omega} |\boldsymbol{u}_t(t)|^2 dx + \frac{\varkappa}{2}\int\limits_{\Omega} |\boldsymbol{Q}\boldsymbol{u}(t)|^2 dx + \nu \int\limits_{\Omega^t} (\boldsymbol{A}\boldsymbol{\varepsilon}(\boldsymbol{u}_{t'})) \cdot \boldsymbol{\varepsilon}(\boldsymbol{u}_{t'}) dx dt' \\
&-\nu \int\limits_{S^t} \boldsymbol{u}_{t'} \cdot (\boldsymbol{A}\boldsymbol{\varepsilon}(\boldsymbol{u}_{t'}))\boldsymbol{n} dS dt' + \varkappa \int\limits_{S^t} \boldsymbol{u}_{t'} \cdot (\boldsymbol{A}\boldsymbol{\varepsilon}(\boldsymbol{Q}\boldsymbol{u}))\boldsymbol{n} dS dt' \\
&-\varkappa \int\limits_{S^t} \boldsymbol{Q}\boldsymbol{u} \cdot (\boldsymbol{A}\boldsymbol{\varepsilon}(\boldsymbol{u}_{t'}))\boldsymbol{n} dS dt' = 0, \quad t \le T.
\end{aligned} \tag{33}$$

We see that the boundary conditions $(28)_1$ and $(28)_2$ imply that all the boundary integrals in (33) vanish. Hence, in view of estimates (15), (16), it follows that

$$\|\boldsymbol{u}_t\|_{L_\infty(0,T;\boldsymbol{L}_2(\Omega))} + \|\boldsymbol{u}\|_{L_\infty(0,T;\boldsymbol{W}_2^2(\Omega))} + \|\boldsymbol{\varepsilon}_t\|_{\boldsymbol{L}_2(\Omega^T)} \le 0,$$

which implies $\boldsymbol{u} = \boldsymbol{0}$ in Ω^T. This concludes the proof. □

4. Auxiliary existence results for parabolic problems of fourth and second order

Let $\Omega \subset \mathbb{R}^3$ be a bounded domain with a sufficiently smooth boundary S. Let us consider the fourth order system

$$\begin{aligned} &\boldsymbol{u}_{tt} - \nu \boldsymbol{Q}\boldsymbol{u}_t + \varkappa \boldsymbol{Q}^2 \boldsymbol{u} = \boldsymbol{f} && \text{in } \Omega^T, \\ &\boldsymbol{u}|_{t=0} = \boldsymbol{u}_0, \quad \boldsymbol{u}_t|_{t=0} = \boldsymbol{u}_1 && \text{in } \Omega, \\ &\boldsymbol{B}(\partial_x)\boldsymbol{u} = \boldsymbol{0} && \text{on } S^T, \end{aligned} \tag{34}$$

where $\boldsymbol{B}(\partial_x)$ is given by $(4)_1$ or $(4)_2$.

Lemma 5: *Let $\boldsymbol{f} \in \boldsymbol{L}_p(\Omega^T)$, $\boldsymbol{u}_0 \in \boldsymbol{W}_p^{4-2/p}(\Omega)$, $\boldsymbol{u}_1 \in \boldsymbol{W}_p^{2-2/p}(\Omega)$, $1 < p < \infty$, $S = \partial\Omega \in C^4$. Then, there exists a unique solution $\boldsymbol{u} \in \boldsymbol{W}_p^{4,2}(\Omega^T)$ of problem (34) such that*

$$\|\boldsymbol{u}\|_{\boldsymbol{W}_p^{4,2}(\Omega^T)} \le c(\|\boldsymbol{f}\|_{\boldsymbol{L}_p(\Omega^T)} + \|\boldsymbol{u}_0\|_{\boldsymbol{W}_p^{4-2/p}(\Omega)} + \|\boldsymbol{u}_1\|_{\boldsymbol{W}_p^{2-2/p}(\Omega)}). \tag{35}$$

Proof: Since the complementarity condition is satisfied (see Lemma 4) we apply the results of [16]. This shows the assertion. □

Let us consider problem

$$\begin{aligned} &\boldsymbol{u}_{tt} - \nu \boldsymbol{Q}\boldsymbol{u}_t + \varkappa \boldsymbol{Q}^2 \boldsymbol{u} = \nabla \cdot \boldsymbol{\sigma} + \boldsymbol{b} && \text{in } \Omega^T, \\ &\boldsymbol{u}|_{t=0} = \boldsymbol{u}_0, \quad \boldsymbol{u}_t|_{t=0} = \boldsymbol{u}_1 && \text{in } \Omega, \\ &\boldsymbol{B}(\partial_x)\boldsymbol{u} = \boldsymbol{0} && \text{on } S^T, \end{aligned} \tag{36}$$

where $\boldsymbol{\sigma} = (\sigma_{ij})_{i,j=1,2,3}$, $\boldsymbol{b} = (b_i)_{i=1,2,3}$, and $\boldsymbol{B}(\partial_x)$ is given by $(4)_1$ or $(4)_2$.

Lemma 6: *Let $\boldsymbol{\sigma} \in \boldsymbol{L}_p(\Omega^T)$, $\boldsymbol{b} \in \boldsymbol{L}_p(\Omega^T)$, $u_0 \in \boldsymbol{W}_p^{3-2/p}(\Omega)$, $\boldsymbol{u}_1 \in W_p^{1-2/p}(\Omega)$, $1 < p < \infty$, $S \in C^3$. Then, the solutions of problem (36) satisfy the inequality*

$$\begin{aligned} &\|\boldsymbol{u}\|_{\boldsymbol{W}_p^{3,3/2}(\Omega^T)} \le c(\|\boldsymbol{\sigma}\|_{\boldsymbol{L}_p(\Omega^T)} \\ &\quad + \|\boldsymbol{b}\|_{\boldsymbol{L}_p(\Omega^T)} + \|\boldsymbol{u}_0\|_{\boldsymbol{W}_p^{3-2/p}(\Omega)} + \|\boldsymbol{u}_1\|_{\boldsymbol{W}_p^{1-2/p}(\Omega)}). \end{aligned} \tag{37}$$

Proof: Let $\boldsymbol{G} = (G_{ij})_{i,j=1,2,3}$ be the Green function of problem (36), with vanishing initial conditions $\boldsymbol{u}_0 = \boldsymbol{u}_1 = \boldsymbol{0}$, on the half-space $x_3 > 0$. Then, $\boldsymbol{u}$ admits the following representation

$$\boldsymbol{u}(\boldsymbol{x}, t) = \int_{\mathbb{R}_+^3 \times \mathbb{R}} \boldsymbol{G}(\boldsymbol{x}, t, \boldsymbol{x}', t')(\nabla_{\boldsymbol{x}'} \cdot \boldsymbol{\sigma} + \boldsymbol{b}) dx' dt'.$$

Since $\boldsymbol{G}|_{x_3=0} = \boldsymbol{0}$, we obtain

$$\boldsymbol{u}(\boldsymbol{x},t) = -\int\limits_{\mathbb{R}^3_+\times\mathbb{R}} \nabla_{\boldsymbol{x}'}\boldsymbol{G}(\boldsymbol{x},t,\boldsymbol{x}',t')\boldsymbol{\sigma} dx'dt' + \int\limits_{\mathbb{R}^3_+\times\mathbb{R}} \boldsymbol{G}\boldsymbol{b} dx'dt', \tag{38}$$

where $(\nabla_{\boldsymbol{x}'}\boldsymbol{G}\boldsymbol{\sigma})_i = \sum_{j,k=1}^3 \partial_{x'_k} G_{ij}\sigma_{jk}$. In view of (38), from [16] it follows that

$$\|\boldsymbol{u}\|_{\boldsymbol{W}_p^{3,3/2}(\mathbb{R}^3_+\times\mathbb{R})} \le c(\|\boldsymbol{\sigma}\|_{\boldsymbol{L}_p(\mathbb{R}^3_+\times\mathbb{R})} + \|\boldsymbol{b}\|_{\boldsymbol{L}_p(\mathbb{R}^3_+\times\mathbb{R})}).$$

Next, by using the regularizer technique and the extension of the initial data, the assertion can be concluded. □

Now, let us consider the parabolic problem

$$\begin{array}{ll} a(\boldsymbol{x},t)\theta_t - \Delta\theta = f & \text{in } \Omega^T, \\ \theta|_{t=0} = \theta_0 & \text{in } \Omega, \\ \boldsymbol{n}\cdot\nabla\theta = 0 & \text{on } S^T. \end{array} \tag{39}$$

Lemma 7: *Let $f \in L_p(\Omega^T)$, $\theta_0 \in W_p^{2-2/p}(\Omega)$, $1 < p < \infty$, $S \in C^2$ and let the coefficient $a \in C^{\alpha,\alpha/2}(\Omega^T)$, $\alpha \in (0,1)$, satisfy $a^* \ge a \ge a_* > 0$ (for two positive constants a_*, a^*); suppose that $a_t \in L_2(0,T;L_2(\Omega))$. Then, there exists a unique solution $\theta \in W_p^{2,1}(\Omega^T)$ of problem (39) such that*

$$\begin{aligned} \|\theta\|_{W_p^{2,1}(\Omega^T)} &\le \varphi\left(\tfrac{1}{\min a}, \max a, \|a\|_{C^{\alpha,\alpha/2}(\Omega^T)}, \|a_t\|_{L_2(\Omega^T)}, T\right) \\ &\cdot[\|f\|_{L_p(\Omega^T)} + \|\theta_0\|_{W_p^{2-2/p}(\Omega)} + \|\theta\|_{V_2^0(\Omega^T)}], \end{aligned} \tag{40}$$

and

$$\|\theta\|_{V_2^0(\Omega^T)} := \operatorname*{ess\ sup}_{t\in(0,T)} \|\theta(t)\|_{L_2(\Omega)} + \|\nabla\theta\|_{\boldsymbol{L}_2(\Omega^T)} \le A(T), \tag{41}$$

where φ is an increasing positive function of its arguments, and $A(T)$ is a positive function depending on the data f, θ_0 and a, $\|a_t\|_{L_2(\Omega^T)}$.

Proof: First, we obtain the energy inequality (41). Next, by applying a partition of unity and using energy inequality (41) we obtain (40). □

5. Outline of the proof of Theorem 1

The idea of the proof is the same as in [10]. It is based on the Leray-Schauder fixed point theorem. In the present proof, an essential role is played by the auxiliary results in Lemmas 5–7. Here we present the main steps, while the details are given in [10].

Step 1. The solution map. We use the Leray-Schauder theorem in the following formulation:

Theorem 8: Let $\mathcal{B}$ be a Banach space. Assume that $T : [0,1] \times \mathcal{B} \to \mathcal{B}$ is a map with the following properties:

(i) for any fixed $\tau \in [0,1]$ the map $T(\tau,\cdot) : \mathcal{B} \to \mathcal{B}$ is completely continuous.
(ii) For every bounded subset $\mathcal{C}$ of $\mathcal{B}$, the family of maps $T(\cdot,\chi) : [0,1] \to \mathcal{B}, \chi \in \mathcal{C}$, is uniformly equicontinuous.
(iii) There is a bounded subset $\mathcal{C}$ of $\mathcal{B}$ such that any fixed point in $\mathcal{B}$ of the map $T(\tau,\cdot), 0 \le \tau \le 1$, is contained in $\mathcal{C}$.
(iv) the map $T(0,\cdot)$ has precisely one fixed point in $\mathcal{B}$.
Then, the map $T(1,\cdot)$ has at least one fixed point in $\mathcal{B}$.

In order to define the corresponding solution map, we extend the definition of $F_1(\boldsymbol{\varepsilon},\theta)$ to all $\theta \in \mathbb{R}$ in such a way that it is of class C^3, and that

$$F_{1,\theta\theta}(\boldsymbol{\varepsilon},\theta) \ge 0 \quad \text{for all} \quad (\boldsymbol{\varepsilon},\theta) \in \boldsymbol{S}^2 \times (-\infty,0).$$

With such extension, the lower bound (18) on $c_0(\boldsymbol{\varepsilon},\theta)$ remains valid for all $(\boldsymbol{\varepsilon},\theta) \in \boldsymbol{S}^2 \times \mathbb{R}$.
The solution space is $\boldsymbol{V}(p,q)$, defined by (23). The solution map

$$T(\tau,\cdot) : (\bar{\boldsymbol{u}},\bar{\theta}) \in \boldsymbol{V}(p,q) \to (\boldsymbol{u},\theta) \in \boldsymbol{V}(p,q), \quad \tau \in [0,1], \tag{42}$$

is defined by means of the following initial-boundary value problems:

$$\begin{aligned} &\boldsymbol{u}_{tt} - \nu \boldsymbol{Q}\boldsymbol{u}_t + \varkappa \boldsymbol{Q}^2\boldsymbol{u} = \tau[\nabla \cdot F_{,\boldsymbol{\varepsilon}}(\bar{\boldsymbol{\varepsilon}},\bar{\theta}) + \boldsymbol{b}] && \text{in } \Omega^T, \\ &\boldsymbol{u}|_{t=0} = \tau\boldsymbol{u}_0, \quad \boldsymbol{u}_t|_{t=0} = \tau\boldsymbol{u}_1 && \text{in } \Omega, \\ &\boldsymbol{B}(\partial_x)\boldsymbol{u} = \boldsymbol{0} && \text{on } S^T, \end{aligned} \tag{43}$$

$$\begin{aligned} &c_0(\boldsymbol{\varepsilon},\bar{\theta},\tau)\theta_t - k_0\Delta\theta = \tau[\bar{\theta}F_{,\theta\boldsymbol{\varepsilon}}(\boldsymbol{\varepsilon},\bar{\theta}) \cdot \boldsymbol{\varepsilon}_t + \nu(\boldsymbol{A}\boldsymbol{\varepsilon}_t)\cdot\boldsymbol{\varepsilon}_t + g] && \text{in } \Omega^T, \\ &\theta|_{t=0} = \tau\theta_0 && \text{in } \Omega, \\ &\boldsymbol{n}\cdot\nabla\theta = 0 && \text{on } S^T, \end{aligned} \tag{44}$$

where

$$c_0(\boldsymbol{\varepsilon},\bar{\theta},\tau) = c_v - \tau\bar{\theta}F_{,\theta\theta}(\boldsymbol{\varepsilon},\bar{\theta}), \quad \bar{\boldsymbol{\varepsilon}} = \boldsymbol{\varepsilon}(\bar{\boldsymbol{u}}).$$

Clearly, any fixed point of $T(1,\cdot)$ in $\boldsymbol{V}(p,q)$ yields a solution $(\boldsymbol{u},\theta)$ of problem (1), (2) in $\boldsymbol{V}(p,q)$. Therefore, the proof is reduced to checking that the solution map $T(\tau,\cdot)$ complies with the properties (i)–(iv).

Step 2. Properties (i), (ii) and (iv) of the solution map. Property (i) follows by showing that, for any fixed $\tau \in [0,1]$, $T(\tau,\cdot)$ maps bounded subsets into precompact subsets in $\boldsymbol{V}(p,q)$. Let $(\bar{\boldsymbol{u}}^n, \bar{\theta}^n)$ be a bounded sequence in $\boldsymbol{V}(p,q)$, such that for $n \to \infty$

$$\bar{\boldsymbol{u}}^n \rightharpoonup \bar{\boldsymbol{u}} \text{ weakly in } \boldsymbol{W}_p^{4,2}(\Omega^T), \quad \bar{\theta}^n \rightharpoonup \bar{\theta} \text{ weakly in } W_q^{2,1}(\Omega^T), \quad 5 < p,q < \infty.$$

By repeating the arguments of [10] and using Lemma 5, we conclude that for the sequence $(\boldsymbol{u}^n, \theta^n) := T(\tau, \bar{\boldsymbol{u}}^n, \bar{\theta}^n)$, the following convergences hold for $n \to \infty$

$$\begin{aligned} &\boldsymbol{u}^n \to \boldsymbol{u} \ \text{ strongly in } \ \boldsymbol{W}_p^{4,2}(\Omega^T), \quad 5 < p < \infty, \\ &\theta^n \to \theta \ \text{ strongly in } \ W_q^{2,1}(\Omega^T), \quad 5 < q < \infty, \end{aligned}$$

where $(\boldsymbol{u},\theta) = T(\tau, \bar{\boldsymbol{u}}, \bar{\theta})$. This shows (i).

Property (ii) follows by direct comparison of two solutions $(\boldsymbol{u},\theta)$ and $(\tilde{\boldsymbol{u}},\tilde{\theta})$, corresponding to parameters τ and $\tilde{\tau}$, respectively. Applying Lemmas 5 and 7, we show that

$$\|\boldsymbol{u} - \tilde{\boldsymbol{u}}\|_{W_p^{4,2}(\Omega^T)}, \quad \|\theta - \tilde{\theta}\|_{W_q^{2,1}(\Omega^T)} \le c|\tau - \tilde{\tau}|, \quad 5 < p,q < \infty.$$

Property (iv) is obvious in view of Lemmas 5, 7 and of the definition of $T(\tau,\cdot)$.

Further steps of the proof concern property (iii) of the solution map.

Step 3. A priori bounds for a fixed point. Without loss of generality, we set $\tau = 1$ and assume that $(\boldsymbol{u},\theta) \in \boldsymbol{V}(p,q)$ is a fixed point of $T(1,\cdot)$. We begin by proving that the temperature is positive. Having this, we establish some energy estimates and then improve them recursively.

Step 3.1. Positivity of temperature.

Lemma 9: *(see [11, Lemma 3.1]). Let*

$$\theta_* \equiv \min_{\Omega} \theta_0 > 0, \quad g \ge 0 \ \ in \ \ \Omega^T,$$

and $(\boldsymbol{u},\theta)$ be a solution to (1), (2) such that $\boldsymbol{\varepsilon}, \boldsymbol{\varepsilon}_t \in \boldsymbol{L}_\infty(\Omega^T)$, $\theta \in L_\infty(\Omega^T)$, $\theta_t \in L_1(0,T;L_q(\Omega))$, $1 < q < \infty$.
Then, there exists a positive finite number ω satisfying

$$[g + \nu(\boldsymbol{A}\boldsymbol{\varepsilon}_t)\cdot\boldsymbol{\varepsilon}_t]\exp(\omega t) + [\omega c_0(\boldsymbol{\varepsilon},\theta) + F_{,\theta\boldsymbol{\varepsilon}}(\boldsymbol{\varepsilon},\theta)\cdot\boldsymbol{\varepsilon}_t]\theta_* \ge 0 \quad in \ \ \Omega^T,$$

such that

$$\theta \ge \theta_* \exp(-\omega t) \quad in \ \ \Omega^T. \tag{45}$$

We point out that the regularity assumptions in Lemma 9 are satisfied for solutions in the space $\boldsymbol{V}(p,q)$.

Step 3.2. Energy estimates.

Lemma 10: *(see [11, Lemma 3.2]). Let*

$$\begin{aligned}&\boldsymbol{u}_0 \in \boldsymbol{W}_2^2(\Omega), \quad \boldsymbol{u}_1 \in \boldsymbol{L}_2(\Omega), \quad \theta_0 \in L_1(\Omega),\\ &(F_1(\boldsymbol{\varepsilon}_0,\theta_0) - \theta_0 F_{1,\theta}(\boldsymbol{\varepsilon}_0,\theta_0)) + F_2(\boldsymbol{\varepsilon}_0) \in L_1(\Omega),\\ &\boldsymbol{b} \in L_1(0,T;\boldsymbol{L}_2(\Omega)), \quad g \in L_1(\Omega^T).\end{aligned}$$

Assume that $\theta \geq 0$ in Ω^T and that the bound (21) holds. Then, a solution $(\boldsymbol{u},\theta)$ to (1), (2) satisfies the estimate

$$\begin{aligned}&\|\theta\|_{L_\infty(0,T;L_1(\Omega))} + \|\boldsymbol{u}_t\|^2_{L_\infty(0,T;\boldsymbol{L}_2(\Omega))} + \|\boldsymbol{Q}\boldsymbol{u}\|^2_{L_\infty(0,T;\boldsymbol{L}_2(\Omega))}\\ &\quad + \|(F_1(\boldsymbol{\varepsilon},\theta) - \theta F_{1,\theta}(\boldsymbol{\varepsilon},\theta)) + F_2(\boldsymbol{\varepsilon})\|_{L_\infty(0,T;L_1(\Omega))} \leq c,\end{aligned} \tag{46}$$

with a constant c only depending on the data.

We indicate the implications of estimate (46) which are important in the next step. Firstly, by property (16) of the operator $\boldsymbol{Q}$,

$$\|\boldsymbol{u}\|_{L_\infty(0,T;\boldsymbol{W}_2^2(\Omega))} \leq c,$$

so that

$$\|\bar{\boldsymbol{\varepsilon}}\|_{L_\infty(0,T;\boldsymbol{W}_2^1(\Omega))\cap L_\infty(0,T;\boldsymbol{L}_6(\Omega))} \leq c. \tag{47}$$

Secondly, (46) implies the bound

$$\|\boldsymbol{u}\|_{\boldsymbol{W}^{2,1}_{2,\infty}(\Omega^T)} \leq c,$$

so that, in view of Sobolev's imbeddings,

$$\|\boldsymbol{\varepsilon}\|_{\boldsymbol{W}^{1,1/2}_{2,\infty}(\Omega^T)\cap \boldsymbol{L}_{10}(\Omega^T)} \leq c. \tag{48}$$

Our aim is to prove the estimates (24). This shall be accomplished with the help of Lemmas 5–7. We point out that, in view of the nonlinearity of the coefficient $c_0(\boldsymbol{\varepsilon},\theta)$, in order to apply Lemma 7 we first have to prove Hölder-norm bounds for $\boldsymbol{\varepsilon}$ and θ. To this aim, we proceed in a number of steps which provide the recursive improvement of estimates for θ and $\boldsymbol{\varepsilon}$.

Step 3.3. The first temperature estimate. According to Lemma 6, we have the following estimate for problem (1):

$$\begin{aligned}&\|\boldsymbol{\varepsilon}\|_{\boldsymbol{W}_p^{2,1}(\Omega^T)} \leq c\|\boldsymbol{u}\|_{\boldsymbol{W}_p^{3,3/2}(\Omega^T)}\\ &\quad \leq c(\|F_{,\boldsymbol{\varepsilon}}(\boldsymbol{\varepsilon},\theta)\|_{\boldsymbol{L}_p(\Omega^T)} + \|\boldsymbol{b}\|_{\boldsymbol{L}_p(\Omega^T)}\\ &\quad + \|\boldsymbol{u}_0\|_{\boldsymbol{W}_p^{3-2/p}(\Omega)} + \|\boldsymbol{u}_1\|_{\boldsymbol{W}_p^{1-2/p}(\Omega)}), \quad 1 < p < \infty.\end{aligned} \tag{49}$$

In view of this estimate, repeating the arguments of Lemma 7 in [10], which rely on multiplying equation $(2)_1$ by θ, integrating over Ω, and applying appropriate imbeddings and interpolation inequalities, we get

Lemma 11: *(see [10, Lemma 4.3]). Suppose that assumption (A3-1) (iv) is satisfied and that* $\|\boldsymbol{\varepsilon}\|_{\boldsymbol{L}_{10}}(\Omega) \leq c(T)$. *Then, there exists a constant* $c(T)$ *depending only on the data and* T^a, $a \in \mathbb{R}_+$, *such that*

$$\|\theta\|_{L_\infty(0,T;L_2(\Omega))} + \|\nabla\theta\|_{\boldsymbol{L}_2(\Omega^T)} \leq c(T). \tag{50}$$

We indicate the implications of (50). By virtue of the imbedding,

$$\|\theta\|_{L_{10/3}(\Omega^T)} \leq c(T). \tag{51}$$

Moreover, using the following estimate derived in the proof of Lemma 11 (see [10, eq. (52)]), namely

$$\|\boldsymbol{\varepsilon}\|_{\boldsymbol{W}^{2,1}_{20/7}(\Omega^T)} \leq c(T)(\|\nabla\theta\|^{3s/4}_{\boldsymbol{L}_2(\Omega^T)} + 1), \tag{52}$$

we have

$$\|\boldsymbol{\varepsilon}\|_{\boldsymbol{W}^{2,1}_{20/7}(\Omega^T)} \leq c(T).$$

Hence, by Sobolev's imbeddings, $\boldsymbol{\varepsilon}$ is Hölder continuous in Ω^T and

$$\|\boldsymbol{\varepsilon}\|_{\boldsymbol{C}^{\alpha_1,\alpha_1/2}(\Omega^T)} \leq c(T), \quad \text{with } \ 0 < \alpha_1 < 1/4. \tag{53}$$

Due to (53), using the growth condition on $F_{1,\boldsymbol{\varepsilon}}$ in (A3-1) (iv) and the estimate (49), we conclude that

$$\|\boldsymbol{\varepsilon}\|_{\boldsymbol{W}^{2,1}_p(\Omega^T)} \leq c(T)(\|\theta\|^s_{L_{ps}(\Omega^T)} + 1) \leq c(T) \quad \text{for } \ p = 10/(3s) > 5. \tag{54}$$

Further, thanks to (53) the bounds (19) imply

$$|c_0(\boldsymbol{\varepsilon},\theta)| + |c_{0,\boldsymbol{\varepsilon}}(\boldsymbol{\varepsilon},\theta)| + |c_{0,\theta}(\boldsymbol{\varepsilon},\theta)| \leq c(T) \quad \text{in } \Omega^T. \tag{55}$$

Step 3.4. The second temperature estimate. Multiplying equation $(2)_1$ by θ_t and integrating over Ω^t we obtain

Lemma 12: *(see [10, Lemma 4.4]). Suppose that*

$$\begin{aligned}
&0 < s < 2/3 \qquad g \in L_2(\Omega^T), \qquad \nabla\theta_0 \in \boldsymbol{L}_2(\Omega), \quad \text{and}\\
&\|\theta\|_{L_{10/3}(\Omega^T)} \leq c(T), \qquad \|\boldsymbol{\varepsilon}\|_{\boldsymbol{C}^{\alpha_1,\alpha_1/2}(\Omega^T)} \leq c(T),\\
&\|\boldsymbol{\varepsilon}_t\|_{\boldsymbol{L}_p(\Omega)} \leq c(T) \quad \text{for} \quad p = 10/(3s).
\end{aligned}$$

Then, there exists a constant $c(T) > 0$ *such that*

$$\|\theta_t\|_{L_2(\Omega^T)} + \|\theta\|_{L_\infty(0,T;W^1_2(\Omega))} \leq c(T). \tag{56}$$

Step 3.5. The third temperature estimate. Writing equation $(2)_1$ in the form

$$-k_0\Delta\theta = -c_0(\varepsilon,\theta)\theta_t + \theta F_{,\theta\varepsilon}(\varepsilon,\theta)\cdot\varepsilon_t + \nu(\boldsymbol{A}\varepsilon_t)\cdot\varepsilon_t + g \tag{57}$$

and using (55) and (56), the right-hand side of (57) can be estimated in $L_2(\Omega^T)$-norm. Consequently, by virtue of the classical elliptic theory,

$$\|\theta\|_{L_2(0,T;W_2^2(\Omega))} \le c(T), \quad \text{so} \quad \|\theta\|_{L_2(0,T;L_\infty(\Omega))} \le c(T). \tag{58}$$

Furthermore, (58) and (56) imply that

$$\|\theta\|_{W_2^{2,1}(\Omega^T)} \le c(T), \quad \|\nabla\theta\|_{\boldsymbol{W}_2^{1,1/2}(\Omega^T)} \le c(T),$$

so, by Sobolev's imbeddings,

$$\|\theta\|_{L_{10}(\Omega^T)} \le c(T), \quad \|\nabla\theta\|_{\boldsymbol{L}_{10/3}(\Omega^T)} \le c(T). \tag{59}$$

Step 3.6. The improvement of the strain estimate. Using $(59)_1$, we repeat the estimate (54) to conclude that

$$\begin{aligned}&\|\varepsilon\|_{\boldsymbol{W}_p^{2,1}(\Omega^T)} \le c(T)(\|\theta\|^s_{L_{ps}(\Omega^T)}+1) \le c(T),\\ &\|\nabla\varepsilon\|_{\boldsymbol{W}_p^{1,1/2}(\Omega^T)} \le c(T)\end{aligned}$$

for $p = 10/s > 15$. Consequently, by (well-known) imbedding results,

$$\|\nabla\varepsilon\|_{\boldsymbol{C}^{\alpha_2,\alpha_2/2}(\Omega^T)} \le c(T) \quad \text{with} \quad 0 < \alpha_2 < 1 - s/2. \tag{60}$$

Now, recalling assumptions (A3-1) (iii), (iv) and (A3-2) (ii), (iii), and using (53), (59), (60), we get

$$\|\nabla\cdot F_{,\varepsilon}(\varepsilon,\theta)\|_{\boldsymbol{L}_{10/3}(\Omega^T)} \le c(T).$$

Hence, by virtue of Lemma 5,

$$\|\boldsymbol{u}\|_{\boldsymbol{W}^{4,2}_{10/3}(\Omega^T)} \le c(T),$$

so that

$$\|\varepsilon_t\|_{\boldsymbol{W}^{1,1/2}_{10/3}(\Omega^T)} \le c(T). \tag{61}$$

In view of the imbedding

$$\boldsymbol{W}^{1,1/2}_{10/3}(\Omega^T) \subset L_q(0,T;\boldsymbol{L}_\infty(\Omega)) \quad \text{for} \quad q < 4,$$

(61) implies

$$\|\varepsilon_t\|_{\boldsymbol{L}_{10}(\Omega^T)} + \|\varepsilon_t\|_{L_q(0,T;\boldsymbol{L}_\infty(\Omega))} \le c(T). \tag{62}$$

This estimate is crucial for obtaining an $L_\infty(\Omega^T)$-norm bound and subsequently Hölder-norm bound on θ. This is done in the next two steps, which follow the outline of [10].

Step 3.7. Pointwise estimate on temperature. Here, we have to impose assumption (A5) on the separable form of $F_1(\varepsilon, \theta)$. Then, we have

Lemma 13: *(see [10, Lemma 5.1]). Suppose (A5) is satisfied, in addition to (A1)-(A4). Moreover, let*

$$\theta_0 \in L_\infty(\Omega^T), \quad g \in L_1(0,T; L_\infty(\Omega)),$$

$$\theta \geq \theta_*(T) = \theta_* \exp(-\omega T),$$

$$\boldsymbol{\varepsilon} \in \boldsymbol{L}_\infty(\Omega^T), \quad \boldsymbol{\varepsilon}_t \in L_2(0,T; \boldsymbol{L}_\infty(\Omega)), \quad \textit{with}$$

$$\|\boldsymbol{\varepsilon}\|_{\boldsymbol{L}_\infty(\Omega^T)} \leq c(T), \quad \|\boldsymbol{\varepsilon}_t\|_{L_2(0,T;\boldsymbol{L}_\infty(\Omega))} \leq c(T),$$

where $c(T)$ is an increasing positive function. Then,

$$\begin{aligned} \|\theta\|_{L_\infty(\Omega^T)} \leq\ & c \exp(c(T)T^{1/2}\|\boldsymbol{\varepsilon}_t\|_{L_2(0,T;\boldsymbol{L}_\infty(\Omega))}) \times \\ & \times \left(\|\boldsymbol{\varepsilon}_t\|^2_{L_2(0,T;\boldsymbol{L}_\infty(\Omega))} + \|g\|_{L_1(0,T;\boldsymbol{L}_\infty(\Omega))} + \|\theta\|_{L_\infty(\Omega)}\right) \leq c(T). \end{aligned} \tag{63}$$

The proof of this lemma is based on multiplying equation $(2)_1$ by θ^r, $r > 1$, integrating over Ω and introducing a specially constructed primitive of the function $-\theta^{r+1}F_{1,\theta\theta}(\boldsymbol{\varepsilon}, \theta)$ with respect to θ. We point out that a similar idea was used in Lemma 11, where $(2)_1$ has been multiplied by θ and a primitive of $-\theta^2 F_{1,\theta\theta}(\boldsymbol{\varepsilon}, \theta)$ with respect to θ has been constructed. As a result we obtain an estimate on θ in the $L_\infty(0,T; L_r(\Omega))$-norm, which, due to bound (62) on $\boldsymbol{\varepsilon}_t$, allows to pass to the limit with $r \to \infty$ and to conclude the assertion.

In view of (63), estimate (54) yields

$$\|\boldsymbol{\varepsilon}\|_{\boldsymbol{W}_p^{2,1}(\Omega^T)} \leq c(T) \quad \text{for } 1 < p < \infty. \tag{64}$$

Step 3.8. Hölder continuity of temperature. To prove the Hölder continuity of θ we apply De Giorgi's method in a way presented in [8]. Namely, we prove that θ is an element of the space $\mathcal{B}_2(\Omega^T, M, \gamma, r, \delta, \varkappa)$, where $M, \gamma, r, \delta, \varkappa$ are positive parameters (for the definition of this space see [8, Chap. II.7]). The key point in the proof is the $L_\infty(\Omega^T)$-norm estimate on θ provided by Lemma 13, as well as the $\boldsymbol{L}_p(\Omega^T)$-norm estimate (64) on $\boldsymbol{\varepsilon}_t$. We have the following

Lemma 14: *(see [10, Lemma 6.1]). Suppose that*

$$\begin{aligned}
&|\varepsilon| \le c(T) \quad in \quad \Omega^T, \quad \|\varepsilon_t\|_{L_p(\Omega^T)} \le c(T), \quad 1 < p < \infty, \\
&\|\theta\|_{W_2^{2,1}(\Omega^T)} \le c(T), \quad \|\theta\|_{L_\infty(\Omega^T)} \le M = c(T), \\
&|c_0(\varepsilon, \theta)| + |c_{0,\varepsilon}(\varepsilon, \theta)| + |c_{0,\theta}(\varepsilon, \theta)| \le c(T) \quad in \quad \Omega^T, \\
&g \in L_p(\Omega^T), \quad 1 < p < \infty, \quad \theta_0 \in C^{\alpha_0}(\Omega), \quad 1 < \alpha_0 < 1.
\end{aligned}$$

Furthermore, let k be a positive number such that

$$k > \sup_{\Omega} \theta_0(x) \quad and \quad M - k < \delta \quad for\ some\ \delta > 0.$$

Then,

$$\theta \in \mathcal{B}_2(\Omega^T, M, \gamma, r, \delta, \varkappa), \tag{65}$$

where

$$r = q = \frac{10}{3}, \quad \varkappa \in \left(0, \frac{2}{3}\right), \quad \gamma = c(T).$$

By virtue of (65), we can apply the imbedding result of [8, Theorem II.7.1] to conclude that θ is Hölder continuous in Ω^T, and that

$$\|\theta\|_{C^{\alpha,\alpha/2}(\Omega^T)} \le c(T), \tag{66}$$

with Hölder exponent $0 < \alpha < 1$ depending on $M = c(T)$, $\gamma = c(T)$, r, δ and $\varkappa$.

Step 3.9. The final estimates. In view of the Hölder continuity of ε and θ as well as bound (64), we can apply Lemmas 5 and 7 to conclude the final estimates (24) and thereby prove property (iii) of the solution map. We have

Lemma 15: *(see [10, Lemma 6.2]). Suppose that ε and θ are Hölder continuous in Ω^T, and*

$$\begin{aligned}
&|\varepsilon| + |\theta| \le c(T) \quad in \quad \Omega^T, \\
&\|\nabla\varepsilon\|_{L_\sigma(\Omega^T)} + \|\varepsilon_t\|_{L_\sigma(\Omega^T)} \le c(T) \quad for \quad 1 < \sigma < \infty.
\end{aligned}$$

Moreover, suppose that the data satisfy (A4). Then

$$\|u\|_{W_p^{4,2}(\Omega^T)} \le c(T), \quad \|\theta\|_{W_q^{2,1}(\Omega^T)} \le c(T), \quad 5 < p, q < \infty. \tag{67}$$

Summarizing, we have shown that the solution map (42) satisfies the assumptions (i)-(iv) of the Leray-Schauder theorem. Thus, $T(1, \cdot)$ has at least one fixed point in $V(p, q)$, which yields a solution $(u, \theta) \in V(p, q)$ to problem (1), (2). Together with the bounds (67) and (45) the proof is completed.

References

[1] R. Abeyaratne and J.K. Knowles, Implications of viscosity and strain-gradient effects for the kinetics of propagating phase boundaries in solids. *SIAM J Appl. Math.* **51**, 1205–1221 (1991).

[2] T. Aiki, Weak solutions for Falk's model of shape memory alloys. *Math. Meth. Appl. Sci.* **23**, 299–319 (2000).

[3] O.V. Besov, V.P. Il'in and S.M. Nikolski, *Integral Representations of Functions and Imbedding Theorems.* Nauka, Moscow, 1975 (in Russian).

[4] M. Brokate and J. Sprekels, *Hysteresis and Phase Transitions.* Springer, New York, 1996.

[5] S.D. Ejdelman, Parabolic equations. In: *Partial Differential Equations VI*, Ed. Yu. V. Egorov and M. A. Shubin (Springer, New York, 1994), p. 205.

[6] F. Falk, Elastic phase transitions and nonconvex energy functions. In: *Free Boundary Problems: Theory and Applications. Vol. I*, Ed. K.-H. Hoffmann and J. Sprekels (Longman Sci. Tech, Harlow, 1990), p. 45.

[7] J. Nečas, *Les Méthodes Directes en Théorie des Équations Elliptiques.* Masson, Paris, 1967.

[8] O.A. Ladyzhenskaya, V.A. Solonnikov and N.N. Ural'tseva, *Linear and Quasilinear Equations of Parabolic Type.* Nauka, Moscow, 1967 (in Russian).

[9] I. Pawłow, Three-dimensional model of thermomechanical evolution of shape memory materials. *Control Cybern.* **29**, 341–365 (2000).

[10] I. Pawłow and W.M. Zajączkowski, Global existence to a three-dimensional non-linear thermoelsticity system arising in shape memory materials. *Math. Methods Appl. Sci.* **28**, 407–442 (2005).

[11] I. Pawłow and W.M. Zajączkowski, Unique global solvability in two-dimensional non-linear thermoelasticity. *Math. Methods Appl. Sci.* **28**, 551–592 (2005).

[12] I. Pawłow and W.M. Zajączkowski, On diffused-interface models of shape memory alloys. *Control Cybern.* **32**, 629–658 (2003).

[13] I. Pawłow and A. Żochowski, Nonlinear thermoelastic system with viscosity and nonlocality. In: *Free Boundary Problems: Theory and Applications. Vol. I*, Ed. N. Kenmochi (Gakkōtosho, Tokyo, 2000), p. 251.

[14] I. Pawłow and A. Żochowski, Existence and uniqueness for a three-dimensional thermoelastic system. *Dissertationes Math.* **406**, pp. 46. (2002).

[15] M. Slemrod, Admissibility criteria for propagating phase boundaries in a van der Waals fluid. *Arch. Rational Mech. Anal.* **81**, 301–315 (1983).

[16] V.A. Solonnikov, On boundary value problems for linear parabolic systems of a general type. *Trudy Mat. Inst. Steklov.* **83**, (1965) (in Russian).

[17] J. Sprekels and S. Zheng, Global solutions to the equations of a Ginzburg-Landau theory for structural phase transitions in shape memeory alloys. *Phys. D* **39**, 56–76 (1989).

[18] S. Yoshikawa, Weak solutions for the Falk model system of shape memory alloys in energy class. *Math. Methods Appl. Sci.* **28**, 1423–1443 (2005).

[19] S. Yoshikawa, Unique global existence for a three-dimensional thermoelastic system of shape memory alloys. *Adv. Math. Sci. Appl.* **15**, 603–627 (2005).
[20] J. Zimmer, Global existence for a nonlinear system in thermoviscoelasticity with nonconvex energy, *J. Math. Anal. Appl.* **292**, (2004), 589–604.

ANALYSIS OF A 1-D THERMOVISCOELASTIC MODEL WITH TEMPERATURE-DEPENDENT VISCOSITY

Robert Peyroux[1] and Ulisse Stefanelli[2]

[1] *Laboratoire de Mécanique et Génie Civil*
CNRS UMR 5508, Université de Montpellier II
place Eugène-Bataillon, F-34095 Montpellier cedex 05, France
E-mail: peyroux@lmgc.univ-montp2.fr

[2] *Istituto di Matematica Applicata e Tecnologie Informatiche – CNR*
via Ferrata 1, I-27100 Pavia, Italy
E-mail: ulisse@imati.cnr.it

This note addresses the analysis of a thermoviscoelastic model with a temperature-dependent viscous modulus. We provide the well-posedness of a related initial and boundary value problem and detail a suitable fully implicit variable time-step discretization. The latter is proved to be conditionally stable and convergent. Moreover, some a priori error estimates of optimal order are established.

1. Introduction

The isothermal evolution of a polymer is generally considered to be viscoelastic. On the other hand, the mechanical behavior of polymers is known to be strongly temperature-dependent, and a certain number of phenomena such as strain localization (shear bands or necking) cannot be modeled in a satisfactory way by means of classic models [18, 24]. Moreover, industrial polymers are known to experience large temperature variations in most (if not all) manufacturing processes[6]. The mathematical description of polymeric materials is therefore forced to take into account thermomechanical couplings. In particular, mechanically induced heat sources, thermal stresses, and strong temperature-dependence of the viscous and elastic moduli have to be considered.

The present analysis is devoted to the discussion of a one-dimensional thermoviscoelastic model of Kelvin-Voigt type. In particular, we shall be concerned with the system

$$c_s\dot{\theta} - k\theta_{xx} = -\alpha c\theta\dot{\varepsilon} + \nu(\theta)\dot{\varepsilon}^2, \tag{1}$$

$$\nu(\theta)\dot{\varepsilon} + c\varepsilon = \sigma + \alpha c(\theta - \theta_c), \tag{2}$$

posed in the space-time domain $Q := \Omega \times (0,T)$ where $\Omega := (0,1)$ and $T > 0$ is a final reference time. In the latter relations, $\theta > 0$ represents the absolute temperature of some thermoviscoelastic wire, ε is its strain, and σ is some given stress. System (1)-(2) arises naturally in the framework of Continuum Thermo-mechanics [8, 9] whenever one prescribes the *free energy density* and the *potential of dissipation* of the medium to be respectively

$$\psi(\theta,\varepsilon) := -c_s\theta\ln\theta - \alpha c(\theta-\theta_c)\varepsilon\frac{c}{2}\varepsilon^2,$$

$$\phi(\theta,\theta_x,\dot{\varepsilon}) := \frac{k}{2\theta}\theta_x^2 + \frac{\nu(\theta)}{2}\dot{\varepsilon}^2.$$

Here, $c_s > 0$ is the specific heat, $k > 0$ is the thermal conductivity, $\alpha \geq 0$ is the thermal dilatation coefficient, $c \geq 0$ is the elastic modulus, $\theta_c > 0$ is some critical temperature, and $\nu : (0,+\infty) \to (0,+\infty)$ is the *temperature-dependent* viscous modulus, which we assume to be bounded, smooth, and uniformly positive. We explicitly observe that the case of a purely viscous material ($c = 0$) is included in the present framework. Relations (1)-(2) arise from the energy balance relation

$$\dot{e} + q_x = \sigma\dot{\varepsilon},$$

where e is the internal energy density, q is the heat flux, and $\sigma\dot{\varepsilon}$ represents the mechanically induced heat sources (note that no external thermal source is considered for simplicity), and from the the classical positions

$$q := \theta R, \quad R := -\frac{\partial\phi}{\partial\theta_x}, \quad \sigma := \sigma_{el} + \sigma_{vis} = \frac{\partial\psi}{\partial\varepsilon} + \frac{\partial\phi}{\partial\dot{\varepsilon}}.$$

Here, R represents the entropy flux and σ_{el} and σ_{vis} are the elastic and the viscous components of the stress, respectively. It may be readily checked that the Second Law of Thermodynamics is fulfilled in the form of the Clausius-Duhem inequality.

We assume the stress $\sigma : Q \to \mathbb{R}$ to be given. This is not very restrictive in one dimension, since the equilibrium equation reads

$$\sigma_x + b = 0, \tag{3}$$

where b is some prescribed body force. Indeed, whenever we consider some traction test $\sigma(1) = g$, one readily computes from the latter relation that $\sigma(x,t) = \int_x^1 b(y,t)\,dy + g(t)$.

Finally, the system is complemented with the initial and boundary conditions

$$\theta(\cdot,0) = \theta^0, \quad \varepsilon(\cdot,0) = \varepsilon^0, \tag{4}$$

$$k\theta_x(0,\cdot) - h(\theta(0,\cdot) - \theta_e) = k\theta(1,\cdot) + h(\theta(1,\cdot) - \theta_e) = 0, \tag{5}$$

where θ^0 and ε^0 are initial data, $h \geq 0$ is a thermal exchange coefficient, and $\theta_e > 0$ is a constant environmental temperature. Note that the case $h = 0$ is included in the analysis. Namely, we are in the position of considering homogeneous Neumann boundary conditions on θ as well.

It is beyond our purposes to provide the reader with a comprehensive survey on the vast mathematical literature on one-dimensional thermoviscoelasticity with constant viscosity. On the other hand, let us at least mention the pioneering papers [4, 5], where the global solvability of the problem was first addressed. Moreover, one-dimensional existence and uniqueness results have been obtained in a variety of different settings [12, 20, 23]. More recently, the modeling of the thermo-mechanical evolution of the so-called shape memory materials has drawn new interest to the coupling of thermal and mechanical effects. As a consequence, some well-posedness results for thermoviscoelasticity can be recovered as special cases in the context of the study of solid-solid phase transformations[3, 10, 11, 15, 16]. Existence and uniqueness models in three space dimensions are comparably less studied[1, 2, 7, 21, 25]. As for the numerical treatment of one-dimensional thermoviscoelasticity, one has to mention some previous results in somehow different settings, see [13, 14, 19].

The main novelty of this paper is that we address the temperature-dependent viscosity situation. Of course, even quite ordinary materials show suitable temperature dependence in both the elastic and the viscous moduli. Our main focus here is, however, on materials which show a strongly distinguished mechanical behavior at different temperature regimes. This is especially the case of polymeric materials, among others. As a first step in the direction of describing the full temperature-dependent situation, we assume here that only the viscous modulus ν depends on the temperature θ while the elastic modulus is constant. Moreover, we impose ν to be uniformly positive for all temperatures. Of course, this prevents the system from degenerating into a purely elastic situation and crucially simplifies the analysis. On the other hand, we shall stress that no monotonicity is assumed on the viscous modulus.

In addition to some (by now quite classical) well-posedness analysis, our main aim consists in providing the detail of some effective fully implicit variable time-step discretization procedure. The latter turns out to be conditionally stable and convergent. Moreover, the discrete approximating temperature remains positive for all times. Finally, we are able to present some a priori error estimates of optimal order. Since no a priori constraints between consecutive time-steps are imposed, our error estimates ensure the possibility of implementing an adaptive time-stepping procedure.

This is the plan of the paper. We recall some notation and state our main assumptions in Section 2. The well-posedness results for the continuous problem are formulated in Section 3, where it is also directly checked that the temperature remains positive for all times. We give a proof of some continuous dependence result in Section 4. Then, our discrete scheme is introduced in Section 5 and proved to be (conditionally) uniquely solvable in Section 6. We establish the conditional stability of the approximation procedure in Section 7 and its convergence in Section 8. Finally, we present our a priori error bounds in Section 9.

2. Notation and assumptions

Let us start by setting some notation. We let

$$H := L^2(\Omega), \quad V := H^1(\Omega), \quad W := H^2(\Omega),$$

endowed with the usual scalar products. The reader is referred to [17] for the definitions and properties of Sobolev spaces. Let also $(\cdot,\cdot)$ denote the scalar product in H, $|\cdot|$ stand for both the norm in H and the modulus in $\mathbb{R}$, and $\|\cdot\|_E$ denote the norm in a generic normed space E.

We can now state our assumptions:

(A1) $\nu : \mathbb{R} \to \mathbb{R}$ is Lipschitz continuous, bounded, and uniformly positive, namely

$$0 < \nu_* \leq \nu(r) \leq \nu^* \quad \text{for all} \quad r \in \mathbb{R},$$

for some given ν_*, ν^*.

(A2) $\sigma \in H^1(0,T;H) \cap L^\infty(0,T;V)$.

(A3) $\theta^0 \in V$, $\theta > 0$ on $[0,1]$, and $\varepsilon^0 \in V$.

Taking into account the above discussion, some sufficient requirements on b and g in view of (A2) are

$$b \in H^1(0,T;L^1(\Omega)) \cap L^\infty(0,T;H), \quad g \in H^1(0,T).$$

3. Continuous problem

The continuous problem (CP) reads as follows:

CP: find $\theta \in H^1(0,T;H) \cap C([0,T],V) \cap L^2(0,T;W)$ and $\varepsilon \in H^2(0,T;H) \cap W^{1,\infty}(0,T;V)$ such that (1)-(2), (4)-(5) are almost everywhere fulfilled.

Indeed, given the above stated regularities, it is straightforward to check that (2) and (4) will actually be fulfilled everywhere. One has the following:

Theorem 1: Under assumptions (A1)-(A3), there exists a unique solution to problem CP.

The proof of this theorem shall follow from the results stated in Section 5 and proved in the subsequent sections.

Lemma 2: *Let $(\sigma_i, \theta_i^0, \varepsilon_i^0)$, $i = 1, 2$, fulfill (A2)-(A3) and $(\theta_i, \varepsilon_i)$ be the respective solutions to CP. Then, there exists a positive constant C, depending on $(\sigma_i, \theta_i^0, \varepsilon_i^0)$ and on data, such that*

$$\|\theta_1 - \theta_2\|_{C([0,T];H)\cap L^2(0,T;V)} + \|\varepsilon_1 - \varepsilon_2\|_{H^1(0,T;H)} \leq C\Big(|\theta_1^0 - \theta_2^0| + |\varepsilon_1^0 - \varepsilon_2^0| + \|\sigma_1 - \sigma_2\|_{L^2(0,T;H)}\Big). \tag{6}$$

Of course, the latter (local) Lipschitz continuous dependence result allows us to consider some weaker notion of solution, corresponding to weaker assumptions of data. On the other hand, let us mention that the uniqueness part of the statement of Theorem 1 will be an easy consequence of Lemma 2.

Moreover, let us point out that the techniques here developed could be easily tailored in order to include in the energy balance equation (1) some distributed heat source term of the form $\ell(x,t,\theta(x,t))$, taking into account, for instance, the Joule effect, radiations, microwaves etc. Here, $\ell : Q\times\mathbb{R} \to \mathbb{R}$ is a Carathéodory function, uniformly Lipschitz continuous with respect to θ, and such that there exists some parameter $\theta_\ell > 0$ such that, for all $\theta \leq \theta_\ell$ one has that $\ell(\cdot,\theta) \geq 0$ almost everywhere in Q. Indeed, with analogous assumptions we could consider also some non-homogeneous boundary conditions of the form

$$k\theta_x(\cdot,0) - h(\theta(\cdot,0) - \theta_e(\cdot)) = g_0(\cdot), \quad k\theta(\cdot,1) + h(\theta(\cdot,1) - \theta_e(\cdot)) = -g_1(\cdot),$$

where g_0, $g_1 \geq 0$ are given heat fluxes, and the variable external temperature $\theta_e(\cdot)$ is uniformly bounded away from zero and suitably behaved. We shall examine the original case of (1)-(5) for the sake of clarity.

Before closing this section, let us explicitly observe that, owing to (A3), the component θ of the solution of Problem CP remains positive for all times. Indeed, the function θ fulfills (4) and

$$c_s\dot{\theta} - k\theta_{xx} \geq -\alpha c\theta\dot{\varepsilon} \quad \text{a.e. in } Q. \tag{7}$$

Letting $\underline{\theta} := \min\{\min_{x\in[0,1]} \theta^0(x), \theta_e\}$, we define the function ζ by

$$\zeta(t) := -\left(\theta(t) - \underline{\theta}\exp\left(-\frac{\alpha c}{c_s}\int_0^t \|\dot{\varepsilon}(s)\|_{L^\infty(\Omega)}ds\right)\right)^- \quad \forall t \in [0,T].$$

Next, we multiply (7) by ζ and take the integral over $\Omega \times (0,t)$ for some $t \in (0,T]$ obtaining that

$$\frac{c_s}{2}|\zeta(t)|^2 + k\int_0^t |\zeta_x|^2 \leq \frac{c_s}{2}|\zeta(0)|^2 + \alpha c\int_0^t \|\dot{\varepsilon}(s)\|_{L^\infty(\Omega)}|\zeta(s)|^2 ds$$

$$+\alpha c\underline{\theta}\int_0^t \left(\|\dot{\varepsilon}(s)\|_{L^\infty(\Omega)} - \dot{\varepsilon}(s)\right)\exp\left(-\frac{\alpha c}{c_s}\int_0^s \|\dot{\varepsilon}(r)\|_{L^\infty(\Omega)}dr\right)\zeta(s)\,ds$$

$$\leq \frac{c_s}{2}|\zeta(0)|^2 + \alpha c\int_0^t \|\dot{\varepsilon}(s)\|_{L^\infty(\Omega)}|\zeta(s)|^2 ds.$$

Hence , since $\zeta(0) = 0$, the Gronwall lemma ensures that $\zeta \equiv 0$ and $\theta(t)$ turns out to be bounded from below by

$$\underline{\theta}\exp\left(-\frac{\alpha c}{c_s}\int_0^t \|\dot{\varepsilon}(s)\|_{L^\infty(\Omega)}ds\right) > 0$$

almost everywhere in Ω, and for all times $t \in [0,T]$.

4. Continuous dependence

We shall be concerned with the proof of Lemma 2. Let us start by fixing a positive constant C_0 such that

$$\|\theta_i\|_{L^\infty(Q)}, \|\dot{\varepsilon}_i\|_{L^\infty(Q)} \leq C_0 \quad \text{for } i = 1,2. \tag{8}$$

Indeed, as it will be clear in the sequel (see Lemma 4), we are entitled to choose C_0 depending on data and on $(\sigma_i, \theta_i^0, \varepsilon_i^0)$, for $i = 1,2$, but independently of $(\theta_i, \varepsilon_i)$.

Next, let us set $\overline{\theta} := \theta_1 - \theta_2$, $\overline{\varepsilon} = \varepsilon_1 - \varepsilon_2$ etc., and denote by C any positive constant, possibly varying from line to line, and depending on data, on C_0, and on $(\sigma_i, \theta_i^0, \varepsilon_i^0)$, for $i = 1,2$. Let us take the difference between (1) written for $(\theta_1, \varepsilon_1)$, and the same relation for $(\theta_2, \varepsilon_2)$. We get

$$c_s\dot{\overline{\theta}} - k\overline{\theta}_{xx} = -\alpha c\dot{\varepsilon}_1\overline{\theta} - \alpha c\theta_2\dot{\overline{\varepsilon}} + \left(\nu(\theta_1) - \nu(\theta_2)\right)\dot{\varepsilon}_1^2 + \nu(\theta_2)(\dot{\varepsilon}_1 + \dot{\varepsilon}_2)\dot{\overline{\varepsilon}}.$$

Hence, by multiplying by $\overline{\theta}$, integrating in space and time, and exploiting (A1) and (8), we readily obtain that

$$\frac{c_s}{2}|\overline{\theta}(t)|^2 + k\int_0^t |\overline{\theta}_x|^2 \leq C\left(|\overline{\theta}^0|^2 + \int_0^t |\overline{\theta}|^2 + \int_0^t |\dot{\overline{\varepsilon}}|^2\right). \tag{9}$$

On the other hand, by taking the difference of the respective relations (2), we may check that $\nu(\theta_1)\dot{\overline{\varepsilon}} = \overline{\sigma} - \big(\nu(\theta_1) - \nu(\theta_2)\big)\dot{\varepsilon}_2 - c\overline{\varepsilon} + \alpha c\overline{\theta}$. We now multiply by $\dot{\overline{\varepsilon}}$, integrate in space and time, and apply Gronwall's lemma, obtaining

$$\int_0^t |\dot{\overline{\varepsilon}}|^2 \leq C\left(|\overline{\varepsilon}^0|^2 + \int_0^t |\overline{\theta}|^2 + \int_0^t |\overline{\sigma}|^2\right).$$

Whence, looking back to (9), the assertion follows from another application of Gronwall's lemma.

5. Approximation

We shall be concerned with a variable time-step discretization of CP. To this aim, let us start by introducing the partition

$$\mathcal{P} := \{0 = t_0 < t_1 < \cdots < t_{N-1} < t_N = T\},$$

with variable time-step $\tau_i := t_i - t_{i-1}$, and let $\tau := \max_{1\leq i\leq N} \tau_i$ denote the diameter of the partition $\mathcal{P}$.

In the forthcoming analysis, the following notation will be extensively used: letting $\{u_i\}_{i=0}^N$ be a vector, we denote by $u_\mathcal{P}$ and $\overline{u}_\mathcal{P}$ two functions of the time interval $[0,T]$ which interpolate the values of the vector $\{u_i\}$ piecewise linearly and backward constantly on the partition $\mathcal{P}$, respectively. Namely,

$$\begin{aligned} u_\mathcal{P}(0) &:= u_0, \quad u_\mathcal{P}(t) := \gamma_i(t)u_i + \big(1-\gamma_i(t)\big)u_{i-1}, \\ \overline{u}_\mathcal{P}(0) &:= u_0, \quad \overline{u}_\mathcal{P}(t) := u_i, \qquad \text{for } t \in (t_{i-1}, t_i], \quad i = 1, \ldots, N, \end{aligned}$$

where $\gamma_i(t) := (t - t_{i-1})/\tau_i$ for $t \in (t_{i-1}, t_i]$, $i = 1, \ldots, N$. Moreover, we shall define a second vector $\{\delta u_i\}_{i=1}^N$ by $\delta u_i := (u_i - u_{i-1})/\tau_i$ (this is nothing but a discrete derivative).

Finally we introduce some approximation of the data. In order to fix ideas (other choices are actually possible), we approximate σ by $\overline{\sigma}_\mathcal{P}$ where $\sigma_i := \sigma(t_i)$ for $i = 0, 1, \ldots, N$. Let us stress that (A2) indeed ensures that

$$\|\dot{\sigma}_\mathcal{P}\|_{L^2(0,T;H)}, \ \|\overline{\sigma}_\mathcal{P}\|_{L^\infty(0,T;V)} \text{ are bounded independently of } \mathcal{P}, \tag{10}$$

$$\sigma_\mathcal{P} \to \sigma \text{ strongly in } H^1(0,T;H) \text{ as } \tau \to 0. \tag{11}$$

Moreover, and for the sake of completeness, we approximate $(\theta^0, \varepsilon^0)$ by some suitable $(\theta^0_{\mathcal{P}}, \varepsilon^0_{\mathcal{P}}) \in W \times V$, strongly converging to $(\theta^0, \varepsilon^0)$ in $V \times V$ as the diameter of the partition goes to zero, and such that $\theta^0_{\mathcal{P}} \geq \min \theta^0$ (this may be achieved by standard singular perturbation procedures).

Finally, let us fix an arbitrary positive parameter θ_* such that $\theta_* \leq \underline{\theta}$, and introduce the function $\varphi : \mathbb{R} \to [0, +\infty)$ by

$$\varphi(r) := \max\{0, \min\{r, 2r - \theta_*\}\}.$$

Of course φ is Lipschitz continuous of constant 2, and it is bounded by $r^+ = \max\{r, 0\}$ independently of the choice of θ_*.

We are interested in the following fully implicit scheme

$$c_s \delta\theta_i - k\theta_{i,xx} = -\alpha c \varphi(\theta_i)\delta\varepsilon_i + \nu(\theta_i)(\delta\varepsilon_i)^2 \quad \text{for } i = 1, \ldots, N, \tag{12}$$

$$\nu(\theta_i)\delta\varepsilon_i + c\varepsilon_i = \sigma_i + \alpha c\big(\varphi(\theta_i) - \theta_c\big) \quad \text{for } i = 1, \ldots, N, \tag{13}$$

$$\theta_0 = \theta^0_{\mathcal{P}}, \quad \varepsilon_0 := \varepsilon^0_{\mathcal{P}}, \tag{14}$$

$$k\theta_{i,x}(0) - h(\theta_i(0) - \theta_e) = k\theta_{i,x}(1) + h(\theta_i(1) - \theta_e) = 0 \quad \text{for } i = 1, \ldots, N. \tag{15}$$

In particular, we shall consider the following discrete problem.

DP: find $\{\theta_i, \varepsilon_i\}_{i=0}^N \in (W \times V)^{N+1}$ such that relations (12)-(15) are almost everywhere fulfilled.

As before, (13)-(15) will actually be fulfilled everywhere. We can prove the following results:

Lemma 3: *Under the above assumptions, problem DP has a unique solution for all τ small enough.*

Lemma 4: *Under the above assumptions, whenever τ is chosen to be small enough, the solution to DP fulfills*

$$\|\theta_{\mathcal{P}}\|_{H^1(0,T;H)\cap C([0,T];V)\cap L^2(0,T;W)} + \|\varepsilon_{\mathcal{P}}\|_{W^{1,\infty}(0,T;V)} + \|e_{\mathcal{P}}\|_{H^1(0,T;H)\cap L^\infty(Q)} \leq C_{stab}, \tag{16}$$

where the positive constant C_{stab} depends just on data and $\overline{e}_{\mathcal{P}} := \dot{\varepsilon}_{\mathcal{P}}$. Moreover, one also has that $\theta_{\mathcal{P}} \geq \theta_*/2$ everywhere.

Lemma 5: *Under the above assumptions and letting the diameter τ of partition $\mathcal{P}$ go to zero, we find (θ, ε) such that the following convergences* *hold:*

$$\theta_{\mathcal{P}} \to \theta \quad \text{weakly star in} \quad H^1(0,T;H) \cap L^\infty(0,T;V) \cap L^2(0,T;W)$$
$$\text{and strongly in} \quad C([0,T];H) \cap L^2(0,T;V), \tag{17}$$
$$\varepsilon_{\mathcal{P}} \to \varepsilon \quad \text{weakly star in} \quad W^{1,\infty}(0,T;V), \tag{18}$$
$$e_{\mathcal{P}} \to \dot{\varepsilon} \quad \text{weakly in} \quad H^1(0,T;H). \tag{19}$$

Moreover, whenever θ_* *is chosen to be small enough, the functions* (θ,ε) *are the unique solution to CP.*

Let us stress that the existence part of the statement of Theorem 1 is indeed a consequence of Lemma 5.

Lemma 6: *Under the above assumptions, whenever* τ *is chosen to be small enough, and letting* (θ,ε) *and* $\{\theta_i,\varepsilon_i\}_{i=0}^N$ *be the solutions to CP and DP, respectively, we have that*

$$\|\theta-\theta_{\mathcal{P}}\|_{L^2(0,T;H)} + \sup_{t\in[0,T]} \left\| \int_0^t (\theta - \overline{\theta}_{\mathcal{P}}) \right\|_V + \|\varepsilon-\varepsilon_{\mathcal{P}}\|_{H^1(0,T;H)}$$
$$\leq C_{err}\left(|\theta^0-\theta^0_{\mathcal{P}}| + |\varepsilon^0-\varepsilon^0_{\mathcal{P}}| + \|\sigma-\overline{\sigma}_{\mathcal{P}}\|_{L^2(0,T;H)} + \tau\right), \tag{20}$$
$$\|\theta-\theta_{\mathcal{P}}\|_{C([0,T];H)\cap L^2(0,T;V)}$$
$$\leq C_{err}\left(|\theta^0-\theta^0_{\mathcal{P}}| + |\varepsilon^0-\varepsilon^0_{\mathcal{P}}| + \|\sigma-\overline{\sigma}_{\mathcal{P}}\|_{L^2(0,T;H)} + \sqrt{\tau}\right) \tag{21}$$

for some positive constant C_{err} *depending on data. Moreover, if* $\nu \in W^{2,\infty}(\mathbb{R})$ *one has that*

$$\|\varepsilon-\varepsilon_{\mathcal{P}}\|_{H^1(0,T;V)} \leq C^*_{err}\left(\|\varepsilon^0-\varepsilon^0_{\mathcal{P}}\|_V + \|\sigma-\overline{\sigma}_{\mathcal{P}}\|_{L^2(0,T;V)} + \sqrt{\tau}\right) \tag{22}$$

for some positive constant C^*_{err} *depending on data and on* $\|\nu''\|_{L^\infty(\mathbb{R})}$.

We point out that the *a priori* estimate (20) is *optimal* with respect to the order of convergence, as we used a first order approximation of time derivatives. Moreover, the proof of Lemma 6 will show that the constants C_{err} and C^*_{err} solely depend on data and, in particular, exponentially on T.

6. Discrete well-posedness

This section is concerned with the proof of Lemma 3. We will proceed by induction. Indeed, we assume to be given the solution $(\theta_j,\varepsilon_j) \in W \times V$ for all $0 \leq j \leq i-1$, and we solve for $(\theta_i,\varepsilon_i) \in W \times V$. To this aim, let us define the set

$$K := \{\tilde{\theta} \in H \ : \ |\tilde{\theta}| \leq \kappa \ \text{ a.e in } \ \Omega\},$$

where the constant $\kappa \geq 1$ satisfies

$$|\sigma_i|,\ |\varepsilon_{i-1}|,\ 2(|\theta_{i-1}| + 2\theta_e),\ \theta_c\ \leq \kappa \ \text{ a.e. in } \ \Omega.$$

For any fixed $\tilde{\theta} \in K$, we may find a unique $\varepsilon \in H$ such that almost everywhere

$$\big(\nu(\tilde{\theta}) + \tau_i c\big)\varepsilon = \tau_i\big(\sigma_i + \alpha c(\varphi(\tilde{\theta}) - \theta_c)\big) + \nu(\tilde{\theta})\varepsilon_{i-1}.$$

Moreover, we readily check that

$$\nu_* \left|\frac{\varepsilon - \varepsilon_{i-1}}{\tau_i}\right| \leq |\sigma_i| + \alpha c(|\tilde{\theta}| + \theta_c) + c\tau_i \left|\frac{\varepsilon - \varepsilon_{i-1}}{\tau_i}\right| + c|\varepsilon_{i-1}|.$$

In particular, for τ small enough, say $\tau \leq \tau_*$, we can easily find a positive constant C_1 such that

$$\left|\frac{\varepsilon - \varepsilon_{i-1}}{\tau_i}\right| \leq C_1\kappa \ \text{ a.e. in } \ \Omega.$$

We have implicitly defined a mapping $T_1 : K \to L^\infty(\Omega)$ by $T_1(\tilde{\theta}) := \varepsilon$.

On the other hand, given $(\tilde{\theta}, \tilde{\varepsilon}) \in H \times L^\infty(\Omega)$, we may find the unique almost everywhere solution $\theta \in W$ to

$$c_s\theta - \tau_i k\theta_{xx} = \tau_i \left(-\alpha c\varphi(\tilde{\theta}) \left(\frac{\tilde{\varepsilon} - \varepsilon_{i-1}}{\tau_i}\right) + \nu(\tilde{\theta}) \left(\frac{\tilde{\varepsilon} - \varepsilon_{i-1}}{\tau_i}\right)^2\right) + c_s\theta_{i-1},$$

$$k\theta_x(0) - h(\theta(0) - \theta_e) = k\theta_x(1) + h(\theta(1) - \theta_e) = 0.$$

In particular, setting $T_2(\tilde{\theta}, \tilde{\varepsilon}) := \theta$, we have defined a mapping $T_2 : H \times L^\infty(\Omega) \to W$. Moreover, let us denote by $f := -\alpha c c_s^{-1}\varphi(\tilde{\theta})(\tilde{\varepsilon} - \varepsilon_{i-1})/\tau_i + c_s^{-1}\nu(\tilde{\theta})((\tilde{\varepsilon} - \varepsilon_{i-1})/\tau_i)^2$, $\mu := c_s^{-1}h$, fix $p > 1$, multiply the latter equation by $|\theta|^{p-2}\theta$ and integrate over Ω. Denoting by p' the conjugate exponent of p (i.e., $1/p + 1/p' = 1$), we readily get that

$$\begin{aligned}
&\int_\Omega |\theta|^p + \mu\tau_i \sum_{j=0}^{1} |\theta(j)|^p \\
&\leq \int_\Omega (\tau_i f + \theta_{i-1})\left(|\theta|^{p-2}\theta\right) + \mu\tau_i \sum_{j=0}^{1} \theta_e |\theta(j)|^{p-2}\theta(j) \\
&\leq \frac{1}{p'}\|\,|\theta|^{p-2}\theta\,\|^{p'}_{L^{p'}(\Omega)} + \frac{\mu\tau_i}{p'}\sum_{j=0}^{1} |\,|\theta(j)|^{p-2}\theta(j)|^{p'} \\
&\quad + \frac{1}{p}\|\tau_i f + \theta_{i-1}\|^p_{L^p(\Omega)} + \frac{2\mu\tau_i}{p}\theta_e^p.
\end{aligned}$$

Hence,

$$\|\theta\|_{L^p(\Omega)} \leq \|\tau_i f + \theta_{i-1}\|_{L^p(\Omega)} + (2\mu\tau_i)^{1/p}\theta_e$$
$$\leq \tau_i\|f\|_{L^\infty(\Omega)} + \|\theta_{i-1}\|_{L^\infty(\Omega)} + (2\mu\tau_i)^{1/p}\theta_e,$$

and, since $r^{1/p} \leq 1+r$ for all $r \geq 0$, $p \geq 1$, there exists a positive constant C_2, just depending on data and, in particular, on C_1, such that

$$\|\theta\|_{L^\infty(\Omega)} \leq \tau_i C_2 \kappa^2 + \|\theta_{i-1}\|_{L^\infty(\Omega)} + \theta_e + 2\mu\tau_i\theta_e.$$

Finally, it suffices to choose $\tau \leq \tau_{**} := \min\{\tau_*, 1/(2C_2\kappa), 1/(2\mu)\}$. In order to conclude that the mapping $T : K \to W$ defined by $T(\tilde{\theta}) := T_2(\tilde{\theta}, T_1(\tilde{\theta})) = \theta$ takes values in the non-empty, closed, and convex set K.

We shall now prove that T is a contraction in H for sufficiently small τ. To this aim, let us fix $\tilde{\theta}_1, \tilde{\theta}_2 \in K$ and define $\varepsilon_j := T_1(\tilde{\theta}_j)$, $\theta_j := T(\tilde{\theta}_j)$ for $j = 1, 2$, and $\overline{\varepsilon} := \varepsilon_1 - \varepsilon_2$, $\overline{\theta} := \theta_1 - \theta_2$, $\overline{\tilde{\theta}} := \tilde{\theta}_1 - \tilde{\theta}_2$. Next, we may write

$$\nu(\tilde{\theta}_1)\varepsilon_2 + \tau_i c \varepsilon_2 = \tau_i\big(\sigma_i + \alpha c(\varphi(\tilde{\theta}_2) - \theta_c)\big) + \big(\nu(\tilde{\theta}_1) - \nu(\tilde{\theta}_2)\big)\varepsilon_2 + \nu(\theta_2)\varepsilon_{i-1},$$

and, by taking the difference between the latter relation and the analogous for $j = 1$, and recalling the boundedness properties of T_1 and the Lipschitz continuity of ν, we readily obtain that

$$\nu_* \left|\frac{\varepsilon_1 - \varepsilon_2}{\tau_i}\right| = \frac{\nu_*}{\tau_i}|\overline{\varepsilon}| \leq \big(2\alpha c + \|\nu'\|_{L^\infty(\mathbb{R})} C_1 \kappa\big)|\overline{\tilde{\theta}}| + c|\overline{\varepsilon}| \quad \text{a.e. in } \Omega.$$

On the other hand, owing to the definition of T_2 we readily check that

$$|\overline{\theta}| \leq \frac{\tau_i}{c_s}\Big(2\alpha c C_1 \kappa |\overline{\tilde{\theta}}| + \alpha c \kappa |\overline{\varepsilon}|/\tau_i + \|\nu'\|_{L^\infty(\mathbb{R})} C_1^2 \kappa^2 |\overline{\tilde{\theta}}| + 2\nu^* C_1 \kappa |\overline{\varepsilon}|/\tau_i\Big)$$

almost everywhere in Ω. Hence, taking into account the above discussion, we may easily find a suitable small τ_{***} such that, for all $\tau \leq \min\{\tau_{**}, \tau_{***}\}$, the mapping T is a contraction in H, and therefore has a unique fixed point $\theta \in W$. In order to conclude the proof, it suffices to observe that, since φ is Lipschitz continuous, one has that $\varepsilon := T_1(\theta) \in V$.

Let us moreover stress that the above constructed solution of DP depends continuously on data in suitable spaces. We omit the proof of this fact for the sake of simplicity.

7. Stability

Let us now turn to the proof of Lemma 4 by providing some a priori bounds on the solution of DP. Throughout this section, the symbol C stands for any positive constant, possibly depending on data but independent of $\mathcal{P}$. Of course C may vary from line to line.

Lower bound. First of all, we shall prove by induction some uniform and positive lower bound on θ_i. In particular, let $\theta_{i-1} \geq \theta_*/2$ on $\overline{\Omega}$, where θ_* is exactly the parameter in the definition of φ. Then, we multiply (12) by the function $-\tau_i(\theta_i - \theta_*/2)^- := \tau_i \min\{\theta - \theta_*/2, 0\} \in V$, integrate over Ω, and obtain

$$c_s|(\theta_i - \theta_*/2)^-|^2 - \tau_i h \sum_{j=0}^{1} (\theta_i(j) - \theta_e)(\theta_i(j) - \theta_*/2)^- \\ \leq \alpha c \tau_i \int_\Omega \varphi(\theta_i)\delta\varepsilon_i(\theta_i - \theta_*/2)^-.$$

The right-hand side in the inequality above vanishes due to the definition of φ, and the assertion follows.

First estimate. Let us sum (12) and (13), multiplied by $\delta\varepsilon_i$. We multiply the result by τ_i, integrate over Ω, and take the sum for $i = 1, \ldots, m$, obtaining

$$c_s \int_\Omega \theta_m + \frac{c}{2}|\varepsilon_m|^2 \\ \leq \sum_{i=1}^{m} \tau_i \sigma_i \delta\varepsilon_i - \alpha c \theta_c \int_\Omega (\varepsilon_m - \varepsilon_{\mathcal{P}}^0) + c_s \int_\Omega \theta_{\mathcal{P}}^0 + \frac{c}{2}|\varepsilon_{\mathcal{P}}^0|^2 + 2hT\theta_e.$$

The first term in the above right-hand side can be controlled by means of a discrete integration by parts as follows

$$\sum_{i=1}^{m} \tau_i \sigma_i \delta\varepsilon_i \leq |\sigma_m|\,|\varepsilon_m| + |\sigma_0|\,|\varepsilon_{\mathcal{P}}^0| + \sum_{i=1}^{m} \tau_i |\delta\sigma_i|\,|\varepsilon_{i-1}|.$$

In particular, it is a standard matter to deduce the bounds

$$\|\overline{\theta}_{\mathcal{P}}\|_{L^\infty(0,T;L^1(\Omega))} + \|\overline{\varepsilon}_{\mathcal{P}}\|_{L^\infty(0,T;H)} \leq C. \tag{23}$$

For later convenience, let us observe that (13) yields in particular

$$\nu_*|\delta\varepsilon_i| \leq C\left(|\sigma_i| + \alpha c(|\theta_i| + \theta_c) + |\varepsilon_{\mathcal{P}}^0|\right) \quad \text{a.e. in } \Omega, \tag{24}$$

by taking τ small enough and exploiting the discrete Gronwall lemma. Hence, (10), (23) and (24) ensure that

$$\|\overline{e}_{\mathcal{P}}\|_{L^\infty(0,T;L^1(\Omega))} \leq C. \tag{25}$$

Second estimate. We multiply (12) by $-\tau_i/\theta_i$ (recall the lower bound on θ_i), integrate over Ω, and take the sum for $i = 1, \ldots, m$. Owing to the concavity of the logarithm, we get

$$-c_s \int_\Omega \ln \theta_i + k \sum_{i=1}^{m} \tau_i |\theta_{i,x}/\theta_i|^2 \leq -c_s \int_\Omega \ln \theta^0_{\mathcal{P}} + \alpha c \sum_{i=1}^{m} \tau^i \int_\Omega |\delta \varepsilon_i| + 2hT.$$

Thus, also using (23)-(25), the latter relation yields

$$\|(\ln \overline{\theta}_{\mathcal{P}})_x\|_{L^2(0,T;H)} \leq C. \tag{26}$$

Let us now recall the continuity of the injection of $W^{1,1}(\Omega)$ into $L^\infty(\Omega)$ and perform the standard computation

$$\begin{aligned} \|\theta_i\|_{L^\infty(\Omega)} &= \|\theta^{1/2}\|^2_{L^\infty(\Omega)} \leq 2\|\theta^{1/2}\|^2_{L^1(\Omega)} + 2\|(\theta^{1/2})_x\|^2_{L^1(\Omega)} \\ \leq 2|\theta^{1/2}|^2 &+ \frac{1}{2}\left(\int_\Omega \frac{(\theta_i)_x}{\theta_i}\theta_i^{1/2}\right)^2 \leq 2\|\theta_i\|_{L^1(\Omega)} + \frac{1}{2}|(\ln\theta_i)_x|^2|\theta^{1/2}|^2 \\ &= 2\|\theta_i\|_{L^1(\Omega)} + \frac{1}{2}|(\ln\theta_i)_x|^2\|\theta_i\|_{L^1(\Omega)}. \end{aligned}$$

Exploiting (23) and (26), we deduce that

$$\|\overline{\theta}_{\mathcal{P}}\|_{L^1(0,T;L^\infty(\Omega))} \leq C. \tag{27}$$

Finally, by elementary interpolation and (23) we conclude that

$$\|\overline{\theta}_{\mathcal{P}}\|_{L^2(0,T;H)} \leq C. \tag{28}$$

Third estimate. This is nothing but the energy estimate. Indeed, we multiply (12) by $\tau_i\theta_i$, integrate over Ω, and take the sum for $i = 1, \ldots, m$. Owing to (10) and (23)-(24), we readily obtain

$$\begin{aligned} &\frac{c_s}{2}|\theta_m|^2 + k\sum_{i=1}^{m} \tau_i|\theta_{i,x}|^2 + h \sum_{\substack{i=1,\ldots,m \\ j=0,1}} \tau_i\theta_i^2(j) \\ &\leq C\sum_{i=1}^{m} \tau_i \int_\Omega \theta_i^3 + h \sum_{\substack{i=1,\ldots,m \\ j=0,1}} \tau_i\theta_e\theta_i(j) + C. \end{aligned}$$

In order to control the cubic term above, we just exploit (23) and the compactness of the injection of V into L^∞. In particular, for all $\delta > 0$ there exists a positive constant C_δ such that

$$\begin{aligned} &\|\overline{\theta}_{\mathcal{P}}\|^3_{L^3(0,T;L^3(\Omega))} \leq \|\overline{\theta}_{\mathcal{P}}\|_{L^\infty(0,T;L^1(\Omega))}\|\overline{\theta}^2_{\mathcal{P}}\|_{L^1(0,T;L^\infty(\Omega))} \\ &\leq C\|\overline{\theta}_{\mathcal{P}}\|^2_{L^2(0,T;L^\infty(\Omega))} \leq \delta\|\overline{\theta}_{\mathcal{P},x}\|^2_{L^2(0,T;H)} + C_\delta\|\overline{\theta}_{\mathcal{P}}\|^2_{L^2(0,T;H)}. \end{aligned}$$

Hence, by choosing $\delta < k$ and τ small enough, and applying the discrete Gronwall lemma we deduce that

$$\|\overline{\theta}_{\mathcal{P}}\|_{L^\infty(0,T;H)\cap L^2(0,T;V)} \leq C. \tag{29}$$

Fourth estimate. We multiply (12) by $\tau_i \delta\theta_i$, integrate over Ω, and take the sum for $i = 1, \ldots, m$, getting

$$\begin{aligned} c_s \sum_{i=1}^{m} \tau_i |\delta\theta_i|^2 + \frac{k}{2}|\theta_{m,x}|^2 + \frac{h}{2}\sum_{j=0}^{1}(\theta_m(j) - \theta_e)^2 \\ \leq \frac{k}{2}|\theta^0_{\mathcal{P},x}|^2 + \frac{h}{2}\sum_{j=0}^{1}(\theta^0_{\mathcal{P}}(j) - \theta_e)^2 \\ + \sum_{i=1}^{m} \tau_i \big(\alpha c \|\theta_i\|_{L^4(\Omega)} \|\delta\varepsilon_i\|_{L^4(\Omega)} + \nu^* |(\delta\varepsilon_i)^2|\big)|\delta\theta_i| \\ \leq C + C\sum_{i=1}^{m}\tau_i \|\theta_i\|^4_{L^4(\Omega)} + \frac{c_s}{2}\sum_{i=1}^{m}\tau_i|\delta\theta_i|^2, \end{aligned}$$

where we also exploited (10) and (24). By observing that, thanks to (29) and some standard interpolation, one has that

$$\sum_{i=1}^{N} \tau_i \|\theta_i\|^4_{L^4(\Omega)} \leq \|\overline{\theta}_{\mathcal{P}}\|^2_{L^\infty(0,T;H)} \|\overline{\theta}_{\mathcal{P}}\|^2_{L^2(0,T;L^\infty(\Omega))} \leq C,$$

and we readily obtain the bound

$$\|\dot{\theta}_{\mathcal{P}}\|_{L^2(0,T;H)} + \|\overline{\theta}_{\mathcal{P}}\|_{L^\infty(0,T;V)} \leq C. \tag{30}$$

In particular, again using (10) and (24) we have

$$\|\overline{\theta}_{\mathcal{P}}\|_{L^\infty(Q)} + \|\dot{\varepsilon}_{\mathcal{P}}\|_{L^\infty(Q)} \leq C, \tag{31}$$

and, by comparison in (12) and standard elliptic estimates,

$$\|\overline{\theta}_{\mathcal{P}}\|_{L^2(0,T;W)} \leq C, \tag{32}$$

(actually much more is true).

Fifth estimate. We define $\delta\varepsilon_0 = e_0 := \big(\sigma_0 + \alpha c(\theta_0 - \theta_c) - c\varepsilon_0\big)/\nu(\theta_0)$, in such a way that relation (13) is fulfilled for $i = 0$ as well. Next, let us take the difference between (13) and the same relation, written for the index $i-1$, multiply by δe_i, integrate over Ω, and take the sum for $i = 1, \ldots, m$. One has

$$\nu_* \sum_{i=1}^{m} \tau_i |\delta e_i|^2$$

$$\leq \sum_{i-1}^{m} \tau_i \int_\Omega \left(\delta\sigma_i \delta e_i - \frac{\nu(\theta_i) - \nu(\theta_{i-1})}{\tau_i} e_{i-1}\delta e_i - c e_i \delta e_i + 2\alpha c|\delta\theta_i|\,|\delta e_i| \right)$$

$$\leq \frac{\nu_*}{2} \sum_{i=1}^{m} \tau_i |\delta e_i|^2 + C,$$

where we have just made use of (10), (30)-(31). In particular, we are entitled to deduce the bound

$$\|\dot{e}_{\mathcal{P}}\|_{L^2(0,T;H)} \leq C. \tag{33}$$

Sixth estimate. By exploiting (10), (30)-(31), and performing a comparison in (13), we readily get that $\|\bar{e}_{\mathcal{P},x}\|_{L^\infty(0,T;H)} \leq C(1 + \|\bar{\varepsilon}_{\mathcal{P},x}\|_{L^\infty(0,T;H)})$. Hence, at least for sufficiently small diameters τ, it suffices to apply the discrete Gronwall lemma in order to conclude the bound

$$\|\bar{e}_{\mathcal{P},x}\|_{L^\infty(0,T;H)} \leq C. \tag{34}$$

8. Convergence

Let us turn to the proof of Lemma 5. For the sake of later convenience, we rewrite (12)-(13) in the more compact form

$$c_s\dot{\theta}_{\mathcal{P}} - k\bar{\theta}_{\mathcal{P},x} = -\alpha c\varphi(\bar{\theta}_{\mathcal{P}})\bar{e}_{\mathcal{P}} + \nu(\bar{\theta}_{\mathcal{P}})(\bar{e}_{\mathcal{P}})^2, \tag{35}$$

$$\nu(\bar{\theta}_{\mathcal{P}})\bar{e}_{\mathcal{P}} + c\bar{\varepsilon}_{\mathcal{P}} = \bar{\sigma}_{\mathcal{P}} + \alpha c\big(\varphi(\bar{\theta}_{\mathcal{P}}) - \theta_c\big). \tag{36}$$

Next, we let the diameter τ of the partition $\mathcal{P}$ go to zero. Taking advantage of the estimate (16) and of well-known compactness results [22], we may actually find a pair (θ, ε) such that (possibly taking not relabeled subsequences) the following convergences hold

$$\theta_{\mathcal{P}} \to \theta \quad \text{weakly star in } \ H^1(0,T;H)\cap L^\infty(0,T;V)\cap L^2(0,T;W)$$
$$\text{and strongly in } \ C([0,T];H)\cap L^2(0,T;V), \tag{37}$$

$$\overline{\theta}_{\mathcal{P}} \to \theta \quad \text{weakly star in } \ L^\infty(0,T;V)\cap L^2(0,T;W)$$
$$\text{and strongly in } \ L^\infty(0,T;H)\cap L^2(0,T;V), \tag{38}$$

$$\varepsilon_{\mathcal{P}} \to \varepsilon \quad \text{weakly star in } \ W^{1,\infty}(0,T;V), \tag{39}$$

$$e_{\mathcal{P}} \to \dot{\varepsilon} \quad \text{weakly in } \ H^1(0,T;H)\cap L^2(0,T;V)$$
$$\text{and strongly in } \ C([0,T];H). \tag{40}$$

$$\overline{e}_{\mathcal{P}} \to \dot{\varepsilon} \quad \text{weakly in } \ L^2(0,T;V)$$
$$\text{and strongly in } \ L^\infty(0,T;H). \tag{41}$$

Hence, also owing to (11), we readily pass to the limit in (35)-(36) and obtain that the pair indeed (θ,ε) solves

$$c_s\dot{\theta} - k\theta_{xx} = -\alpha c\varphi(\theta)\dot{\varepsilon} + \nu(\theta)\dot{\varepsilon}^2, \tag{42}$$
$$\nu(\theta)\dot{\varepsilon} + c\varepsilon = \sigma + \alpha c\big(\varphi(\theta) - \theta_c\big), \tag{43}$$

as well as (4)-(5), at least almost everywhere. The latter problem may be proved to have unique solutions by arguing exactly as in the proof of Lemma 2. Hence, the above stated convergences turn out to hold for the whole sequences and not only for some extracted subsequences.

We shall conclude the proof of Lemma 5 by providing a suitable θ_* such that any solution (θ,ε) of the problem (42)-(43) fulfills $\theta \geq \theta_*$ everywhere. Then, owing to the definition of φ, in this case (1)-(2) and (42)-(43) coincide. Namely, we need to establish an a priori positive lower bound for θ. This actually follows along the very same lines of Section 3, i.e., upon suitably choosing the test function ζ as

$$\zeta(t) := -\left(\theta(t) - \underline{\theta}\exp\left(-\frac{\alpha c}{c_s}\int_0^t \frac{\varphi(\theta(s))}{\theta(s)}\|\dot{\varepsilon}(s)\|_{L^\infty(\Omega)}\,ds\right)\right)^- .$$

Omitting the details for the sake of brevity, it suffices to let

$$\theta_* := \underline{\theta}\,\exp\left(-\frac{\alpha c}{c_s}\int_0^T \frac{\varphi(\theta(s))}{\theta(s)}\|\dot{\varepsilon}(s)\|_{L^\infty(\Omega)}\,ds\right)$$

in order to prove that the above established solution to problem (42)-(43) fulfills $\theta \geq \theta_*$, and thus solves problem CP as well.

9. Error control

Let us now turn to the proof of Lemma 6. Henceforth, C will stand for any positive constant depending on data and, in particular, on $(\sigma, \theta^0, \varepsilon^0)$ and $(\overline{\sigma}_{\mathcal{P}}, \theta^0_{\mathcal{P}}, \varepsilon^0_{\mathcal{P}})$. To be more precise, let $C_3 > 0$ be such that

$$\|\sigma\|_{L^2(0,T;V)} + \|\overline{\sigma}_{\mathcal{P}}\|_{L^2(0,T;V)} + |\theta^0| + |\theta^0_{\mathcal{P}}| + \|\varepsilon^0\|_V + \|\varepsilon^0_{\mathcal{P}}\|_V \leq C_3.$$

Then, all the forthcoming constants indicated by C will possibly depend on $c_s, k, h, \alpha, c, \nu, T$, and C_3, but not on $\mathcal{P}$.

First of all, we shall observe that Lemma 4 entails in particular that

$$\|\theta_{\mathcal{P}} - \overline{\theta}_{\mathcal{P}}\|_{L^2(0,T;H)} + \|\varepsilon_{\mathcal{P}} - \overline{\varepsilon}_{\mathcal{P}}\|_{L^2(0,T;V)} \leq C\tau. \tag{44}$$

Let us now take the difference of equations (1) and (35), take the integral over $(0, s)$ for $s \in (0, T]$, multiply by the function $\theta - \overline{\theta}_{\mathcal{P}}$, and take the integral again over $\Omega \times (0, t)$ for $t \in (0, T]$. One has that

$$c_s \int_0^t |\theta - \theta_{\mathcal{P}}|^2 + \frac{k}{2} \left| \int_0^t (\theta - \overline{\theta}_{\mathcal{P}})_x \right|^2 + \frac{h}{2} \sum_{j=0}^{1} \left(\int_0^t (\theta - \overline{\theta}_{\mathcal{P}})(j) \right)^2 = \sum_{i=1}^{6} I_i(t), \tag{45}$$

where, letting $e := \dot{\varepsilon}$,

$$I_1(t) := c_s \int_0^t (\theta^0 - \theta^0_{\mathcal{P}}, \theta - \overline{\theta}_{\mathcal{P}}), \qquad I_2(t) := c_s \int_0^t (\theta - \theta_{\mathcal{P}}, \overline{\theta}_{\mathcal{P}} - \theta_{\mathcal{P}}),$$

$$I_3(t) := -\alpha c \int_0^t \left(\int_0^s (\theta - \overline{\theta}_{\mathcal{P}}) e, \theta - \overline{\theta}_{\mathcal{P}} \right),$$

$$I_4(t) := -\alpha c \int_0^t \left(\int_0^s (e - \overline{e}_{\mathcal{P}}) \overline{\theta}_{\mathcal{P}}, \theta - \overline{\theta}_{\mathcal{P}} \right),$$

$$I_5(t) := \int_0^t \left(\int_0^s (\nu(\theta) - \nu(\overline{\theta}_{\mathcal{P}})) e^2, \theta - \overline{\theta}_{\mathcal{P}} \right),$$

$$I_6(t) := \int_0^t \left(\int_0^s \nu(\overline{\theta}_{\mathcal{P}})(e + \overline{e}_{\mathcal{P}})(e - \overline{e}_{\mathcal{P}}), \theta - \overline{\theta}_{\mathcal{P}} \right).$$

Next, also owing to (16) and (44), it is a standard matter to check that

$$
\begin{aligned}
I_1(t) &\le \frac{c_s}{5}\int_0^t |\theta-\theta_{\mathcal{P}}|^2 + C\left(|\theta^0-\theta^0_{\mathcal{P}}|^2+\tau^2\right),\\
I_2(t) &\le \frac{c_s}{5}\int_0^t |\theta-\theta_{\mathcal{P}}|^2 + C\tau^2,\\
I_3(t)+I_5(t) &\le \frac{c_s}{5}\int_0^t |\theta-\theta_{\mathcal{P}}|^2 + C\left(\int_0^t \|\theta-\theta_{\mathcal{P}}\|^2_{L^2(0,s;H)}+\tau^2\right),\\
I_4(t)+I_6(t) &\le \frac{c_s}{5}\int_0^t |\theta-\theta_{\mathcal{P}}|^2 + C\left(\int_0^t \|e-\overline{e}_{\mathcal{P}}\|^2_{L^2(0,s;H)}+\tau^2\right),
\end{aligned}
$$

so that (45) reduces to

$$
\begin{aligned}
\int_0^t |\theta-\theta_{\mathcal{P}}|^2 + \left|\int_0^t (\theta-\overline{\theta}_{\mathcal{P}})_x\right|^2 + \sum_{j=0}^1\left(\int_0^t (\theta-\overline{\theta}_{\mathcal{P}})(j)\right)^2 \\
\le C\left(|\theta^0-\theta^0_{\mathcal{P}}|^2 + \int_0^t \|\theta-\theta_{\mathcal{P}}\|^2_{L^2(0,s;H)}\right)\\
+C\left(\int_0^t \|e-\overline{e}_{\mathcal{P}}\|^2_{L^2(0,s;H)}+\tau^2\right). \qquad (46)
\end{aligned}
$$

On the other hand, one easily computes that, for all $t\in(0,T]$,

$$
|(\varepsilon-\overline{\varepsilon}_{\mathcal{P}})(t)| \le C\left(|(\varepsilon-\varepsilon_{\mathcal{P}})(t)| + |\varepsilon^0-\varepsilon^0_{\mathcal{P}}| + \int_0^t |e-\overline{e}_{\mathcal{P}}|\right).
$$

Hence, using (44), we get that

$$
\int_0^t |\varepsilon-\overline{\varepsilon}_{\mathcal{P}}|^2 \le C\left(|\varepsilon^0-\varepsilon^0_{\mathcal{P}}|^2 + \int_0^t \|e-\overline{e}_{\mathcal{P}}\|^2_{L^2(0,s;H)}+\tau^2\right). \qquad (47)
$$

Let us now take the difference between (2) and (36), multiply it by $e-\overline{e}_{\mathcal{P}}$, and take the integral over $\Omega\times(0,t)$ for $t\in(0,T]$. We obtain

$$
\begin{aligned}
\int_0^t (\nu(\theta)(e-\overline{e}_{\mathcal{P}}), e-\overline{e}_{\mathcal{P}}) = \int_0^t (\sigma-\overline{\sigma}_{\mathcal{P}}, e-\overline{e}_{\mathcal{P}}) + \alpha c\int_0^t (\theta-\overline{\theta}_{\mathcal{P}}, e-\overline{e}_{\mathcal{P}})\\
-c\int_0^t (\varepsilon-\overline{\varepsilon}_{\mathcal{P}}, e-\overline{e}_{\mathcal{P}}) + \int_0^t ((\nu(\overline{\theta}_{\mathcal{P}})-\nu(\theta))\overline{e}_{\mathcal{P}}, e-\overline{e}_{\mathcal{P}}).
\end{aligned}
$$

Hence, again owing to (16), (44), and (47), one can handle the latter right-hand side in order to get that

$$\frac{\nu_*}{2}\int_0^t |e-\overline{e}_{\mathcal{P}}|^2 \le C\left(\int_0^t \left(|\sigma-\overline{\sigma}_{\mathcal{P}}|^2+|\theta-\theta_{\mathcal{P}}|^2+|\varepsilon-\overline{\varepsilon}_{\mathcal{P}}|^2\right)+\tau^2\right)$$
$$\le C\int_0^t \left(|\sigma-\overline{\sigma}_{\mathcal{P}}|^2+|\theta-\theta_{\mathcal{P}}|^2+\|e-\overline{e}_{\mathcal{P}}\|^2_{L^2(0,s;H)}\right)$$
$$+C\left(|\varepsilon^0-\varepsilon^0_{\mathcal{P}}|^2+\tau^2\right). \tag{48}$$

Combining now (46) and (48), we deduce that

$$\int_0^t |\theta-\theta_{\mathcal{P}}|^2+\left|\int_0^t(\theta-\overline{\theta}_{\mathcal{P}})_x\right|^2+\sum_{j=0}^{1}\left(\int_0^t(\theta-\overline{\theta}_{\mathcal{P}})(j)\right)^2+\int_0^t|e-\overline{e}_{\mathcal{P}}|^2$$
$$\le C\left(|\theta^0-\theta^0_{\mathcal{P}}|^2+|\varepsilon^0-\varepsilon^0_{\mathcal{P}}|^2+\int_0^t|\sigma-\overline{\sigma}_{\mathcal{P}}|^2\right)$$
$$+C\left(\int_0^t\|\theta-\theta_{\mathcal{P}}\|^2_{L^2(0,s;H)}+\int_0^t\|e-\overline{e}_{\mathcal{P}}\|^2_{L^2(0,s;H)}+\tau^2\right),$$

and (20) follows from an application of Gronwall's lemma.

As for (21), we simply take the difference of equations (1) and (35), multiply it by $\theta-\overline{\theta}_{\mathcal{P}}$, and take the integral over $\Omega\times(0,t)$ for $t\in(0,T]$, obtaining

$$\frac{c_s}{2}|(\theta-\theta_{\mathcal{P}})(t)|^2+k\int_0^t|(\theta-\overline{\theta}_{\mathcal{P}})_x|^2+\sum_{j=0}^{1}(\theta-\overline{\theta}_{\mathcal{P}})^2(j)$$
$$=\frac{c_s}{2}|\theta^0-\theta^0_{\mathcal{P}}|^2+\sum_{i=7}^{11}I_i(t), \tag{49}$$

where

$$I_7(t):=c_s\int_0^t(\dot\theta-\dot\theta_{\mathcal{P}},\overline{\theta}_{\mathcal{P}}-\theta_{\mathcal{P}}),\qquad I_8(t):=-\alpha c\int_0^t((\theta-\overline{\theta}_{\mathcal{P}})e,\theta-\overline{\theta}_{\mathcal{P}}),$$
$$I_9(t):=-\alpha c\int_0^t(\overline{\theta}_{\mathcal{P}}(e-\overline{e}_{\mathcal{P}}),\theta-\overline{\theta}_{\mathcal{P}}),$$
$$I_{10}(t):=\int_0^t((\nu(\theta)-\nu(\overline{\theta}_{\mathcal{P}}))e^2,\theta-\overline{\theta}_{\mathcal{P}}),$$
$$I_{11}(t):=\int_0^t(\nu(\overline{\theta}_{\mathcal{P}})(e+\overline{e}_{\mathcal{P}})(e-\overline{e}_{\mathcal{P}}),\theta-\overline{\theta}_{\mathcal{P}}).$$

Next, taking into account (16), (44), and the established (20), we control

$$I_7(t) \leq C\left(\int_0^t |\theta_{\mathcal{P}} - \overline{\theta}_{\mathcal{P}}|^2\right)^{1/2} \leq C\tau,$$

$$I_8(t) + I_{10}(t) \leq C\int_0^t |\theta - \overline{\theta}_{\mathcal{P}}|^2,$$

$$I_9(t) + I_{11}(t) \leq C\int_0^t \left(|\theta - \overline{\theta}_{\mathcal{P}}|^2 + |e - \overline{e}_{\mathcal{P}}|^2\right).$$

Finally, relation (49) yields

$$|(\theta - \theta_{\mathcal{P}})(t)|^2 + \int_0^t |(\theta - \overline{\theta}_{\mathcal{P}})_x|^2 + \sum_{j=0}^{1}(\theta - \overline{\theta}_{\mathcal{P}})^2(j)$$
$$\leq C\left(|\theta^0 - \theta^0_{\mathcal{P}}|^2 + |\varepsilon^0 - \varepsilon^0_{\mathcal{P}}|^2 + \int_0^t |\sigma - \overline{\sigma}_{\mathcal{P}}|^2 + \tau^2\right) + C\tau,$$

and estimate (21) follows.

Let us now check relation (22). To this aim we take the difference between (2) and (36), differentiate it with respect to space, multiply by $(e - \overline{e}_{\mathcal{P}})_x$, and take the integral over $\Omega \times (0,t)$ for $t \in (0,T]$, obtaining

$$\int_0^t (\nu(\theta)(e - \overline{e}_{\mathcal{P}})_x, (e - \overline{e}_{\mathcal{P}})_x) = \sum_{i=12}^{18} I_i(t), \tag{50}$$

where

$$I_{12} := \int_0^t ((\sigma - \overline{\sigma}_{\mathcal{P}})_x, (e - \overline{e}_{\mathcal{P}})_x), \qquad I_{13} := \alpha c\int_0^t ((\theta - \overline{\theta}_{\mathcal{P}})_x, (e - \overline{e}_{\mathcal{P}})_x),$$

$$I_{14} := -c\int_0^t ((\varepsilon - \overline{\varepsilon}_{\mathcal{P}})_x, (e - \overline{e}_{\mathcal{P}})_x),$$

$$I_{15} := \int_0^t ((\nu(\overline{\theta}_{\mathcal{P}}) - \nu(\theta))\overline{e}_{\mathcal{P},x}, (e - \overline{e}_{\mathcal{P}})_x),$$

$$I_{16} := \int_0^t ((\nu'(\overline{\theta}_{\mathcal{P}}) - \nu'(\theta))\overline{\theta}_{\mathcal{P},x}\overline{e}_{\mathcal{P}}, (e - \overline{e}_{\mathcal{P}})_x),$$

$$I_{17} := \int_0^t (\nu'(\theta)(\overline{\theta}_{\mathcal{P}} - \theta)_x\overline{e}_{\mathcal{P}}, (e - \overline{e}_{\mathcal{P}})_x),$$

$$I_{18} := \int_0^t (\nu'(\theta)\theta_x(\overline{e}_{\mathcal{P}} - e), (e - \overline{e}_{\mathcal{P}})_x).$$

Once more, owing to (16) and assuming $\nu \in W^{2,\infty}(\mathbb{R})$, one has that

$$I_{12} \leq \frac{\nu_*}{7}\int_0^t |(e - \overline{e}_{\mathcal{P}})_x|^2 + C\int_0^t |(\sigma - \overline{\sigma}_{\mathcal{P}})_x|^2,$$

$$
\begin{aligned}
I_{13} + I_{17}(t) &\leq \frac{\nu_*}{7} \int_0^t |(e - \bar{e}_{\mathcal{P}})_x|^2 + C \int_0^t |(\theta - \bar{\theta}_{\mathcal{P}})_x|^2, \\
I_{14}(t) &\leq \frac{\nu_*}{7} \int_0^t |(e - \bar{e}_{\mathcal{P}})_x|^2 + C \int_0^t |(\varepsilon - \bar{\varepsilon}_{\mathcal{P}})_x|^2, \\
I_{15}(t) &\leq \frac{\nu_*}{7} \int_0^t |(e - \bar{e}_{\mathcal{P}})_x|^2 + C \|\theta - \bar{\theta}_{\mathcal{P}}\|^2_{L^2(0,T;L^\infty(\Omega))}, \\
I_{16}(t) &\leq \frac{\nu_*}{7} \int_0^t |(e - \bar{e}_{\mathcal{P}})_x|^2 + C \|\theta - \bar{\theta}_{\mathcal{P}}\|^2_{L^\infty(0,T;H)}, \\
I_{18}(t) &\leq \frac{\nu_*}{7} \int_0^t |(e - \bar{e}_{\mathcal{P}})_x|^2 + C \int_0^t |e - \bar{e}_{\mathcal{P}}|^2.
\end{aligned}
$$

Finally, looking back to (50) and exploiting (20)-(21), one readily concludes (22).

Acknowledgments

The main part of this research was performed during a visit to France under the partial sponsorship of the MIUR-COFIN 2002 research program on *Free Boundary Problems in Applied Sciences.* The kind hospitality of the *Laboratoire de Mécanique et Génie Civil de Montpellier* is gratefully acknowledged.

References

[1] D. Blanchard and O. Guibé, Existence of a solution for a nonlinear system in thermoviscoelasticity. *Adv. Differential Equations* **5**, 1221–1252 (2000).

[2] E. Bonetti and G. Bonfanti, Existence and uniqueness of the solution to a 3D thermoviscoelastic system. *Electron. J. Differential Equations* (electronic), No. **50**, 15 pp. (2003).

[3] M. Brokate and J. Sprekels, *Hysteresis and Phase Transitions.* Springer, New York, 1996.

[4] C.M. Dafermos, Global smooth solutions to the initial-boundary value problem for the equations of one-dimensional nonlinear thermoviscoelasticity. *SIAM J. Math. Anal.* **13**, 397–408 (1982).

[5] C.M. Dafermos and L. Hsiao, Global smooth thermomechanical processes in one-dimensional nonlinear thermoviscoelasticity. *Nonlinear Anal.* **6**, 435–454 (1982).

[6] A.D. Drozdov, A constitutive model in linear thermoviscoelasticity of polymers based on the concept of cooperative relaxation. *Contin. Mech. Thermodyn.* **11**, 193–216 (1999).

[7] G.A. Francfort and P.M. Suquet, Homogenization and mechanical dissipation in thermoviscoelasticity. *Arch. Rational Mech. Anal.* **96**, 265–293 (1986).

[8] M. Frémond, *Non-smooth thermomechanics.* Springer-Verlag, Berlin, 2002.
[9] P. Germain, *Mécanique des milieux continus.* Masson, Paris, 1973.
[10] K.-H. Hoffmann, M. Niezgódka and S. Zheng, Existence and uniqueness to an extended model of the dynamical developments in shape memory alloys. *Nonlinear Anal.* **15**, 977–990 (1990).
[11] K.-H. Hoffmann and S. Zheng, Uniqueness for structural phase transitions in shape memory alloys. *Math. Methods Appl. Sci.* **10**, 145–1511 (1988).
[12] K.-H. Hoffmann and A. Żochowski, Existence of solutions of some nonlinear thermoelastic systems with viscosity. *Math. Methods Appl. Sci.* **15**, 187–204 (1992).
[13] S. Jiang, On a finite element method for the equations of one-dimensional nonlinear thermoviscoelasticity. *Computing* **40**, 111–124 (1988).
[14] S. Jiang, A C^1 finite element collocation method for the equations of one-dimensional nonlinear thermoviscoelasticity. *Math. Comput. Simulation* **31**, 227–239 (1989).
[15] P. Krejčí, J. Sprekels and U. Stefanelli, Phase-field models with hysteresis in one dimensional thermo-visco-plasticity. *SIAM J. Math. Anal.* **34**, 409–434 (2002).
[16] P. Krejčí, J. Sprekels, and U. Stefanelli, One-dimensional thermo-visco-plastic processes with hysteresis and phase transitions. *Adv. Math. Sci. Appl.* **13**, 695–712 (2003).
[17] J.-L. Lions and E. Magenes, *Non-homogeneus boundary value problems and applications. Vol.I.* Springer-Verlag, New York-Heidelberg, 1972.
[18] F. Meissonnier, R. Peyroux and A. Chrysochoos, Finite element analysis of strain localization induced by thermomechanical couplings in a kelvin-voigt behaviour model. In: *ESDA'96, ASME International Conference, Montpellier, 1-4.7.96*, Ed. R. Ohayon, A. Faghri and A. Chrysochoos (1996), p. 207.
[19] J. T. Oden and W.H. Armstrong, Analysis of nonlinear, dynamic coupled thermoviscoelasticity problems by the finite element method. *Computers & Structures* **1**, 603–621 (1971).
[20] R. Racke and S. Zheng, Global existence and asymptotic behavior in nonlinear thermoviscoelasticity. *J. Differential Equations* **134**, 46–67 (1997).
[21] Y. Shibata, Global in time existence of small solutions of nonlinear thermoviscoelastic equations. *Math. Methods Appl. Sci.* **18**, 871–895 (1995).
[22] J. Simon, Compact sets in the space $L^p(0,T;B)$. *Ann. Mat. Pura Appl. (4)* **146**, 65–96 (1987).
[23] S.J. Watson, Unique global solvability for initial-boundary value problems in one-dimensional nonlinear thermoviscoelasticity. *Arch. Ration. Mech. Anal.* **153**, 1–37 (2000).
[24] B. Wattrisse, J.M. Muracciole and A. Chrysochoos, Thermomechanical effects accompanying the localized necking of semi-crystalline polymers. *Int. J. Therm. Sci.* **41**, 422–427 (2002).
[25] J. Zimmer, Global existence for a nonlinear system in thermoviscoelasticity with nonconvex energy. *J. Math. Anal. Appl.* **292**, 589–604 (2004).

GLOBAL ATTRACTOR FOR THE WEAK SOLUTIONS OF A CLASS OF VISCOUS CAHN-HILLIARD EQUATIONS

Riccarda Rossi

Dipartimento di Matematica, Università di Brescia
Via Valotti 9, I-25133 Brescia, Italy
e-mail: riccarda.rossi@ing.unibs.it

We address the long-time behaviour of a class of viscous Cahn-Hilliard equations, modelling phase separation in mixtures and alloys. Specifically, we prove the existence of (a suitable notion of) the global attractor for the *weak solutions* of the so-called *generalized* viscous Cahn-Hilliard equation.

1. Introduction

This paper is concerned with the analysis of the long-time behavior of the (weak) solutions of the following fourth-order equation

$$\partial_t\chi - \Delta(\alpha(\partial_t\chi - \Delta\chi + \chi^3 - \chi)) = 0 \quad \text{in } \Omega \times (0, +\infty). \tag{1}$$

Here, Ω a bounded, connected domain in $\mathbb{R}^N$, $N = 1, 2, 3$, with smooth boundary Γ; $\alpha : D(\alpha) \subset \mathbb{R} \to \mathbb{R}$ is a (strictly) increasing, differentiable function, while the term $\chi^3 - \chi$ is the derivative of the double-well potential

$$\mathcal{W}(x) = \frac{(x^2-1)^2}{4}, \quad x \in \mathbb{R}. \tag{2}$$

In fact, (1) models the evolution of a phase separation process, to which a two-phase material (for instance, a binary alloy or a mixture), occupying the domain Ω, is subject. In this connection, the variable χ, usually referred to as *order parameter*, stands for the local concentration of one of the two components.

Equation (1) is indeed a *generalized* viscous Cahn-Hilliard equation: in fact, the viscous Cahn-Hilliard equation

$$\partial_t\chi - \kappa\Delta(\partial_t\chi - \Delta\chi + \chi^3 - \chi) = 0 \quad \text{in } \Omega \times (0, +\infty) \tag{3}$$

can be retrieved from (1) by choosing $\alpha(r) := \kappa r$ for every $r \in \mathbb{R}$, $\kappa > 0$ being the *mobility coefficient.*

On the other hand, (3) is itself a viscous regularization of the well-known Cahn-Hilliard equation

$$\partial_t \chi - \kappa\Delta(-\Delta\chi + \chi^3 - \chi) = 0 \quad \text{in } \Omega \times (0, +\infty), \tag{4}$$

originally introduced in the paper [6], dating back to 1961, for modelling phase separation phenomena driven by a quick quenching. Almost thirty years later, (3) was proposed in [16] to account for viscosity effects in the phase separation of polymeric systems. We refer to the survey [17] for a detailed overview of several analytical results on equations on (4) and (3), especially concerning well-posedness issues, and the dynamics of pattern formation. The asymptotic behavior as $t \to +\infty$ of the solutions to (suitable boundary value problems for) (4) and (3) has also been extensively tackled. In particular, we mention the seminal papers [9, 15] (see also Chap. III in [25]), for (4), and [8] for (3).

The generalized viscous Cahn-Hilliard equation (1) has instead been proposed in the much more recent paper [11], in the framework of a new approach to the modeling of phase separation phenomena. The key idea in [11] is that the work of the internal microforces, accompanying the changes of χ, should be taken into account. On these grounds, a new, unified derivation (also expounded in [12, 13]) of (4) and (3) is developed, leading to the *generalized Cahn-Hilliard equation*

$$\partial_t \chi - \operatorname{div}(\mathbf{M}(Z)\nabla(\delta\partial_t\chi - \Delta\chi + \mathcal{W}'(\chi))) = 0 \quad \text{in } \Omega \times (0, +\infty). \tag{5}$$

Here, $\delta \geq 0$ is a positive constant ($\delta > 0$ in the viscous case), and Z denotes the set of the independent constitutive variables the mobility tensor $\mathbf{M}$ (in general, a positive definite $N \times N$ matrix), is allowed to depend on. We refer to [12, 13], to the survey [14], to [19], and the references therein, for a detailed account of the results on well-posedness and long-time behavior of several kinds of Cahn-Hilliard equations derived from (5).

As far as (1) is concerned, its well-posedness has been indeed first tackled in [19], where it has also been shown that (1) can be obtained as a particular case of (5). Indeed, it is sufficient to choose a mobility tensor $\mathbf{M}$ only depending on the *modified* chemical potential

$$w = \partial_t\chi - \Delta\chi + \chi^3 - \chi,$$

through the formula $\mathbf{M} = \mathbf{M}(w) := \alpha'(w)\mathbf{I}$. Note that in [19] (1) has been formulated in the variables χ and $u := \alpha(w)$, which splits (1) into the system

$$\begin{aligned} \partial_t \chi - \Delta u = 0 \quad \text{in } \Omega \times (0, +\infty), \\ \partial_t \chi - \Delta \chi + \chi^3 - \chi = \rho(u) \quad \text{in } \Omega \times (0, +\infty), \end{aligned} \tag{6}$$

where ρ is the inverse of α. Further, two different choices for the nonlinearity α and, accordingly, two different choices for the boundary conditions on χ and u, have been considered therein. Indeed, the well-posedness of the initial-boundary value problem for (6) (supplemented with a source term f in the first equation), on a *finite time interval* $(0, T)$, has been first tackled in the case of

$$\text{a } \textit{bi-Lipschitz}, \text{ strictly increasing function } \alpha_1 : \mathbb{R} \to \mathbb{R}, \tag{7}$$

coupled with *Neumann* boundary conditions

$$\partial_n \chi = 0 \quad \text{and} \quad \partial_n u = 0 \quad \text{in } \Gamma \times (0, T), \tag{8}$$

on χ and u (non-homogeneous Neumann boundary conditions on u have also been considered). Secondly, the well-posedness of (6) on $(0, T)$ has been addressed in the case of a

$$\begin{aligned} &\text{a strictly increasing function } \alpha_2 : (a, +\infty) \to \mathbb{R}, \lim_{r \downarrow a} \alpha_2(r) = -\infty, \\ &\quad \text{with inverse } \rho_2 := \alpha_2^{-1} \text{ Lipschitz (but } \textit{not bi-Lipschitz}). \end{aligned} \tag{9}$$

In the latter case, (6) has been supplemented with a homogeneous Neumann boundary condition on χ, and with a third type (or Robin) condition on u, i.e. (here ω denotes a positive constant)

$$\partial_n \chi = 0 \quad \text{and} \quad -\partial_n u = \omega u \quad \text{in } \Gamma \times (0, T). \tag{10}$$

We refer to [19] for further modeling details, also motivating the choices (7) and (9) of the nonlinearities. Instead, we want to stress the main difficulties connected with the analysis of (6), both in the case of the choices (7)-(8), and of (9)-(10). Roughly speaking, due to the presence of the nonlinearity ρ in the second equation of (6), the *a priori* bounds on (suitable norms of) u (needed to pass to the limit in, possibly, an approximation procedure), can only be obtained through careful estimates. In the case of (7)-(8), these computations substantially rely on the fact that ρ_1 is bi-Lipschitz. On the other hand, in the case of (9)-(10) - when only the Lipschitz continuity of ρ is assumed -, the third type condition (10) on u plays a key role. Thus, well-posedness and, in the case of (9)-(10), also regularity results have been obtained in [19] for both boundary-value problems, supplemented with an initial datum $\chi_0 \in H^1(\Omega)$. In view of such results, the related solution operators form a *strongly continuous* semigroup on the phase space $H^1(\Omega)$. The existence of the universal attractor in this regularity framework shall be investigated in the paper [20].

Here, paralleling the analysis of [15] (see also [25]), we shall instead focus on the existence and the long-time behaviour of a class of *weak solutions* to (6). Such solutions in fact emanate from initial data for χ in the *bigger* phase space $L^4(\Omega)$ - note that such a choice is linked to the natural domain of the nonlinearly $\mathcal{W}$ (2), as in the approach of [18].

In this weaker regularity setting, the analytical drawbacks due to the nonlinearity ρ become more patent, and are less easy to overcome. Indeed (see Theorem 9 later on), we have been able to prove the existence of solutions to the initial-boundary value problem for (6), with an initial datum $\chi_0 \in L^4(\Omega)$ only when with the *same* kind of boundary condition on χ and u is considered. We shall denote by J the operator realizing $-\Delta$ in the first and in the second of (6) with such a boundary condition. Hence, the key assumption is that J be coercive w.r.t. the $H^1(\Omega)$-norm. For example, we may tackle within this framework homogeneous Dirichlet conditions on χ and u

$$\chi = u = 0 \quad \text{in } \Gamma \times (0, +\infty) \tag{11}$$

or third type boundary conditions on χ and u, namely

$$\partial_n \chi + \omega \chi = 0 \quad \text{and} \quad \partial_n u + \omega u = 0 \quad \text{in } \Gamma \times (0, +\infty), \tag{12}$$

ω being a positive constant.

Note that homogeneous *Neumann* boundary conditions on u and χ are more usual in the framework of the Cahn-Hilliard equations. Nonetheless (cf. [13]), Dirichlet conditions on u might for instance be considered in the case in which (6) models the propagation of a solidification front in a medium at rest with respect to the front; furthermore, the homogeneous Dirichlet conditions (11) have been already considered in [8] and [26].

Still, even in the setting of these easier-to-handle conditions, no uniqueness/continuous dependence results are available: (6) does not generate a semigroup, and the standard theory of [25] cannot be directly applied. Thus, we have exploited the new theory of *generalized semiflows*, recently proposed by J. M. Ball in [1, 2] for the study of the long-time behavior of solutions of differential problems lacking uniqueness. Let us stress that this theory is also related to the approach of *trajectory attractors* developed in [7] (see also [21, 22]). For instance, Ball's approach has been successfully applied to the study of the long-time behavior of the solutions of the

Navier-Stokes equation [1], and of the semilinear damped wave equation [3].

In fact, we have shown (cf. Theorem 11) that the set of the *weak solutions* has the structure of a generalized semiflow on the phase space $L^4(\Omega)$, according to the definition given in [1] (see Section 2). Hence, we have proved in Theorem 12 that the generalized semiflow of the *weak solutions* possesses a unique *global attractor*.

1.1. *Plan of the paper*

In Section 2, for the reader's convenience we briefly recall the main definitions and results on *generalized semiflows* and their long-time behavior developed in the papers [1, 2], of which we closely follow the outline. In Section 3, we introduce the definition of weak solution to (1) by means of a suitable variational formulation, which in fact subsumes several kinds of "coercive" boundary conditions (like (11) and (12)), see Remarks 4 and 6. Hence, we state our main theorems, whose proof is developed in Section 4.

2. Preliminaries: generalized semiflows

2.1. *Definition of generalized semiflow*

Let $(\mathcal{X}, d_{\mathcal{X}})$ be a (not necessarily complete) metric space; we recall that the *Hausdorff semidistance* $e(A,B)$ of two subsets $A,\ B \subset \mathcal{X}$ is given by $e(A,B) := \sup_{a\in A} \inf_{b\in B} d_{\mathcal{X}}(a,b)$, while the *Hausdorff distance* $\operatorname{dist}(A,B)$ of A and B is defined by $\operatorname{dist}(A,B) := \max\{e(A,B), e(B,A)\}$.

Definition 1: A *generalized semiflow* $\mathcal{S}$ on $\mathcal{X}$ is a family of maps $u : [0,+\infty) \to \mathcal{X}$ (referred to as "solutions"), satisfying:

(H1) *(Existence)* for any $v \in \mathcal{X}$ there exists at least one $u \in \mathcal{S}$ with $u(0) = v$;

(H2) *(Translates of solutions are solutions)* for any $u \in \mathcal{S}$ and $\tau \geq 0$, the map $u^\tau(t) := u(t+\tau)$, $t \in [0,+\infty)$, is in $\mathcal{S}$;

(H3) *(Concatenation)* for any u, $w \in \mathcal{S}$ and $t \geq 0$ with $w(0) = u(t)$, then $z \in \mathcal{S}$, z being the map defined by

$$z(\tau) := \begin{cases} u(\tau) & \text{if } 0 \leq \tau \leq t, \\ w(\tau - t) & \text{if } t < \tau; \end{cases}$$

(H4) *(Upper-semicontinuity w.r.t. initial data)* if $\{u_n\} \subset \mathcal{S}$ and $u_n(0) \to v$, then there exists a subsequence $\{u_{n_k}\}$ of $\{u_n\}$, and $u \in \mathcal{S}$, such that $u(0) = v$ and $u_{n_k}(t) \to u(t)$ for all $t \geq 0$.

2.2. *Continuity properties*

A generalized semiflow may enjoy the following properties:

(C1) Each $u \in \mathcal{S}$ is continuous from $(0, +\infty) \to \mathcal{X}$;
(C2) for any $\{u_n\} \subset \mathcal{S}$ with $u_n(0) \to v$, there exists a subsequence $\{u_{n_k}\}$ of $\{u_n\}$, and $u \in \mathcal{S}$, such that $u(0) = v$ and $u_{n_k} \to u$ uniformly on the compact subsets of $(0, +\infty)$;
(C3) Each $u \in \mathcal{S}$ is continuous from $[0, +\infty) \to \mathcal{X}$;
(C4) for any $\{u_n\} \subset \mathcal{S}$ with $u_n(0) \to v$, there exists a subsequence $\{u_{n_k}\}$ of $\{u_n\}$, and $u \in \mathcal{S}$, such that $u(0) = v$ and $u_{n_k} \to u$ uniformly on the compact subsets of $[0, +\infty)$.

Of course, (C3) $\Rightarrow$ (C1) and (C4) $\Rightarrow$ (C2).

In addition, the notions of *(positive) orbit* and *ω-limit* (both of a solution $u \in \mathcal{S}$ and of a subset $E \subset \mathcal{X}$), which are classical within the theory of universal attractors for dynamical systems (cf. [25]), can be extended to this multivalued setting. In the same way, the *attraction* and *invariance* properties can be suitably introduced. We refer to [1, Sec. 3] for all the precise definitions, which eventually lead to the definition of *global attractor*. Quoting [1], we say that a set $\mathcal{A} \subset \mathcal{X}$ is a *global attractor* for $\mathcal{S}$ if $\mathcal{A}$ is compact, invariant, and attracts all the bounded sets of $\mathcal{X}$.

2.3. *Compactness and dissipativity*

By definition, a generalized semiflow $\mathcal{S}$ is *eventually bounded* if for every bounded $B \subset \mathcal{X}$ there exists $\tau \geq 0$ such that $\gamma^{\tau}(B)$ is bounded; *point dissipative* if there exists a bounded set $B_0 \subset \mathcal{X}$ such that for any $u \in \mathcal{S}$ there exists $\tau \geq 0$ such that $u(t) \in B_0$ for all $t \geq \tau$; *asymptotically compact* if for any sequence $\{u_n\} \subset \mathcal{S}$ such that $\{u_n(0)\}$ is bounded and for any sequence $t_n \uparrow \infty$, the sequence $\{u_n(t_n)\}$ admits a convergent subsequence; *compact* if for any sequence $\{u_n\} \subset \mathcal{S}$ with $\{u_n(0)\}$ bounded there exists a subsequence $\{u_{n_k}\}$ such that $\{u_{n_k}(t)\}$ is convergent for any $t > 0$.

Remark 2: It has also been shown (cf. Prop. 3.2 in [2]), that if $\mathcal{S}$ is eventually bounded and compact, then it is also asymptotically compact.

Finally, we say that a global attractor $\mathcal{A}$ for $\mathcal{S}$ is *Lyapunov stable* if for any $\varepsilon > 0$ there exists $\delta > 0$ such that for any $E \subset \mathcal{S}$ with $e(E, \mathcal{A}) < \delta$, then $e(T(t)E, \mathcal{A}) < \varepsilon$ for all $t \geq 0$.

2.4. *Existence of the attractor*

In the end, we recall a criterion (cf. Thms. 3.3 and 6.1 in [1]) for the existence of a global attractor $\mathcal{A}$ for $\mathcal{S}$.

Theorem 3: A generalized semiflow $\mathcal{S}$ has a global attractor if and only if it is point dissipative and asymptotically compact; in that case, the attractor $\mathcal{A}$ is unique, it is the maximal compact invariant subset of $\mathcal{X}$, and it can be characterized as

$$\mathcal{A} = \cup\{\omega(B) \ : \ B \subset \mathcal{X}, \text{ bounded}\} = \omega(\mathcal{X}).$$

Moreover, if $\mathcal{S}$ complies with (C1) and (C4), $\mathcal{A}$ is Lyapunov stable.

3. Main results

3.1. *Notation*

We denote by H the space $L^2(\Omega)$, $((\cdot,\cdot)_H$ will be the scalar product and $\|\cdot\|_H$ the norm of $H)$, while V will denote a Hilbert space $V \subset H^1(\Omega)$, endowed with the norm $\|\cdot\|_V$ of $H^1(\Omega)$; $\|\cdot\|_{V'}$ will denote the norm on V', and $\langle\cdot,\cdot\rangle$ both the duality pairing between V' and V, and between $(H^1(\Omega))'$ and $H^1(\Omega)$. Hence,

$$\begin{gathered}(V, H, V') \text{ is a Hilbert triplet, and} \\ V \subset H \equiv H' \subset V' \quad \text{with dense and compact embeddings.}\end{gathered} \tag{13}$$

We also consider a continuous and symmetric bilinear form j on $H^1(\Omega) \times H^1(\Omega)$, and the associated operator $J : H^1(\Omega) \to (H^1(\Omega))'$. We will denote by the same symbol J also its restriction to V. We will suppose that

$$\begin{gathered}J \text{ is bounded, linear, symmetric, and coercive on } V, \text{ i.e.,} \\ \exists \gamma > 0 \ : \ \langle Ju, u\rangle \geq \gamma \|u\|_V^2 \ \forall u \in V.\end{gathered} \tag{14}$$

In the sequel, we will also assume that, for any differentiable non decreasing function $h : \mathbb{R} \to \mathbb{R}$, with $h(0) = 0$,

$$\langle J(h(v)), v\rangle \geq 0 \quad \forall v \in D_V(h) := \left\{v \in V \ : \ h(v) \in H^1(\Omega)\right\}. \tag{15}$$

Remark 4: For example, we may choose as V the space

$$H^1_{\Gamma_0}(\Omega) := \left\{v \in H^1(\Omega) \ : \ v|_{\Gamma_0} = 0\right\} \tag{16}$$

(where Γ_0 is a measurable subset of Γ, with positive measure, and $v|_{\Gamma_0}$ is the trace of v on Γ_0). Hence, an admissible choice for J is

$$\langle Ju, v\rangle := \int_\Omega \nabla u \nabla v \, dx \quad \forall u, v \in H^1_{\Gamma_0}(\Omega). \tag{17}$$

Another possibility is to choose

$$V := H^1(\Omega) \quad \text{and} \quad \langle Ju, v\rangle := \int_\Omega \nabla u \cdot \nabla v + \omega\langle u, v\rangle_\Gamma \quad \forall u, v \in V, \tag{18}$$

with $\omega > 0$, where $\langle\cdot,\cdot\rangle_\Gamma$ is the duality pairing between $H^{-1/2}(\Gamma)$ and $H^{1/2}(\Gamma)$. In fact, in both cases the coercivity (14) follows from Poincaré's inequality, and (15) trivially holds.

It follows from (14) J has a bounded inverse $J^{-1} : V' \to V$; henceforth, will consider on the space V (V', respectively), the scalar product $(\!(v_1, v_2)\!) := j(v_1, v_2) = \langle Jv_1, v_2\rangle =$ for every v_1, $v_2 \in V$ (the scalar product $(\!(w_1, w_2)\!)_* := j(J^{-1}(w_1), J^{-1}(w_2)) = \langle w_1, J^{-1}(w_2)\rangle$ for every w_1, $w_2 \in V'$, resp.). Accordingly, we will endow V and V' with the norms

$$\|v\|_V^2 := \langle Jv, v\rangle \quad \forall v \in V, \quad \|w\|_{V'}^2 := \langle w, J^{-1}(w)\rangle \quad \forall w \in V', \tag{19}$$

which are equivalent to the standard norms of $H^1(\Omega)$ and of $(H^1(\Omega))'$. In particular, $J : V \to V'$ is an isometry. We will also make use of the relation

$$\langle Jw, J^{-1}(v)\rangle = (w, v)_H \quad \forall w \in V, v \in H. \tag{20}$$

Finally, given a Banach space Y, $C^0_w([0,T];Y)$ will denote the space of the weakly continuous Y-valued functions on $[0,T]$.

3.2. *Assumptions on the data*

We suppose that $\alpha : (a, +\infty) \to \mathbb{R}$, $a \in \mathbb{R}$, is a differentiable and strictly increasing function, with

$$m := \inf_{r>a} \alpha'(r) > 0, \quad \text{and} \quad \lim_{r\downarrow a^+} \alpha(r) = -\infty; \tag{21}$$

Hence, its inverse function ρ is defined on the whole real line (up to a translation, we can suppose $\rho(0) = 0$), and

$$\begin{gathered}\rho : \mathbb{R} \to \mathbb{R} \text{ is strictly increasing, differentiable and Lipschitz,} \\ \text{with Lipschitz constant } \frac{1}{m}.\end{gathered} \tag{22}$$

3.3. *Statement of the problem*

First of all, we give a variational formulation of the initial-boundary value problem for (1) on the finite time interval $(0, T)$. We focus on the autonomous case, since we are interested in the analysis of the long-time behavior of the solutions. In this connection, towards the construction of a *generalized semiflow*, we shall also introduce the notion of *weak solution* to (6) on the half-line $(0, +\infty)$.

Problem 5: *Given* $\chi_0 \in H$*, find* $\chi \in H^1(0,T;V') \cap L^2(0,T;V) \subset C^0([0,T];H)$ *and* $u \in L^2(0,T;V)$ *fulfilling*

$$\partial_t \chi + Ju = 0 \quad \text{in } V', \quad \text{for a.e. } t \in (0,T), \tag{23}$$

$$\partial_t \chi + J\chi + \chi^3 - \chi = \rho(u) \quad \text{in } V', \quad \text{for a.e. } t \in (0,T), \tag{24}$$

$$\chi(x,0) = \chi_0(x) \quad \text{for a.e. } x \in \Omega. \tag{25}$$

Remark 6: Let us stress that the variational formulation of Problem 5 entails the same boundary conditions on χ and u, depending on the choice of the space V and of the operator J. For example, the choices (16) for V and $J := A|_V$ yield the homogeneous Dirichlet conditions (11); the choice (18) yields the third type conditions (12).

Now, for any pair of functions $\chi : [0, +\infty) \to L^4(\Omega)$ and $u : [0, +\infty) \to V$, let us set for a.e. $t \in (0, +\infty)$

$$\begin{aligned} \mathcal{V}(\chi, u)(t) &:= \frac{1}{4}\|\chi(t)\|^4_{L^4(\Omega)} \\ &+ \int_0^t \left(\|\chi^3(\tau)\|^2_H - \|\chi(\tau)\|^4_{L^4(\Omega)} - \left(\rho(u(\tau)), \chi^3(\tau)\right)_H \right) d\tau. \end{aligned} \tag{26}$$

We are now in the position to give the following

Definition 7: A function $\chi : [0, +\infty) \to L^4(\Omega)$ is called a *weak solution* to (6) if there exists $u : [0, +\infty) \to V$ such that for any $T > 0$

$$\begin{cases} \chi \in H^1(0,T;V') \cap L^2(0,T;V) \cap C^0_w([0,T];L^4(\Omega)) \cap L^6(0,T;L^6(\Omega)) \\ u \in L^2(0,T;V), \end{cases} \tag{27}$$

the pair (χ, u) fulfils

$$\partial_t \chi + Ju = 0 \quad \text{in } V', \quad \text{for a.e. } t \in (0, +\infty), \tag{28}$$

$$\partial_t \chi + J\chi + \chi^3 - \chi = \rho(u) \quad \text{in } V', \quad \text{for a.e. } t \in (0, +\infty), \tag{29}$$

and there exists a negligible set $\mathcal{N} \subset (0, T]$ such that (χ, u) complies with

$$\mathcal{V}(\chi, u)(t) \leq \mathcal{V}(\chi, u)(s) \quad \forall t \in [0,T], \quad \forall s \in [0,t] \setminus \mathcal{N}. \tag{30}$$

We denote by $\mathcal{S}_w$ the set of all weak solutions.

Note that for any weak solution (χ, u), the map $t \mapsto \mathcal{V}(u,\chi)(t)$ is in $L^\infty_{loc}(0,+\infty)$ thanks to the regularity (27). In the sequel (see e.g. the statement of Theorem 9), with a slight abuse of notation we shall also call *weak solution* any solution to Problem 5, on the finite-time interval $[0,T]$, having the regularity (27) and complying with the inequality (30).

Remark 8: Definition 7 highlights the auxiliary role of the variable u. For simplicity, in the sequel we will sometimes happen to call the *pair* (χ, u) weak/strong solution. Nonetheless, the most relevant solution component, also in view of the long-time behavior analysis, is in fact χ.

3.4. *Existence of weak solutions*

Theorem 9: Assume (13), (14), (15), (21), and

$$m > \frac{1}{3}. \tag{31}$$

Then, for any

$$\chi_0 \in L^4(\Omega) \tag{32}$$

Problem 5 admits a *weak solution* χ (in the sense of Definition 7), enjoying the *further regularity*

$$\chi \in C^0([0,T]; L^4(\Omega)). \tag{33}$$

Remark 10: Trivial changes in the proof of Theorem 9 yield the existence of solutions (in the sense of Definition 7) also when (23) is supplemented with a forcing term $F \in L^2(0,T;V')$ on the right-hand side.

3.5. *Generalized semiflow and long-time behavior of the weak solutions*

The long-time dynamics of the weak solutions to (6) will be analyzed in the phase space $\mathcal{X}_w := L^4(\Omega)$ (endowed with the metric of $\|\cdot\|_{L^4(\Omega)}$).

Theorem 11: Assume (13)-(14), (21), and (31). Then, the set of the weak solutions $\mathcal{S}_w$ is a generalized semiflow on $\mathcal{X}_w$, enjoying the continuity properties **(C3)** and **(C4)**.

Theorem 12: Assume (13)-(14), (21), and (31). Then, the generalized semiflow $\mathcal{S}_w$ possesses a unique global attractor $\mathcal{A}_w$ on $\mathcal{X}_w$, given by $\mathcal{A}_w := \omega(\mathcal{X}_w)$. Moreover, $\mathcal{A}_w$ is Lyapunov stable.

Henceforth, we adopt the convention of denoting by the same symbol C, whose meaning may vary even within the same line, (almost) all the constants occurring in the estimates.

4. Generalized semiflow and Global Attractor of the *weak* solutions

4.1. *Proof of Theorem 9*

We fix a sequence $\{\chi_0^k\}_k \subset V$ (which we shall suppose to suitably approximate the initial datum χ_0 of Problem 5), and for any $k \in \mathbb{N}$ we consider the following Cauchy problem:

Problem $\mathbf{P_k}$: *Find* $\chi_k \in H^1(0,T;H) \cap C^0([0,T];V) \cap L^2(0,T;W)$ *and* $u_k \in L^2(0,T;V)$ *fulfilling* (23), $\chi(0) = \chi_0^k$, *and*

$$\partial_t \chi_k + J\chi_k + \chi_k^3 - \chi_k = \rho(u_k) \quad \text{in } H, \quad \text{for a.e. } t \in (0,T). \tag{34}$$

A straightforward adaptation of the argument in the proof of Thm. 2.2 in [19] ensures that Problem $\mathbf{P_k}$ admits a unique solution (χ_k, u_k) for any $k \in \mathbb{N}$.

The following result yields our existence Theorem 9.

Proposition 13: *Under the assumptions* (13)-(15), (21), (31)-(32), *suppose that* $\{\chi_0^k\}_k \subset V$ *, and*

$$\chi_0^k \to \chi_0 \quad \text{in } L^4(\Omega). \quad \text{as } k \uparrow \infty. \tag{35}$$

Let $\{(\chi_k, u_k)\}$ *be the sequence of the solutions to* $\mathbf{P_k}$, *supplemented with the data* $\{\chi_0^k\}$.
Then, there exist $\chi \in H^1(0,T;V') \cap C^0([0,T];H) \cap L^2(0,T;V) \cap C^0_w([0,T];L^4(\Omega)) \cap L^6(0,T;L^6(\Omega))$, *and* $u \in L^2(0,T;V)$, *such that, up to the extraction of a subsequence, the following convergences hold as* $k \uparrow \infty$

$$\chi_k \rightharpoonup^* \chi \quad \text{in } H^1(0,T;V') \cap L^\infty(0,T;L^4(\Omega)) \cap L^2(0,T;V), \tag{36}$$

$$\chi_k \to \chi \quad \text{in } C^0([0,T];V') \cap L^2(0,T;L^{6-\varepsilon}(\Omega)) \cap L^p(0,T;L^4(\Omega)) \quad \text{for any } 0 < \varepsilon \le 5 \text{ and } 1 \le p < \infty, \tag{37}$$

$$\begin{cases} \chi_k(t) \to \chi(t) & \text{in } L^4(\Omega)) \quad \text{for a.e. } t \in (0,T), \\ \chi_k(t) \rightharpoonup \chi(t) & \text{in } L^4(\Omega)) \quad \text{for all } t \in [0,T], \end{cases} \tag{38}$$

$$u_k \rightharpoonup u \quad \text{in } L^2(0,T;V), \tag{39}$$

and (χ, u) *is a solution to Problem 5. Moreover, the pair* (χ, u) *fulfils* (30), *and* χ *has the additional regularity* (33).

Proof: First, we provide *a priori* estimates for the norms of the approximate solutions $\{(\chi_k, u_k)\}$ in suitable function spaces.

***A priori* estimates.** Let us test (23) by $J^{-1}(\rho(u_k))$, (34) by $J^{-1}(\partial_t\chi_k) + \chi_k^3$, add the resulting equations and integrate on $(0,t)$, for $t \in (0,T]$. Two terms cancel out; taking into account (19) and (20) (which for example yields

$$\langle J\chi_k(t), J^{-1}(\partial_t\chi_k(t))\rangle = (\partial_t\chi_k(t), \chi_k(t))_H = \frac{1}{2}\frac{d}{dt}\|\chi_k\|_H^2(t)$$

for a.e. $t \in (0,T)$), we obtain

$$\begin{aligned}
&\int_0^t (u_k(s), \rho(u_k(s)))_H\, ds + \int_0^t \|\partial_t\chi_k(s)\|_{V'}^2 ds\, ds \\
&+ \int_0^t \langle \chi_k^3(s), J^{-1}(\partial_t\chi_k(s))\rangle + \frac{1}{2}\|\chi_k(t)\|_H^2 + \frac{1}{4}\|\chi_k(t)\|_{L^4(\Omega)}^4 \\
&\quad + \int_0^t \langle J\chi_k(s), \chi_k^3(s)\rangle\, ds + \int_0^t \|\chi_k^3(s)\|_H^2 ds \\
&\quad = \frac{1}{2}\|\chi_0^k\|_H^2 + \frac{1}{4}\|\chi_0^k\|_{L^4(\Omega)}^4 \\
&+ \int_0^t \langle \chi_k(s), J^{-1}(\partial_t\chi_k(s))\rangle\, ds + \int_0^t \|\chi_k(s)\|_{L^4(\Omega)}^4\, ds \\
&\quad + \int_0^t (\rho(u_k(s)), \chi_k^3(s))_H ds.
\end{aligned} \tag{40}$$

Note that, in view of (21), $\int_0^t (u_k(s), \rho(u_k(s)))_H\, ds \geq m \int_0^t \|\rho(u_k(s))\|_H^2 ds$, while the sixth term on the left-hand side of (40) is positive thanks to (15).

On the other hand, by (31) we can fix a constant $\varepsilon_1 > 0$ such that $m > 1/3 + \varepsilon_1$, which guarantees, by elementary computations, that there exists a constant $\sigma_1 > 1/4$ such that $1 - (4(m-\varepsilon_1))^{-1} > \sigma_1$. Hence, we combine $\left|\int_0^t (\rho(u_k(s)), \chi_k^3(s))_H\, ds\right| \leq (m - \varepsilon_1)\int_0^t \|\rho(u_k(s))\|_H^2\, ds$ $+(4(m-\varepsilon_1))^{-1}\int_0^t \|\chi_k^3(s)\|_H^2\, ds$ with the following estimates:

$$\left|\int_0^t \langle \chi_k^3(s), J^{-1}(\partial_t\chi_k(s))\rangle\, ds\right| \leq \sigma_1 \int_0^t \|\chi_k^3(s)\|_{V'}^2\, ds$$

$$+\tfrac{1}{4\sigma_1}\int_0^t \|J^{-1}(\partial_t\chi_k(s))\|_V^2\, ds \leq \sigma_1 \int_0^t \|\chi_k^3(s)\|_H^2\, ds + \tfrac{1}{4\sigma_1}\int_0^t \|\partial_t\chi_k(s)\|_{V'}^2\, ds;$$

$$\left|\int_0^t \langle \chi_k(s), J^{-1}(\partial_t\chi_k(s))\rangle\, ds\right| \leq \kappa_1 \int_0^t \|\partial_t\chi_k(s)\|_{V'}^2\, ds + \tfrac{1}{4\kappa_1}\int_0^t \|\chi_k(s)\|_H^2\, ds,$$

where we have chosen $0 < \kappa_1 < 1 - 1/(4\sigma_1)$ (which is possible since $\sigma_1 > 1/4$).

We thus deduce from (40) and the above estimates that there exist three constants $c_1, c_2, c_3 > 0$ such that

$$\begin{aligned} c_1 \int_0^t \|\rho(u_k(s))\|_H^2 \, ds + c_2 \int_0^t \|\partial_t \chi_k(s)\|_{V'}^2 \, ds \\ + c_3 \int_0^t \|\chi_k^3(s)\|_H^2 \, ds + \frac{1}{2}\|\chi_k(t)\|_H^2 + \frac{1}{4}\|\chi_k(t)\|_{L^4(\Omega)}^4 \\ \leq \frac{1}{2}\|\chi_0^k\|_H^2 + \frac{1}{4}\|\chi_0^k\|_{L^4(\Omega)}^4 \\ + \frac{1}{4\kappa_1} \int_0^t \|\chi_k(s)\|_H^2 \, ds + \int_0^t \|\chi_k(s)\|_{L^4(\Omega)}^4 \, ds. \end{aligned} \tag{41}$$

Recalling (35), an easy application of the Gronwall's Lemma yields that there exists a constant $C \geq 0$, independent of $k \in \mathbb{N}$, such that

$$\|\chi_k\|_{L^\infty(0,T;L^4(\Omega)) \cap H^1(0,T;V')} + \|\chi_k^3\|_{L^2(0,T;H)} + \|\rho(u_k)\|_{L^2(0,T;H)} \leq C. \tag{42}$$

By comparison in (23), we get $\|J(u_k)\|_{L^2(0,T;V')} \leq C$, hence

$$\|u_k\|_{L^2(0,T;V)} + \|\rho(u_k)\|_{L^2(0,T;H^1(\Omega))} \leq C \quad \forall k \in \mathbb{N}, \tag{43}$$

the second estimate following from the Lipschitz continuity of ρ. Finally, testing (34) by χ_k and integrating on the interval $(0, t)$, we find

$$\begin{aligned} \frac{1}{2}\|\chi_k(t)\|_H^2 + \int_0^t \langle J(\chi_k(s)), \chi_k(s)\rangle \, ds + \int_0^t \|\chi_k(s)\|_{L^4(\Omega)}^4 \leq \\ \frac{1}{2}\|\chi_0^k\|_H^2 + \frac{1}{2}\int_0^t \|\rho(u_k(s))\|_{L^2(0,T;H)}^2 \, ds + \frac{3}{2}\int_0^t \|\chi_k(s)\|_H^2 \, ds. \end{aligned}$$

In view of (19) and (42)-(43), this entails

$$\|\chi_k\|_{L^2(0,T;V)} \leq C \quad \forall k \in \mathbb{N}. \tag{44}$$

Compactness for the approximate solutions. The *a priori estimates* (42) and (44) and the Lions-Aubin's theorem (see Thm.4, Cor.5 in [23]) yield that $\{\chi_k\}$ is relatively compact for the strong topologies of $C^0([0,T]; V')$ and of $L^2(0,T; L^{6-\varepsilon}(\Omega))$, for any $0 < \varepsilon \leq 5$. Note that the latter compactness property is due to the continuity of the embedding $V \subset H^1(\Omega)$, and to the compactness of $H^1(\Omega) \subset\subset L^{6-\varepsilon}(\Omega)$. On the other hand, $\{\chi_k\}$ is relatively weakly (weakly star) compact in $H^1(0,T;V') \cap L^\infty(0,T;L^4(\Omega)) \cap L^2(0,T;V)$. Analogously, thanks to (43) $\{u_k\}$ and $\{\rho(u_k)\}$ are relatively weakly compact in $L^2(0,T;V)$, and so is $\{\chi_k^3\}$ in $L^2(0,T;H)$.

Thus, there exists a subsequence, which we do not relabel, and a triplet (χ, u, ζ), such that $u \in L^2(0,T;V)$, $\zeta \in L^2(0,T;H^1(\Omega))$,

$\chi \in (H^1(0,T;V') \cap L^2(0,T;V) \cap L^\infty(0,T;L^4(\Omega))) \subset C^0([0,T];H) \cap C^0_w([0,T];L^4(\Omega))$ (the last inclusion following from Lemma III.1.4 in [24]), and the convergences (36), (37), and (39) hold. Combining the first of (37) with (35), we have $\chi(0) = \chi_0$. Besides,

$$\rho(u_k) \rightharpoonup \zeta \quad \text{in } L^2(0,T;H^1(\Omega)) \quad \text{as } k \uparrow \infty. \tag{45}$$

As for (38), the a.e. pointwise convergence in $L^4(\Omega)$ of χ_k is, upon the extraction of a subsequence, a consequence of the strong convergence of χ_k to χ in $L^2(0,T;L^4(\Omega))$. Note that this pointwise convergence and the bound for χ_k in $L^\infty(0,T;L^4(\Omega)$ imply the last of (37) via the dominated convergence theorem. On the other hand, the second of (38) can be proved by considering on the separable reflexive space $L^4(\Omega)$ the norm $|||\cdot|||$ which induces the weak convergence on the bounded subsets of $L^4(\Omega)$ (cf. e.g. Chap. III in [5] for the definition of $|||\cdot|||$). Hence, the estimates (42) and an argument of Ascoli-Arzelà type (cf. Cor. 5 in [23]) yield (38). Along the same subsequence, we also have

$$\begin{gathered} \chi_k^3 \to \chi^3 \quad \text{in } L^2(0,T;V'), \\ \text{whence } \chi^3 \in L^2(0,T;H), \text{ and } \quad \chi_k^3 \rightharpoonup \chi^3 \quad \text{in } L^2(0,T;H). \end{gathered} \tag{46}$$

(the latter conclusion follows from the fact that $\{\chi_k^3\}$ is weakly relatively compact in $L^2(0,T;H)$). In particular, $\chi \in L^6(0,T;L^6(\Omega))$. Indeed, to check (46), let us fix a constant $0 < \tau < 1/2$: then, elementary computations and Hölder's inequality yield that there exists a constant $0 < \varepsilon_\tau \leq 5$ such that for a.e. $t \in (0,T)$

$$\begin{aligned} &\left|\langle \chi_k^3(t) - \chi^3(t), v\rangle\right| \\ &\leq \|v\|_{L^6(\Omega)} \|\chi_k(t) - \chi(t)\|_{L^{6-\varepsilon_\tau}(\Omega)} \|\chi_k^2(t) + \chi^2(t)\|_{L^{\tau+3/2}(\Omega)} \quad \forall v \in V. \end{aligned}$$

Hence, we easily obtain that $\int_0^T \|\chi_k^3(t) - \chi^3(t)\|^2_{V'}\,dt$ is

$\leq \int_0^T \|\chi_k(t) - \chi(t)\|^2_{L^{6-\varepsilon_\tau}(\Omega)} \left(\|\chi_k^2(t)\|^2_{L^{\tau+3/2}(\Omega)} + \|\chi^2(t)\|^2_{L^{\tau+3/2}(\Omega)}\right) dt$

$\leq \int_0^T \|\chi_k(t) - \chi(t)\|^2_{L^{6-\varepsilon_\tau}(\Omega)} \left(\|\chi_k(t)\|^4_{L^{3+2\tau}(\Omega)} + \|\chi(t)\|^4_{L^{3+2\tau}(\Omega)}\right) dt$

$\leq \int_0^T \left(\|\chi_k\|^4_{L^\infty(0,T;L^4(\Omega))} + \|\chi\|^4_{L^\infty(0,T;L^4(\Omega))}\right) \int_0^T \|\chi_k(t) - \chi(t)\|^2_{L^{6-\varepsilon_\tau}(\Omega)}\,dt,$

which yields (46) thanks to (37) and (42).

Passage to the limit and proof of existence. It follows from (35) and (37) that $\chi(0) = \chi_0$; moreover, (36) and (39) also yield that (χ, u) solves (23). On the other hand, the convergences (36)-(37) and (45)-(46)

allow to pass to the limit in (34), thus obtaining

$$\partial_t \chi + J\chi + \chi^3 - \chi = \zeta \quad \text{in } V', \quad \text{for a.e. } t \in (0,T). \tag{47}$$

Hence, to conclude that (χ, u) is a solution to Problem 5, it is sufficient to prove that $\zeta = \rho(u)$. Note that ρ induces a maximal monotone graph on $L^2(0,T;H)$: by the theory of maximal monotone operators (cf. Prop.1.1, p.42, in [4]), it is then sufficient to show that

$$\limsup_{k\uparrow\infty} \int_0^t (\rho(u_k(s)), u_k(s))_H \, ds \le \int_0^t (\zeta(s), u_k(s))_H \, ds. \tag{48}$$

In fact, testing (23) by $J^{-1}(\rho(u_k))$ and (34) by $J^{-1}(\partial_t \chi_k)$, we obtain

$$\begin{aligned}
&\limsup_{k\uparrow\infty} \int_0^t (\rho(u_k(s)), u_k(s))_H \, ds \\
&= \limsup_{k\uparrow\infty} \left(- \int_0^t \|\partial_t \chi_k(s)\|_{V'}^2 \, ds - \frac{1}{2}\|\chi_k(t)\|_H^2 \right) + \frac{1}{2} \lim_{k\uparrow\infty} \|\chi_0^k\|_H^2 \\
&- \lim_{k\uparrow\infty} \int_0^t \left(\langle \chi_k^3(s), J^{-1}(\partial_t \chi_k(s)) \rangle - \langle \chi_k(s), J^{-1}(\partial_t \chi_k(s)) \rangle \right) \, ds \\
&\le - \int_0^t \|\partial_t \chi(s)\|_{V'}^2 \, ds - \frac{1}{2}\|\chi(t)\|_H^2 + \frac{1}{2}\|\chi_0\|_H^2 \\
&- \int_0^t \left(\langle \chi^3(s), J^{-1}(\partial_t \chi(s)) \rangle - \langle \chi(s), J^{-1}(\partial_t \chi(s)) \rangle \right) \, ds \\
&= - \int_0^t \langle \zeta(s), J^{-1}(\partial_t \chi_k(s)) \rangle = \int_0^t \langle J(u(s)), J^{-1}(\zeta(s)) \rangle \, ds,
\end{aligned} \tag{49}$$

whence (48) is a consequence of (20). Indeed, the intermediate inequality in (49) follows from (35) (note that this is the first point where we use the strong convergence of χ_0^k to χ_0 in H), and from the weak and strong convergences (36)-(38), also combined with (46). In particular, (38) yields $\liminf_{k\uparrow\infty} \|\chi_k(t)\|_H^2 \ge \|\chi(t)\|_H^2$ for all $t \in [0,T]$. The final identities in (49) are due to (47) and to (23). Thus, (45) becomes

$$\rho(u_k) \rightharpoonup \rho(u) \quad \text{in } L^2(0,T;V) \text{ as } k \uparrow \infty. \tag{50}$$

In the end, let us point out that, since χ fulfils (24) and $\rho(u) \in L^2(0,T;H)$, by standard regularity results for parabolic equations (see e.g. [25, Chap.3]), χ has the further regularity

$$\chi \in C^0([\delta,T]; L^4(\Omega)) \quad \forall\, \delta > 0. \tag{51}$$

Proof of (30) **and of** (33)**.** Due to (35), χ_k converges to χ in $L^4(\Omega)$ for $t = 0$; in general, by (38), there exists a negligible set $\overline{\mathcal{N}} \subset (0,T]$ outside which χ_k strongly converges pointwisely to χ in $L^4(\Omega)$. Hence, let us prove (30) for any $t \in (0,T]$ and $s \in (0,t] \setminus \overline{\mathcal{N}}$. Indeed, by (35), (38) and the weak lower semicontinuity of the norm, for any such t and s we have

$$\liminf_{k\uparrow\infty} \|\chi_k(t)\|^4_{L^4(\Omega)} \geq \frac{1}{4}\|\chi(t)\|^4_{L^4(\Omega)}, \quad \chi_k(s) \to \chi(s) \quad \text{in } L^4(\Omega). \tag{52}$$

Let us then test (34) by χ_k^3 and integrate on the interval (s,t). Taking the $\liminf_{k\uparrow\infty}$ of the resulting relation and developing the same computations as for (40), we obtain

$$\begin{aligned} 0 \geq \frac{1}{4} \liminf_{k\uparrow\infty} \|\chi_k(t)\|^4_{L^4(\Omega)} - \frac{1}{4}\lim_{k\uparrow\infty} \|\chi_k(s)\|^4_{L^4(\Omega)} \\ + \liminf_{k\uparrow\infty} \int_s^t \langle J\chi_k(\tau), \chi_k^3(\tau)\rangle d\tau \\ + \liminf_{k\uparrow\infty} \int_s^t \left(\|\chi_k^3(\tau)\|^2_H - \|\chi_k(\tau)\|^4_{L^4(\Omega)}\right) d\tau \\ - \lim_{k\uparrow\infty} \int_s^t (\rho(u_k(\tau)), \chi_k^3(\tau))_H d\tau \\ \geq \frac{1}{4}\|\chi(t)\|^4_{L^4(\Omega)} - \frac{1}{4}\|\chi(s)\|^4_{L^4(\Omega)} \\ + \int_s^t \left(\|\chi^3(\tau)\|^2_H + \|\chi(\tau)\|^4_{L^4(\Omega)} - (\rho(u(\tau)), \chi^3(\tau))_H\right) d\tau, \end{aligned} \tag{53}$$

which trivially yields (30). Note that (53) follows from (52), (15), (37), and by combining the strong convergence (46) and the weak one (50).

Owing to (51), in order to conclude (33) it is sufficient to show that

$$\forall \{t_n\} \subset (0,T] \text{ with } t_n \downarrow 0, \quad \chi(t_n) \to \chi_0 \text{ in } L^4(\Omega) \text{ as } n \uparrow \infty. \tag{54}$$

By (27), we have $\liminf_{n\uparrow\infty} \|\chi(t_n)\|^4_{L^4(\Omega)} \geq \|\chi_0\|^4_{L^4(\Omega)}$. On the other hand, (30) reads, for $t = t_n$ and $s = 0$,

$$\begin{aligned} \frac{1}{4}\|\chi(t_n)\|^4_{L^4(\Omega)} + \int_0^{t_n} \Big(\|\chi^3(\tau)\|^2_H - \|\chi(\tau)\|^4_{L^4(\Omega)} \\ . - \left(\rho(u(\tau)), \chi^3(\tau)\right)\big)_H\Big) d\tau \leq \frac{1}{4}\|\chi_0\|^4_{L^4(\Omega)}. \end{aligned} \tag{55}$$

Taking the lim sup as $n \uparrow \infty$ of (55), we deduce $\limsup_{n\uparrow\infty} \|\chi(t_n)\|^4_{L^4(\Omega)} \leq \|\chi_0\|^4_{L^4(\Omega)}$, whence (54). □

Remark 14: Let us stress that, while the continuity on $[\delta, T]$, for all $\delta > 0$, of the above weak solution χ is a consequence of the structure of equation (24), the continuity in $t = 0$ is obtained by combining the weak continuity

$\chi \in C^0_w([0,T]; L^4(\Omega))$ with the energy inequality (30). As a consequence, we conclude that

$$\text{any } \chi \in \mathcal{S}_w \text{ has the regularity } \quad \chi \in C^0([0,T]; L^4(\Omega)) \ \forall T > 0. \tag{56}$$

4.2. *Proof of Theorem 11*

Proof: It follows from Theorem 9 that for any $\chi_0 \in \mathcal{X}_w$ there exists a weak solution $\chi \in \mathcal{S}_w$ fulfilling $\chi(0) = \chi_0$, so that $\mathcal{S}_w$ complies with **(H1)**; **(C3)** holds thanks to Remark 14.

It can be readily checked that $\mathcal{S}_w$ fulfils **(H3)**; further, in view of (56) and of (30), the map $t \mapsto \mathcal{V}(u,\chi)(t)$ is continuous and non-increasing, so that **(H2)** holds.

We shall verify **(C4)**, which trivially entails **(H4)**. To this aim, we fix a sequence

$$\chi_0^n \text{ converging to } \chi^0 \text{ in } L^4(\Omega), \tag{57}$$

and a sequence $\{(\chi_n, u_n)\} \subset \mathcal{S}_w$ of weak solutions fulfilling $\chi_n(0) = \chi_0^n$. Like in the proof of Proposition 13, we test (28) by $J^{-1}(\rho(u_n))$, (29) by $J^{-1}(\partial_t \chi_n) + \chi_n^3$, integrate on $(0,t)$ for any $t > 0$, and add the resulting relations. Hence, we obtain the estimates (42)-(44) for the sequence $\{(\chi_n, u_n)\}$. Note that, on this level, testing (29) by χ_n^3 is only a formal estimate, since the regularity (27) for (χ_n, u_n) does not guarantee $\chi_n^3 \in V$. Nonetheless, $\chi_n^3 \in H$, and, by comparison in (29), we see that $\partial_t \chi_n + J\chi_n$ is in H as well. Replacing the test function χ_n^3 by its Yosida regularization, we can thus make the whole procedure rigorous.

Arguing exactly as in the same way as in the proof of Proposition 13, we conclude from the estimates (42)-(44) that there exists a limit pair (χ, u) with the regularity (27), and a subsequence $\{(\chi_{n_k}, u_{n_k})\}$ (extracted by a diagonal argument), along which the convergences (36)-(39) and (45)-(46) hold for any $T > 0$. This entails that $\chi(0) = \chi^0$, and that (χ, u) fulfils (28) and (29). Moreover, since (χ_{n_k}, u_{n_k}) complies with (30), passing to the limit we conclude that (χ, u) fulfils (30) as well (cf. with (53)), and it is thus a weak solution in the sense of Definition 7.

Therefore, thanks to (56), the map $t \mapsto \mathcal{V}(\chi, u)(t)$ is continuous on $[0, +\infty)$ and non increasing on $(0, +\infty)$.

In view of (57) and of the pointwise weak convergence (38), **(H4)** follows once we prove that

$$\limsup_{k \uparrow \infty} \|\chi_{n_k}(t)\|_{L^4(\Omega)} \le \|\chi(t)\|_{L^4(\Omega)} \quad \forall t > 0, \tag{58}$$

which can be obtained arguing in the same way as in the proof of Prop.7.4 in [1]. Namely, the convergences (36)-(38) and (46) and (50) yield

$$\mathcal{V}(\chi_{n_k}(t), u_{n_k}(t)) \to \mathcal{V}(\chi(t), u(t)) \quad \text{for a.e. } t \in [0, +\infty).$$

However, since both maps $t \mapsto \mathcal{V}(\chi_{n_k}(t), u_{n_k}(t))$ and $t \mapsto \mathcal{V}(\chi(t), u(t))$ are continuous and non increasing on $[0, +\infty)$ thanks to Remark 14, the latter convergence necessarily holds *for all* $t \in [0, +\infty)$, i.e.

$$\begin{gathered}\int_0^t \Big(\|\chi_{n_k}^3(\tau)\|_H^2 - \|\chi_{n_k}(\tau)\|_{L^4(\Omega)}^4 \\ - \big(\rho(u_{n_k}(\tau)), \chi_{n_k}^3(\tau)\big)_H \Big) \, d\tau + \frac{1}{4} \|\chi_{n_k}(t)\|_{L^4(\Omega)}^4 \\ \downarrow \\ \int_0^t \Big(\|\chi^3(\tau)\|_H^2 - \|\chi(\tau)\|_{L^4(\Omega)}^4 \\ - \big(\rho(u(\tau)), \chi^3(\tau)\big)_H \Big) \, d\tau + \frac{1}{4} \|\chi(t)\|_{L^4(\Omega)}^4\end{gathered} \tag{59}$$

$\forall\, t \in [0, +\infty)$. By a lower semicontinuity argument, (59) yields (58), as well as $\limsup_{k \uparrow \infty} \int_0^t \|\chi_{n_k}^3(\tau)\|_H^2 d\tau \leq \int_0^t \|\chi^3(\tau)\|_H^2 d\tau$, so that

$$\chi_{n_k}^3 \to \chi^3 \quad \textit{strongly} \text{ in } L^2(0, T; H). \tag{60}$$

Finally, (59), combined with (60) and (50) also entails that $\chi_{n_k}(t)$ converges to $\chi(t)$ in $L^4(\Omega)$ uniformly on the compact subsets of $[0, +\infty)$, which gives **(C4)**. □

Remark 15: Indeed, arguing along the same lines as in the proof of (33) in Theorem 9 and of the property continuity **(C4)**, it is possible to obtain the following characterization, which is the analogue of Prop. 7.4 in [1]. Namely, that the following conditions are equivalent:

i). $\mathcal{S}_w$ is generalized semiflow on $\mathcal{X}_w$;
ii). each $\chi \in \mathcal{S}_w$ is continuous from $(0, +\infty)$ to $\mathcal{X}_w$;
iii). each $\chi \in \mathcal{S}_w$ is continuous from $[0, +\infty)$ to $\mathcal{X}_w$.

4.3. *Proof of Theorem 12*

Proof: We shall check that $\mathcal{S}_w$ complies with the two necessary and sufficient (by Theorem 3) conditions for the existence of the global attractor:

namely, that

$$\mathcal{S}_w \text{ is asymptotically compact,} \tag{61}$$

$$\mathcal{S}_w \text{ is point dissipative.} \tag{62}$$

The Lyapunov stability of the attractor $\mathcal{A}_w$ shall then follow from the continuity properties **(C3)** and **(C4)** of $\mathcal{S}_w$.

Ad (62). (62) ensues from suitable *a priori* estimates on the weak solutions. Indeed, let us fix an element $(\chi, u) \in \mathcal{S}_w$, and let us test (28) by $J^{-1}(\rho(u))$, (29) by $J^{-1}(\partial\chi_t) + \chi^3$, and add the resulting relations. Thus, developing the same computations as for concluding (41) in the proof of Proposition 13, we deduce that for a.e. $t \in (0, +\infty)$

$$\begin{aligned} &c_1\|\rho(u(t))\|_H^2 + c_2\|\partial_t\chi(t)\|_{V'}^2 + c_3\|\chi^3(t)\|_H^2 + \frac{1}{2}\frac{d}{dt}\|\chi\|_H^2(t) \\ &+\frac{1}{4}\frac{d}{dt}\|\chi\|_{L^4(\Omega)}^4(t) \le \frac{1}{4\kappa_1}\|\chi(t)\|_H^2 + \|\chi(t)\|_{L^4(\Omega)}^4, \end{aligned} \tag{63}$$

with $c_1, c_2, c_3 > 0$ and $0 < \kappa_1 < 1$ as in (41), *independent* of t. Again, note that, on this level, such computations are only formal: on the other hand, they could be made rigorous by performing the regularization procedure hinted in the proof of Theorem 11.

Next, we test (29) by χ. Also taking into account (19), we get

$$\begin{aligned} &\frac{1}{2}\frac{d}{dt}\|\chi\|_H^2(t) + \|\chi(t)\|_V^2 + \|\chi(t)\|_{L^4(\Omega)}^4 \\ &\le \frac{c_1}{4}\|\rho(u(t))\|_H^2 + \left(\frac{1}{c_1} + 1\right)\|\chi(t)\|_H^2. \end{aligned} \tag{64}$$

We multiply (64) by 2 and add it to (63): easy computations entail

$$\begin{aligned} \frac{3}{2}\frac{d}{dt}\|\chi\|_H^2(t) + \frac{1}{4}\frac{d}{dt}\|\chi\|_{L^4(\Omega)}^4(t) + \|\chi(t)\|_{L^4(\Omega)}^4 &\le \left(\frac{2}{c_1} + 2 + \frac{1}{4\kappa_1}\right)\|\chi(t)\|_H^2 \\ &\le \frac{3}{4}\|\chi(t)\|_{L^4(\Omega)}^4 + K_1 - \frac{3}{2}\|\chi(t)\|_H^2, \end{aligned}$$

where we have also used Hölder's inequality, yielding that for any constants $r, R > 0$ there exists a constant $K = K(r, R, |\Omega|)$ such that $R\|v\|_H^2 \le r\|v\|_{L^4(\Omega)}^4 + K$ for any $v \in L^4(\Omega)$. Therefore, a differential form of the Gronwall Lemma (see e.g. Lemma 2.5 in [10]) gives for all $t > 0$

$$\Phi(\chi(t)) \le \Phi(\chi(0))\exp(-t) + K_1, \quad \Phi(\chi) := \frac{3}{2}\|\chi\|_H^2 + \frac{1}{4}\|\chi\|_{L^4(\Omega)}^4. \tag{65}$$

Defining $\mathcal{B}_0 := \{\chi \in \mathcal{X}_w \ : \ \|\chi\|_{L^4(\Omega)} \le 2K_1\}$, we deduce from (65) that for any weak solution (χ, u) there exists $\tau > 0$, depending on K_1 and $\chi(0)$, such that $\chi(t) \in \mathcal{B}_0$ for $t \ge \tau$.

Ad (61). It can be readily checked by (65) that $\mathcal{S}_w$ is eventually bounded. Owing to Remark 2, (61) follows once we prove that $\mathcal{S}_w$ is compact. To this aim, we fix a bounded sequence $\{\chi_0^n\} \subset L^4(\Omega)$, and denote by $\{(\chi_n, u_n)\}$ the sequence of the associated weak solutions. We may suppose that χ_0^n weakly converges, possibly on a subsequence, to some χ_∞^0 in $L^4(\Omega)$. Exactly as in the proof of the upper-semicontinuity property **(H4)** (cf. the proof of Theorem 11), upon regularizing we may repeat the estimates in the proof of Proposition 13 (which in fact only rely on the *boundedness* of the approximate initial data in $L^4(\Omega)$).

Thus, there exists a limit triplet $(\chi_\infty, u_\infty, \zeta_\infty)$, with (χ_∞, u_∞) as in (27) and $\zeta_\infty \in L^2(0,T;V)$, and a diagonal subsequence (which we do not relabel) of $\{(\chi_n, u_n)\}$, along which the convergences (36)-(39) and (45)-(46) hold for any $T > 0$. We aim at proving that, up to the extraction of a subsequence,

$$\chi_n(t) \to \chi_\infty(t) \quad \text{in } L^4(\Omega) \quad \forall t > 0. \tag{66}$$

Again, $\chi_\infty(0) = \chi_\infty^0$, the pair (χ_∞, u_∞) complies with (28). However, we cannot conclude anymore that (χ_∞, u_∞) fulfils (29) via (49), due to the only *weak* convergence of $\{\chi_0^n\}$: indeed, we just conclude that

$$\partial_t \chi_\infty + J\chi_\infty + \chi_\infty^3 - \chi_\infty = \zeta_\infty \quad \text{a.e. in } (0, +\infty). \tag{67}$$

The next step consists in showing that

$$\chi_n(t) \to \chi_\infty(t) \quad \text{in } H \quad \forall t > 0. \tag{68}$$

Indeed, (38) guarantees that for any $t > 0$, there exists $0 < \tau < t$ such that

$$\chi_n(\tau) \to \chi_\infty(\tau) \quad \text{in } H. \tag{69}$$

Hence, testing (29) (written for χ_n) by χ_n and integrating on the interval (τ, t), we get

$$\begin{aligned}
\limsup_{n\uparrow\infty} \frac{1}{2}\|\chi_n(t)\|_H^2 &\leq \limsup_{n\uparrow\infty} \frac{1}{2}\|\chi_n(\tau)\|_H^2 \\
+\int_\tau^t \Big(- \|\chi_n(r)\|_V^2 &- \|\chi_n(r)\|_{L^4(\Omega)}^4 \\
&+\|\chi_n(r)\|_H^2 + \langle \rho(u_n(r)), \chi_n(r)\rangle \Big) dr \\
&\leq \frac{1}{2}\|\chi_\infty(\tau)\|_H^2 + \int_\tau^t \langle \zeta_\infty(r), \chi_\infty(r)\rangle \, dr \\
-\int_\tau^t \Big(\|\chi_\infty(r)\|_V^2 &+ \|\chi_\infty(r)\|_{L^4(\Omega)}^4 - \|\chi_\infty(r)\|_H^2 \Big) dr \\
&= \frac{1}{2}\|\chi_\infty(t)\|_H^2
\end{aligned} \tag{70}$$

where the second passage follows from the convergences (36)-(37), (45), and (69), whereas the final equality is a consequence of (67). In view of the pointwise weak convergence (38), we deduce (68) from (70).

Now, we test (28) by $J^{-1}(\rho(u_n))$, (29) for $J^{-1}(\partial_t(\chi_n))$, add the resulting equations and integrate on (s,t), for *any* $0 < s \leq t < +\infty$. Thanks to (68), we can repeat the estimate (49) and get that

$$\limsup_{n\uparrow\infty} \int_s^t (\rho(u_n)(r), u_n(r))_H dr \leq \int_s^t (\zeta_\infty(r), u_\infty(r))_H dr \quad \forall\, 0 < s \leq t < \infty,$$

which yields that $\zeta_\infty(t) = \rho(u_\infty(t))$ for a.e. $t \in (0,+\infty)$, and the pair (χ_∞, u_∞) also fulfils (29).

Arguing as in the proof of Theorem 11, we manage to pass to the limit in (30) for (χ_n, u_n): thus, there exists a negligible set $\mathcal{N} \subset [0,+\infty)$ such that (χ_∞, u_∞) complies with (30) for all $t \in (0,+\infty)$ and for $s \in (0,t] \setminus \mathcal{N}$. Therefore, χ is a weak solution in the sense of Definition 7 on any half-line $[\delta,+\infty)$, $\delta > 0$. Arguing as in the proof of Theorem 11, we thus conclude that

$$\chi_n(t) \to \chi(t) \quad \text{in } L^4(\Omega) \quad \forall\, t \geq \delta,\ \forall\, \delta > 0,$$

whence (66). □

References

[1] J. Ball, Continuity properties and global attractors of generalized semiflows and the Navier-Stokes equations. *J. Nonlinear Sci.* **7**, 475–502 (1997).

[2] J. Ball, Erratum: "Continuity properties and global attractors of generalized semiflows and the Navier-Stokes equations". *J. Nonlinear Sci.* **8**, 233 (1998).

[3] J. Ball, Global attractors for damped semilinear wave equations. *Discrete Contin. Dyn. Syst.* **10**, 31–52 (2004).

[4] V. Barbu, *Nonlinear semigroups and differential equations in Banach spaces.* Noordhoff, Leyden, 1976.

[5] H. Brezis, *Analyse fonctionnelle.* Masson, Paris, 1983.

[6] J.W. Cahn, On spinodal decomposition. *Acta Metall.* **9**, 795–801 (1961).

[7] V.V. Chepyzhov and M.I. Vishik, Trajectory attractors for evolution equations. *C. R. Acad. Sci. Paris Sér. I Math.* **321**, 1309–1314 (1995).

[8] C.M. Elliott and A.M. Stuart, Viscous Cahn-Hilliard equation II. Analysis. *J. Differential Equations* **128**, 387–414 (1996).

[9] C.M. Elliott and S. Zheng, On the Cahn-Hilliard equations. *Arch. Rational Mech. Anal.* **96**, 339–357 (1986).

[10] C. Giorgi, M. Grasselli and V. Pata, Uniform attractors for a phase-field model with memory and quadratic nonlinearity. *Indiana Univ. Math. J.* **48**, 1395–1445 (1999).

[11] M.E. Gurtin, Generalized Ginzburg-Landau and Cahn-Hilliard equations based on a microforce balance. *Phys. D* **92**, 178–192 (1996).
[12] A. Miranville, Some generalizations of the Cahn-Hilliard equation. *Asymptot. Anal.* **22**, 235–259 (2000).
[13] A. Miranville, Generalized Cahn-Hilliard equations based on a microforce balance. *J. Appl. Math.* **4**, 165–185 (2003).
[14] A. Miranville, Some models of phase separation based on a microforce balance. (2004) (to appear).
[15] B. Nikolaenko, B. Scheurer, and R. Temam, Some global dynamical properties of a class of pattern formation equations. *Comm. Partial Differential Equations* **14**, 245–297 (1989).
[16] A. Novick-Cohen, *On the viscous Cahn-Hilliard equation.* In: *Material instability in continuum mechanics and related mathematical problems*, Ed. J. Ball, (Oxford Univ. Press, New York, 1988), p. 329.
[17] A. Novick-Cohen, The Cahn-Hilliard equation: mathematical and modeling perspectives. *Adv. Math. Sci. Appl.* **8**, 965-985 (1998).
[18] E. Rocca and G. Schimperna, Universal attractor for some singular phase transition systems. *Phys. D* **192**, 279–307 (2004).
[19] R. Rossi, On two classes of generalized viscous Cahn-Hilliard equations. *Commun. Pure Appl. Anal.* **4**, 405–430 (2005).
[20] R. Rossi, On the long-time behaviour of a class of viscous Cahn-Hilliard equations. (in preparation).
[21] G.R. Sell, Differential equations without uniqueness and classical topological dynamics. *J. Differential Equations* **14**, 42–56 (1973).
[22] G.R. Sell, Global attractors for the three-dimensional Navier-Stokes equations. *J. Dynam. Differential Equations* **8**, 1–33 (1996).
[23] J. Simon, Compact Sets in the space $L^p(0,T;B)$. *Ann. Mat. Pura Appl. (4)* **146**, 65-96 (1987).
[24] R. Temam, *Navier-Stokes equations. Theory and numerical analysis. Third Edition.* North-Holland, Amsterdam, 1984.
[25] R. Temam, *Infinite dimensional mechanical systems in mechanics and physics.* Springer, New York, 1988.
[26] S. Zheng and A. Milani, Global attractors for singular perturbations of the Cahn-Hilliard equations. *J. Differential Equations* **209**, 101–139 (2005).

STABILITY FOR PHASE FIELD SYSTEMS INVOLVING INDEFINITE SURFACE TENSION COEFFICIENTS

Ken Shirakawa

Department of Applied Mathematics,
Faculty of Engineering, Kobe University
1-1 Rokkodai, Nada, Kobe, 657-8501, Japan
E-mail: sirakawa@cs.kobe-u.ac.jp

In this paper, we shall consider a mathematical model of solid-liquid phase transition, which is described as a coupled system of two evolution equations. Here, the first equation is a kinetic equation of heat exchanges, and the second one is a type of Allen-Cahn equation, involving a spatially indefinite surface tension coefficient. Recently, in a known setting of the temperature, the author of [14] has studied the geometry of pattern formations which give some stability in the dynamical system generated by the single Allen-Cahn equation. In the main theorem of this paper, the result of [14] will be somehow extended to the framework of our model, in which the temperature is unknown.

1. Introduction

Let $\Omega \subset \mathbb{R}^2$ be a two dimensional bounded domain with a boundary $\Gamma := \partial\Omega$ given by a smooth Jordan curve, and let $\sigma : \overline{\Omega} \longrightarrow [0, +\infty)$ be a nonnegative and Lipschitz continuous function. In this paper, we shall deal with the following coupled system of evolution equations, of the form:

$$\begin{cases} (\theta + w)_t - \Delta(\theta + \mu\theta_t) = 0 \ \text{ in } Q := (0, +\infty) \times \Omega, \\ \theta = \theta_* \ \text{ on } \Sigma := (0, +\infty) \times \Gamma, \\ \theta(0, \cdot) = \theta_0 \ \text{ in } \Omega; \end{cases} \tag{1}$$

$$\begin{cases} w_t(t) + \partial\Phi_\sigma(w(t)) \ni w(t) + \theta(t) \ \text{ in } L^2(\Omega), \ t > 0, \\ w(0) = w_0 \ \text{ in } L^2(\Omega). \end{cases} \tag{2}$$

Let us denote by (S) the system $\{(1),(2)\}$, for simplicity. This system is a mathematical model for the dynamics of solid-liquid phase transitions. In the context, $\theta = \theta(t, x)$ is the relative temperature, and $w = w(t, x)$ is the nonconserved order parameter, describing the phase of the system in a point x at a time t. Indeed, w always takes a value on the closed interval $[-1, 1]$, and the threshold value 1 (resp. -1) indicates the physical state of pure liquid (resp. pure solid).

As it is well-known, solid-liquid phase transitions are observed on reaching a certain degree of temperature. Such a degree is often called "critical temperature": in this model, the degree is settled as zero for simplicity.

The first equation is the heat equation subject to Dirichlet boundary condition, where μ is a (small) positive constant satisfying $0 < \mu < 1$, and θ_* is a constant boundary source. In this equation, the heat flux is given by the Fourier law, with an additional (viscosity) time-relaxation term $\mu\theta_t$. This type of heat flux has been treated in [13], and it has been shown that the viscosity makes the temperature θ belong to a more regular class than the one in the heat equation.

The second equation (2) is a type of Allen-Cahn equation. In general, Allen-Cahn equation is known as a kinetic equation of phase field dynamics, that is derived as a gradient flow of an appropriate functional, called "free energy". In the case of our mathematical model, the free energy is given as the following functional:

$$w \in L^2(\Omega) \mapsto V_\sigma(w) + \int_\Omega \left\{ I_{[-1,1]}(w) - \frac{1}{2}|w + \theta|^2 \right\} dx; \tag{3}$$

where V_σ is a proper l.s.c. and convex function on $L^2(\Omega)$, defined as:

$$V_\sigma(z) := \inf \left\{ \liminf_{i \to +\infty} \int_\Omega \sigma |\nabla \zeta_i| \, dx \;\middle|\; \begin{array}{l} \{\zeta_i\} \subset W^{1,1}(\Omega), \\ \zeta_i \to z \text{ in } L^2(\Omega) \\ \text{as } i \to +\infty \end{array} \right\}, \quad z \in L^2(\Omega), \tag{4}$$

and $I_{[-1,1]}$ is the indicator function on the closed interval $[-1, 1]$. Additionally, we denote by Φ_σ the convex part of the free energy, more precisely:

$$\Phi_\sigma(z) := V_\sigma(z) + \int_\Omega I_{[-1,1]}(z) \, dx, \quad \text{for any } z \in L^2(\Omega), \tag{5}$$

and denote by $\partial\Phi_\sigma$ the subdifferential of Φ_σ in the topology of $L^2(\Omega)$.

The indicator function $I_{[-1,1]}$ is inserted to force the range of the parameter into the closed interval $[-1, 1]$. However, let us further note that the indicator function makes the density:

$$w \in \mathbb{R} \mapsto I_{[-1,1]}(w) - \frac{1}{2}|w + \theta|^2 \quad \text{as in (3)}$$

have a graph of double-well type. It implies that the indicator function also contributes to the bipolarization of stable states around the critical temperature.

On the other hand, the functional V_σ characterizes the geometry of interfaces (interfacial energy). In our model, the interfacial energy is given by a generalized form of the total variation energy, proposed in Chapter VII of [16] and studied in [7, 8, 9, 13, 14, 15, 12]. Such a generalization is motivated to represent some effects of geometrical constraints for masses in pattern formations. But, unfortunately, the expression of this interfacial energy would be too simple to represent the so-called "anisotropic effect". Therefore, it could be said that the mathematical model (S) is just at transitional level of the research.

In the first part of this paper, we shall study the key properties for solutions of our system (S). Then, assuming appropriate conditions for the indefinite coefficient σ, it will be concluded that:

(a) $\theta(t) \to \theta_*$, uniformly on $\overline{\Omega}$, as $t \to +\infty$;
(b) any cluster point (ω-limit point) w_* of the orbit $\{w(t) \,|\, t \geq 0\}$ (as $t \to +\infty$), satisfies the following inclusion

$$\partial \Phi_\sigma(w_*) \ni w_* + \theta_* \quad \text{in } L^2(\Omega). \tag{6}$$

The above items (a) and (b) imply that any stationary point in the dynamical system generated by (S) is given by a pair $\{\theta_*, w_*\}$ fulfilling the inclusion (6). In view of this, we call (6) the steady-state problem, throughout this paper.

In the second part of this paper, we will focus on the analysis of stability (stability analysis) for each stationary point $\{\theta_*, w_*\}$ in the dynamical system generated by the system (S). Then, since the asymptotic profile of the temperature is perfectly characterized by (a), our interest is naturally to know the concrete geometric pattern represented by solutions of the steady-state problem (6) (steady-state pattern). To this end, we first refer to the results obtained in [14]. Although the stability analysis of [14] has been performed in the framework of the single Allen-Cahn equation (2) (i.e., in a known setting of the temperature), the author of [14] has specified the expression of the coefficient σ to exemplify the possible profile of the stable steady-state pattern (stable pattern) in the corresponding dynamical system. In the main theorem of this paper, the result of [14] will be somehow extended to the framework of the system (S), in which the temperature is unknown. Consequently, it will be concluded in the main theorem that the stable patterns reported in [14] will also show some stability in the dynamical system generated by the system (S).

Notation. In general, for an abstract Banach space X, we denote by $|\cdot|_X$ the norm of X. Here, let H be an abstract Hilbert space. Then, we denote by $(\cdot,\cdot)_H$ the inner product in H. Furthermore, for a proper l.s.c. and convex function Φ defined on H, we use the notation $D(\Phi)$, $\partial\Phi$ and $D(\partial\Phi)$ to indicate the effective domain of Φ, the subdifferential of Φ and the domain of $\partial\Phi$, respectively.

2. Preliminaries

For any $m \in \mathbb{N}$, we denote by $\mathcal{L}^m$ the m-dimensional Lebesgue measure; in each dimension, we will use this measure when it is specified nothing particular. Also, for any $m \in \mathbb{N}$ we denote by $\mathcal{H}^m$ the m-dimensional Hausdorff measure.

Throughout this paper, let $\Omega \subset \mathbb{R}^2$ be a two-dimensional bounded domain with boundary $\Gamma := \partial\Omega$ given by a smooth Jordan curve. Let $\mathcal{B}(\Omega)$ be the class of all Borel subsets in Ω, and for any $B \in \mathcal{B}(\Omega)$, let χ_B be the characteristic function of B.

For any open subset $D \subset \Omega$, let us denote by D^{ex} the external open part $\Omega \setminus \overline{D}$ of D, and for any (small) $\rho \geq 0$ let us put

$$\partial D[\rho] := \left\{ x \in \Omega \;\middle|\; \mathrm{dist}\,(x, \partial D \cap \Omega) \leq \rho \right\};$$

and we set $\partial D[\rho] = \emptyset$, if $\partial D \cap \Omega = \emptyset$.

Let $\sigma : \overline{\Omega} \longrightarrow [0, +\infty)$ be a given nonnegative and Lipschitz continuous function, and let V_σ be the functional on $L^2(\Omega)$ defined by (4). The functional V_σ is known to be the lower-semicontinuous extension of the following functional:

$$z \in W^{1,1}(\Omega) \mapsto \int_\Omega \sigma(x)|\nabla z(x)|\, dx$$

onto the space $L^2(\Omega)$. Therefore, we easily check that V_σ is proper l.s.c. and convex on $L^2(\Omega)$.

In particular, when $\sigma \equiv 1$ on $\overline{\Omega}$ the functional V_1 coincides with the L^2-restriction of the total variation functional V, which is defined on $L^1(\Omega)$ by:

$$V(z) := \sup \left\{ \int_\Omega z \,\mathrm{div}\, \varphi \, dx \;\middle|\; \begin{array}{l} \varphi \in C^1(\Omega)^2 \text{ with} \\ \text{a compact support,} \\ \text{and } |\varphi| \leq 1 \text{ on } \overline{\Omega} \end{array} \right\}, \quad z \in L^1(\Omega).$$

In this context, any function $z \in L^1(\Omega)$ satisfying $V(z) < +\infty$ is called a bounded variation function, or simply BV-function, and the space of all BV-functions is denoted by $BV(\Omega)$. Also, the space $BV(\Omega)$ is known to be a Banach space with the following norm:

$$|z|_{BV(\Omega)} := |z|_{L^1(\Omega)} + V(z) \quad \text{for any } z \in BV(\Omega).$$

Moreover, $BV(\Omega)$ is compactly embedded into the space $L^1(\Omega)$. Further, for any $z \in BV(\Omega)$ there exists a vectorial Radon measure $\nabla z : \mathcal{B}(\Omega) \longrightarrow \mathbb{R}^2$, called "variation measure", such that the value $V(z)$ coincides with the integration value $\int_\Omega |\nabla z|$ (of constant 1), with respect to the total variation measure $|\nabla z|$ of ∇z (cf. Chapter 3 of [1] or Chapter 5 of [3]).

Similar properties also hold in the more general setting of the function σ, and such generalized results have been reported in several papers, see e.g. [11]. Let us recall some of them in the following remark.

Remark 1: (c.f. sections 2∼3 of [11])
(I) For any $z \in D(V_\sigma)$ there exists a vectorial Radon measure $D_\sigma z : \mathcal{B}(\Omega) \longrightarrow \mathbb{R}^2$, such that:

(i) $\int_\Omega z \operatorname{div}(\sigma\hat{\varphi})\, dx = -\int_\Omega \hat{\varphi} \cdot D_\sigma z$ for any Lipschitz continuous function $\hat{\varphi} \in C(\Omega)^2$ with a compact support;

(ii) $$V_\sigma(z) \geq \int_\Omega |D_\sigma z| = \sup\left\{ \int_\Omega z \operatorname{div}(\sigma\varphi)\, dx \;\middle|\; \begin{array}{l} \varphi \in C^1(\Omega)^2 \text{ with compact} \\ \text{support, and } |\varphi| \leq 1 \text{ on } \overline{\Omega} \end{array} \right\},$$
where $|D_\sigma z|$ is the total variation measure of $D_\sigma z$;

(iii) if $z \in BV(\Omega)$, then
$$D_\sigma z(B) = \int_B \sigma(x)\, \nabla z \quad \text{and} \quad |D_\sigma z|(B) = \int_B \sigma(x)\, |\nabla z|$$
for any $B \in \mathcal{B}(\Omega)$, moreover $V_\sigma(z) = \int_\Omega |D_\sigma z| = \int_\Omega \sigma(x)|\nabla z|$.

(II) (Approximation by smooth functions) For any $z \in D(V_\sigma)$ there exists a sequence $\{\zeta_i\} \subset C^\infty(\overline{\Omega})$ such that $\zeta_i \to z$ in $L^2(\Omega)$ and $V_\sigma(\zeta_i) \to V_\sigma(z)$ as $i \to +\infty$.

Next, let us fix any function $\bar{\theta} \in L^2_{loc}([0, +\infty); L^2(\Omega))$ and consider the following evolution equation, of the form:

$$(AC_\sigma; \bar{\theta}) \quad u_t(t) + \partial\Phi_\sigma(u(t)) \ni u(t) + \bar{\theta}(t) \ \text{ in } L^2(\Omega), \ \ t > 0;$$

where Φ_σ is the proper l.s.c. and convex function given in (5).

For any $\bar{\theta} \in L^2_{loc}([0,+\infty); L^2(\Omega))$, a function $u : [0,+\infty) \longrightarrow L^2(\Omega)$ is called a solution of $(AC_\sigma; \bar{\theta})$, if $u \in W^{1,2}_{loc}([0,+\infty); L^2(\Omega))$, $\Phi_\sigma(u) \in L^1_{loc}[0,+\infty)$, and there exists a function $u^* \in L^2_{loc}([0,+\infty); L^2(\Omega))$ such that $u(t) + \bar{\theta}(t) - u_t(t) \in \Phi_\sigma(u(t))$ in $L^2(\Omega)$ a.e. $t > 0$.

Each equation $(AC_\sigma; \bar{\theta})$ ($\bar{\theta} \in L^2_{loc}([0,+\infty); L^2(\Omega))$) is a type of Allen-Cahn equation, namely it is a kinetic equation of phase field dynamics in a known setting of the temperature. In fact, equation $(AC_\sigma; \bar{\theta})$ turns out to be a gradient flow of the free energy defined by (3), with $\theta = \bar{\theta}$ in Q. Note that the subdifferential $\partial\Phi_\sigma$ of the convex function Φ_σ is associated with the first variation of the interfacial energy V_σ.

Nowadays, several mathematicians have established various abstract theories for evolution equations governed by subdifferentials, such as $(AC_\sigma; \bar{\theta})$ (cf. [2, 4, 12, 6]). So, referring to some of the theories (e.g. [2]), we immediately have the following propositions.

Proposition 2:

(I) (Solvability) Let $\bar{\theta} \in L^2_{loc}([0,+\infty); L^2(\Omega))$ be a given function. Then, for any $u_0 \in D(\Phi_\sigma)$ there exists a unique solution u of the equation $(AC_\sigma; \bar{\theta})$, satisfying $u(0) = u_0$ in $L^2(\Omega)$. Moreover, $\Phi_\sigma(u(\cdot))$ is absolutely continuous on any compact interval in $[0,+\infty)$.

(II) (Boundedness) Let us take $\bar{\theta} \in L^2_{loc}([0,+\infty); L^2(\Omega))$ satisfying

$$S(\bar{\theta}) := \sup_{t \geq 0} |\bar{\theta}|_{L^2(t,t+1;L^2(\Omega))} < +\infty. \tag{7}$$

Then, for any solution u of the equation $(AC_\sigma; \bar{\theta})$ there exists a positive constant R_0 such that:

$$\sup_{t \geq 0} \int_t^{t+1} |u_t(\tau)|^2_{L^2(\Omega)}\, d\tau + \sup_{t \geq 0} \Phi_\sigma(u(t)) \leq R_0 \left(1 + \Phi_\sigma(u_0) + S(\bar{\theta})^2\right). \tag{8}$$

Proposition 3: *(Comparison principle) Let $\bar{\theta}_i \in L^2_{loc}([0,+\infty); L^2(\Omega))$ ($i = 1, 2$) be two functions, satisfying $\bar{\theta}_1 \leq \bar{\theta}_2$ a.e. in Q, and, respectively, let $u_i \in C([0,+\infty); L^2(\Omega))$ be solutions of $(AC_\sigma; \bar{\theta}_i)$ ($i = 1, 2$) such that $u_1(0) \leq u_2(0)$ a.e. in Ω. Then, $u_1(t) \leq u_2(t)$ a.e. in Ω, for any $t \geq 0$.*

Furthermore, the author of [14] has reported the following criterion for the solution of equation $(AC_\sigma; \bar{\theta})$ ($\bar{\theta} \in L^2_{loc}([0,+\infty); L^2(\Omega))$).

Proposition 4: *(Criterion for solutions) Let $\bar{\theta} \in L^2_{loc}([0,+\infty); L^2(\Omega))$ be any given function. Then, a function $u : [0,+\infty) \longrightarrow D(\Phi_\sigma)$ is a solution* *of the equation* $(AC_\sigma;\bar{\theta})$ *if* $u \in W^{1,2}_{loc}([0,+\infty);L^2(\Omega))$ *and there exists a function* $\boldsymbol{\nu}_u \in L^\infty([0,+\infty);L^\infty(\Omega)^2)$ *such that:*

(a) $|\boldsymbol{\nu}_u(t)| \leq 1$ *a.e. in* Ω, *for a.e.* $t \geq 0$;

(b) *for a.e.* $t \geq 0$ *the function* $\sigma\boldsymbol{\nu}_u(t)$ *is Lipschitz continuous with compact support in* Ω;

(c) $\Phi_\sigma(u(t)) = -\int_\Omega \mathrm{div}\,(\sigma\boldsymbol{\nu}_u(t))u(t)\,dx$, *for a.e.* $t \geq 0$;

(d) *for a.e.* $(t,x) \in Q$,

$$-\mathrm{div}\,(\sigma\boldsymbol{\nu}_u)(t,x) \begin{cases} \leq (u+\bar{\theta}-u_t)(t,x), & \text{if } u(t,x)=1, \\ = (u+\bar{\theta}-u_t)(t,x), & \text{if } -1<u(t,x)<1, \\ \geq (u+\bar{\theta}-u_t)(t,x), & \text{if } u(t,x)=-1. \end{cases}$$

Finally, let us consider the large-time behavior of solutions. For any $r \geq 0$, let us denote by $\mathrm{Lv}(r;\Phi_\sigma)$ the so-called sublevel set of Φ_σ, namely let us set:

$$\mathrm{Lv}(r;\Phi_\sigma) := \left\{ z \in D(\Phi_\sigma) \,\middle|\, \Phi_\sigma(z) \leq r \right\} \ (r \geq 0).$$

Then, for any function $\bar{\theta}$ satisfying (7) and any solution u of equation $(AC_\sigma;\bar{\theta})$ with initial condition $u(0) = u_0 \in D(\Phi_\sigma)$, we see from (8) that $\left\{ u(t) \,\middle|\, t \geq 0 \right\} \subset \mathrm{Lv}(\Phi_\sigma;r)$ for any $r \geq R_0\left(1+\Phi_\sigma(u_0)+S(\bar{\theta})^2\right)$. This implies that the compactness of the sublevel sets $\mathrm{Lv}(r;\Phi_\sigma)$, $r \geq 0$, will play an important role in the argument of the large-time behavior of solutions. But, in general, the current assumption for σ is not enough to recover the compactness, because σ may degenerate on $\overline{\Omega}$. Hence, we add the following assumption on σ:

(s1) $\mathcal{L}^2\left(\sigma^{-1}(0)\right) = 0$, where $\sigma^{-1}(0) := \left\{ x \in \overline{\Omega} \,\middle|\, \sigma(x) = 0 \right\}$.

The above condition is one of typical assumptions to guarantee the compactness of the sublevel sets. In fact, it has been proved in Corollary 3.1 of [11] that

$$\text{the sublevel set } \mathrm{Lv}(r;\Phi_\sigma) \text{ is compact in } L^2(\Omega) \text{ for any } r \geq 0 \qquad (9)$$

whenever σ satisfies the condition (s1).

Now, under the additional condition (9), the large-time behavior of solutions has been characterized as follows.

Proposition 5: *(cf. [6]) For any* $\bar{\theta} \in L^2_{loc}([0,+\infty);L^2(\Omega))$ *and any solution* u *of* $(AC_\sigma;\bar{\theta})$, *let us denote by* $\omega(u;\bar{\theta})$ *the so-called* ω*-limit set of* u, *that is defined as:*

$$\omega(u;\bar\theta) := \left\{ w \in L^2(\Omega) \;\middle|\; \begin{array}{l} u(t_i) \to w \text{ in } L^2(\Omega) \text{ as } i \to +\infty \\ \text{for some } \{t_i\} \subset (0,+\infty), \text{ satis-} \\ \text{fying } t_i \nearrow +\infty \text{ as } i \to +\infty \end{array} \right\}. \tag{10}$$

If the Lipschitz continuous function σ satisfies the condition (s1), and if there exists a constant $\theta_* \in \mathbb{R}$ such that:

$$\bar\theta - \theta_* \in L^2(0,+\infty; L^2(\Omega)) \text{ and } \bar\theta(t) \to \theta_* \text{ in } L^2(\Omega) \text{ as } t \to +\infty, \tag{11}$$

then the ω-limit set $\omega(u;\bar\theta)$ is nonempty, connected, and compact in $L^2(\Omega)$, and it coincides with the set of all the solutions of the inclusion (6).

In view of Proposition 5, inclusion (6) can be called the steady-state problem for Allen-Cahn equations $(AC_\sigma;\bar\theta)$ under the assumptions (s1) and (11). Furthermore, it will be proved in the next section that inclusion (6) is also linked to the asymptotic behavior of solutions of the system (S) anew.

3. Key properties for the system (S)

In this section, we shall discuss about key properties for the system (S), such as the existence, the uniqueness, the boundedness and the large-time behavior of solutions.

Let X_0 be the Sobolev space $H_0^1(\Omega)$ with the following inner product:

$$(z_1, z_2)_{X_0} := \int_\Omega \nabla z_1 \cdot \nabla z_2 \, dx, \; z_i \in X_0 \; (i = 1, 2).$$

The space X_0 is compactly embedded into the space $L^2(\Omega)$, so there exists a positive constant C_P such that:

$$|z|_{L^2(\Omega)} \le C_P |z|_{X_0} \text{ for any } z \in X_0. \tag{12}$$

The above inequality is known as Poincaré's inequality. Additionally, since Ω is a two-dimensional domain, the Sobolev space $H^2(\Omega)$ is continuously embedded into the Banach space $C(\overline{\Omega})$, and there exists a positive constant $\bar C$ such that:

$$|z|_{C(\overline{\Omega})} \le \bar C |z|_{X_0}^{1/2} |z|_{H^2(\Omega)}^{1/2} \le \bar C |z|_{H^2(\Omega)} \text{ for any } z \in X_0 \cap H^2(\Omega). \tag{13}$$

Let $\Delta_0 : D(\Delta_0) \subset L^2(\Omega) \longrightarrow L^2(\Omega)$ be the operator defined by

$$D(\Delta_0) := X_0 \cap H^2(\Omega) \text{ and } \Delta_0 z := \Delta z \text{ for any } z \in D(\Delta_0).$$

Remark 6: (Fundamental properties of $-\Delta_0$) It is well-known (cf. [2, 10]) that the operator $-\Delta_0$ is a single-valued and maximal monotone operator from $D(\Delta_0)$ into $L^2(\Omega)$, such that

$$(-\Delta_0\, z_1, z_2)_{L^2(\Omega)} = (z_1, -\Delta_0\, z_2)_{L^2(\Omega)} = (z_1, z_2)_{X_0} \quad \text{for all } z_i \in D(\Delta_0)\ (i = 1, 2). \tag{14}$$

Moreover, for any $f \in L^2(\Omega)$, the equation $-\Delta_0\, u = f$ in $L^2(\Omega)$ has a unique solution u, and we find a positive constant C_0 such that

$$|u|_{H^2(\Omega)} \le C_0 |\Delta_0 u|_{L^2(\Omega)} = C_0 |f|_{L^2(\Omega)}. \tag{15}$$

Now, the solutions of the system (S) are defined as follows.

Definition 7: (Definition of solutions) Let $\theta_* \in \mathbb{R}$ be any constant. Then, a pair $\{\theta, w\}$ of functions is called a solution of the system (S), if:

(s1) $\theta \in W^{1,2}_{loc}([0, +\infty); H^2(\Omega))$, $w \in W^{1,2}_{loc}([0, +\infty); L^2(\Omega))$, $\Phi_\sigma(w) \in L^1_{loc}[0, +\infty)$, and the pair $\{\theta_0, w_0\}$ of initial values satisfies $\theta_0 - \theta_* \in D(\Delta_0)$ and $w_0 \in D(\Phi_\sigma)$;

(s2) $\theta(t) - \theta_* \in D(\Delta_0)$, $\theta_t(t) \in D(\Delta_0)$, and

$$(\theta + w)_t(t) - \Delta_0[(\theta(t) - \theta_*) + \mu\theta_t(t)] = 0 \ \text{ in } L^2(\Omega), \tag{16}$$

for a.e. $t > 0$;

(s3) there exists a function $w^* \in L^2_{loc}([0, +\infty); L^2(\Omega))$ such that

$$w^*(t) \in \partial\Phi_\sigma(w(t)) \ \text{ and } \ w_t(t) + w^*(t) = w(t) + \theta(t) \ \text{ in } L^2(\Omega), \tag{17}$$

for a.e. $t > 0$.

First, let us check the solvability of the system (S).

Theorem 8: (Solvability) Let $\theta_* \in \mathbb{R}$ be any constant. Then, the system (S) admits a unique solution $\{\theta, w\}$.

Proof: We can prove this theorem in a similar way as in the proof of Theorem 2.1 of [13]. In fact, the difference between the system (S) and the mathematical model treated in [13] just consists of the expression of the convex part of free energy, and, in the proof of the result of [13], the convex part is only required to be a proper l.s.c. and convex function. Therefore, the argument for [13, Thm. 2.1] can be adapted to the proof of this theorem with slight modifications. □

Next, for any constant $\theta_* \in \mathbb{R}$ let us denote by $\mathcal{F}_{\theta_*}$ the functional on $L^2(\Omega)$ defined by

$$\mathcal{F}_{\theta_*}(z) := \Phi_\sigma(z) - \frac{1}{2}\int_\Omega |z+\theta_*|^2\, dx \quad \text{for any } z \in L^2(\Omega). \tag{18}$$

Note that the functional $\mathcal{F}_{\theta_*}$ coincides with the free energy given in (3) in the case of $\theta \equiv \theta_*$ (in Q). Then, we can prove the following global estimates of solutions, associated with the energy functional $\mathcal{F}_{\theta_*}$.

Theorem 9: (Global estimates) Let $\theta_* \in \mathbb{R}$ be any constant, let $\{\theta_0, w_0\}$ be a pair of initial data, and let $\{\theta, w\}$ be a solution of the system (S). Then, the following statements hold.

(I) (Property of energy-decay) Let us set:

$$J(t) := \frac{1}{2}|\theta(t)-\theta_*|^2_{L^2(\Omega)} + \frac{\mu}{2}|\theta(t)-\theta_*|^2_{X_0} + \mathcal{F}_{\theta_*}(w(t));$$

for any $t \geq 0$. Then:

$$J(t) \leq J(s) \quad \text{for all } 0 \leq s \leq t < +\infty. \tag{19}$$

(II) (H^2-estimate of θ) There exists a positive constant R_1 such that:

$$\begin{aligned}\sup_{t\geq 0}|\theta(t)-\theta_*|^2_{H^2(\Omega)} + \int_0^{+\infty}|\theta(t)-\theta_*|^2_{H^2(\Omega)}\,dt + \int_0^{+\infty}|w_t(t)|^2_{L^2(\Omega)}\,dt \\ \leq R_1\left(1+|\theta_*|^2+|\theta_0-\theta_*|^2_{H^2(\Omega)}+\Phi_\sigma(w_0)\right).\end{aligned} \tag{20}$$

Proof: First, multiplying both sides of (16) by $(\theta(t)-\theta_*)$, multiplying the both sides of (17) by $w_t(t)$, and taking the sum of the results, we compute that

$$\frac{d}{dt}J(t) + |\theta(t)-\theta_*|^2_{X_0} + |w_t(t)|^2_{L^2(\Omega)} = 0, \quad \text{a.e. } t>0. \tag{21}$$

Thus, we conclude the assertion (I), since (21) implies that the function $J(t)$ is non-increasing.

Secondly, for any $t > 0$, let us integrate the both sides of (21) over $[0, t]$. Then, letting $t \nearrow +\infty$ yields

$$\begin{aligned}\int_0^{+\infty}|\theta(t)-\theta_*|^2_{X_0}\,dt + \int_0^{+\infty}|w_t(t)|^2_{L^2(\Omega)}\,dt \\ \leq \frac{1}{2}|\theta_0-\theta_*|^2_{L^2(\Omega)} + \frac{\mu}{2}|\theta_0-\theta_*|^2_{X_0} + \Phi_\sigma(w_0) + (1+|\theta_*|^2)\mathcal{L}^2(\Omega).\end{aligned} \tag{22}$$

Finally, let us multiply both sides of (16) by $(-\Delta_0)(\theta(t)-\theta_*)$. Then,

$$\begin{aligned}&\frac{1}{2}\frac{d}{dt}\left(|\theta(t)-\theta_*|^2_{X_0}+\mu|\Delta_0(\theta(t)-\theta_*)|^2_{L^2(\Omega)}\right)+|\Delta_0(\theta(t)-\theta_*)|^2_{L^2(\Omega)}\\&\quad\le\frac{1}{2}|w_t(t)|^2_{L^2(\Omega)}+\frac{1}{2}|\Delta_0(\theta(t)-\theta_*)|^2_{L^2(\Omega)}\quad\text{for a.e. } t>0.\end{aligned}\tag{23}$$

Integrating both sides of (23) and using (15) and (22), we find the positive constant R_1 required in the assertion (II). □

Remark 10: Note that the constant R_1 found in the proof of Theorem 9, actually depends on some constants, e.g. μ, C_0 and $\mathcal{L}^2(\Omega)$. Yet, we can neglect such a dependence, since those constants are set as fixed factors.

Additionally, let us assume condition (s1) to guarantee the compactness (9) of sublevel sets. Then, we conclude the following theorem, concerned with the large-time behavior of the solutions.

Theorem 11: (Large-time behavior) Let us take any solution $\{\theta, w\}$ of the system (S). If the Lipschitz continuous function σ satisfies condition (s1), then the following statements hold.

(A) (Convergence of θ) $\theta(t)\to\theta_*$ in $H^2(\Omega)$, therefore, on account of (13), $\theta(t)\to\theta_*$ uniformly on $\overline{\Omega}$ as $t\to+\infty$.

(B) (ω-limit points of w) Noting that w is a solution of the Allen-Cahn equation $(AC_\sigma;\theta)$, let us denote by $\omega(w;\theta)$ the ω-limit set defined by (10), in the case of $u=w$ and $\bar{\theta}=\theta$. Then, the ω-limit set $\omega(w;\theta)$ is nonempty, connected and compact in $L^2(\Omega)$, and it coincides with the class of all solutions of the steady-state problem (6).

Proof: The proof is a simplified version of that of Theorem 2.3 of [13]. In view of (20), we find a sequence $\{t_i\}\subset(0,+\infty)$, such that

$$t_i\ge i,\quad |\theta(t_i)-\theta_*|^2_{H^2(\Omega)}<\frac{1}{i},\quad \int_{t_i}^{+\infty}|w_t(t)|^2_{L^2(\Omega)}\,dt<\frac{1}{i},\quad (i=1,2,3,\cdots).$$

Hence, for any $i\in\mathbb{N}$ and any $t\ge t_i$, let us integrate both sides of (23) over $[t_i,t]$. Then, by (15),

$$
\begin{aligned}
|\theta(t) - \theta_*|^2_{H^2(\Omega)} &\leq \frac{C_0^2}{\mu}\left\{\mu|\Delta_0(\theta(t) - \theta_*)|^2_{L^2(\Omega)}\right\} \\
&\leq \frac{2C_0^2}{\mu}\left(|\theta(t_i) - \theta_*|^2_{H^2(\Omega)} + \int_{t_i}^{+\infty} |w_t(\tau)|^2_{L^2(\Omega)}\, dt\right) \\
&< \frac{4C_0^2}{\mu} \cdot \frac{1}{i}, \quad \text{for any } t \geq t_i,\ (i = 1, 2, 3, \cdots).
\end{aligned}
$$

Thus, we conclude assertion (A).

Now, the assertion (B) immediately follows from assertion (A) and Proposition 5 (see section 5 of [13] for details). □

4. Steady-state patterns

Let $\theta_* \in \mathbb{R}$ be any constant. In this section, we shall focus on the concrete profiles of the solutions of the steady-state problem (6). On account of Theorem 11, each solution of (6) is supposed to represent the ultimate pattern weaved by solid-liquid phases in the steady-state (steady-state pattern). Therefore, our main interest will naturally be to know the geometry of steady-state patterns, represented by the nonconstant solutions of (6).

However, in case of $|\theta_*| \geq 1$, we easily see that any solution of the steady-state problem (6) is constant in Ω. More precisely, when $\theta_* > 1$ (resp. $\theta_* < -1$), the steady-state problem (6) has just one solution, and it is the constant 1 (resp. -1). Also, when $\theta_* = 1$ or $\theta_* = -1$, (6) has just two solutions, and they are the constants 1 and -1. Hence, in the rest of this paper, we will assume that $|\theta_*| < 1$, to enable the observation of nonconstant solutions.

On the other hand, for the geometrical observation of the steady-state patterns, it will be necessary to know the profile of the indefinite coefficient σ, too. In view of this, we introduce the following class of nonnegative and Lipschitz continuous functions, as an example of possible profiles of the coefficient σ.

Definition 12: (Class of indefinite coefficients) Let us assume that $\Omega \subset \mathbb{R}^2$, and for every positive numbers κ and r, let us denote by $\sigma_{\kappa,r}$ a nonnegative and Lipschitz continuous function on $\overline{\Omega}$, given by

$$
\sigma_{\kappa,r}(x) := \frac{\kappa}{r}\,\mathrm{dist}(x, \Sigma_r) \quad \text{for any } x \in \overline{\Omega},
$$

$$
\text{where } \Sigma_r := \left\{ \xi = (\xi_1, \xi_2) \in \mathbb{R}^2 \;\middle|\; \begin{array}{l} |\xi_1 - 6kr| = \sqrt{3}|\xi_2| \\ \text{or } \xi_1 = 3kr,\ k \in \mathbb{Z} \end{array} \right\}.
$$

Then, we define a class $\mathcal{S}$ of nonnegative and Lipschitz continuous functions by putting $\mathcal{S} := \{ \sigma_{\kappa,r} \mid \kappa > 0,\ r > 0 \}$.

Remark 13: For every $\kappa > 0$ and $r > 0$, the set Σ_r as in Definition 12 actually coincides with the set $\sigma_{\kappa,r}^{-1}(0)$ of zero points of the function $\sigma_{\kappa,r}$, and it makes a web of triangular lattice in $\mathbb{R}^2$ (see Fig. 1). Thus, it is easily checked that all functions $\sigma \in \mathcal{S}$ satisfy condition (s1) for the compactness (9) of the sublevel sets of the functional Φ_σ.

When σ belongs to the class $\mathcal{S}$, namely:

$$\sigma = \sigma_{\kappa,r} \text{ on } \overline{\Omega}, \text{ for some } \sigma_{\kappa,r} \in \mathcal{S} \text{ with positive numbers } \kappa \text{ and } r; \tag{24}$$

it is relatively easy to find concrete examples of solutions of the steady-state problem (6). As one such example, let us recall a class of piecewise constant functions, which have been reported in Theorems 4.1 and 4.2 of [11].

Example 14: Let θ_* be a constant satisfying $|\theta_*| < 1$, and let σ be any function satisfying (24). Here, let us set a class of subdomains in Ω by:

$$\mathcal{D}_\sigma := \left\{ D \subset \Omega \;\middle|\; \begin{array}{l} D \text{ is a domain with a Lipschitz continuous} \\ \text{boundary } \partial D, \text{ such that } \partial D \text{ is a disjoint union} \\ \text{of at most a finite number of Jordan curves,} \\ \partial D \cap \Omega = \partial D^{ex} \cap \Omega = \partial D \cap \partial D^{ex},\ \partial D \cap \Omega \subset \Sigma_r, \\ \text{and } \inf_{x \in \Gamma} \operatorname{dist}(x, \partial D \cap \Omega) \geq r \text{ whenever } \partial D \cap \Omega \neq \emptyset \end{array} \right\};$$

and let us assume that $\mathcal{D}_\sigma \neq \emptyset$. Let us define a class $\mathcal{M}_\sigma$ of piecewise constant functions, by putting:

$$\mathcal{M}_\sigma := \left\{ w \in D(\Phi_\sigma) \cap BV(\Omega) \;\middle|\; \begin{array}{l} w = \chi_D - \chi_{D^{ex}} \text{ a.e. in } \Omega, \\ \text{for some } D \in \mathcal{D}_\sigma \end{array} \right\}.$$

Then, any function $w_* \in \mathcal{M}_\sigma$ solves the steady-state problem (6). Moreover, when $\theta_* = 0$, any solution $w_* \in \mathcal{M}_\sigma$ also minimizes the functional $\mathcal{F}_0$, that is the free energy $\mathcal{F}_{\theta_*}$ given in (18) in the case of $\theta_* = 0$.

Remark 15: (Supplements for Example 14) From Remark 13, we see that the class $\mathcal{D}_\sigma$ of each $\sigma \in \mathcal{S}$ includes various domains having piecewise linear boundaries. Here, let us notice that the domain illustrated as in Fig. 1 can be one example of such domains.

Furthermore, in the case when the temperature is known, the author of [14] has recently studied the stability that steady-state patterns, as in Example 14, show in the dynamical system generated by the single Allen-Cahn

Fig. 14.1.

equations $(AC_\sigma; \bar{\theta})$ $(\bar{\theta} \in L^2_{loc}([0, +\infty); L^2(\Omega)))$. In general, the terminology "stability" means some kind of restoring force for oscillations of parameters. From this viewpoint, the author of [14] has introduced the following mathematical tools to characterize the range of oscillations.

Definition 16: For any $\rho \geq 0$ and any open set $D \subset \Omega$ $(\subset \mathbb{R}^2)$, let us define a subset D_ρ and a relatively closed set C_ρ, by putting:

$$\begin{cases} \bullet\ D_\rho := \bigcup\limits_{\substack{x \in \Omega \\ B_\rho[x] \subset D \setminus \partial D[\rho]}} B_\rho[x] \ \text{ and } \ C_\rho := \Omega \setminus (D_\rho \cup (D^{ex})_\rho), \text{ if } \partial D \cap \Omega \neq \emptyset, \\ \bullet\ D_\rho := D \ \text{ and } \ C_\rho := \emptyset, \text{ if } \partial D \cap \Omega = \emptyset; \end{cases}$$

where $B_\rho[x] := \{ y \in \Omega \mid |y - x| \leq \rho \}$ for any $x \in \mathbb{R}^2$.

Remark 17: Let us consider the case of $D \in \mathcal{D}_\sigma$ for some function σ satisfying (24). Then, for any $0 < \rho < r/2$ we notice that the set C_ρ is a modified version of the set $\partial D[\rho]$ (ρ-neighborhood of ∂D), such that each corner on the boundary of $\partial D[\rho]$ is regularized by a sector with radius ρ (see Fig. 2). Thus in this case, both sets D_ρ and $(D^{ex})_\rho$ have Hölder continuous boundaries with $C^{1,1}$-regularity.

Now, referring to the arguments developed in section 5 of [14], we conclude the following proposition.

Proposition 18: *Let θ_* be a constant satisfying $|\theta_*| < 1$, and let σ be a Lipschitz continuous function satisfying (24). Let us take any solution*

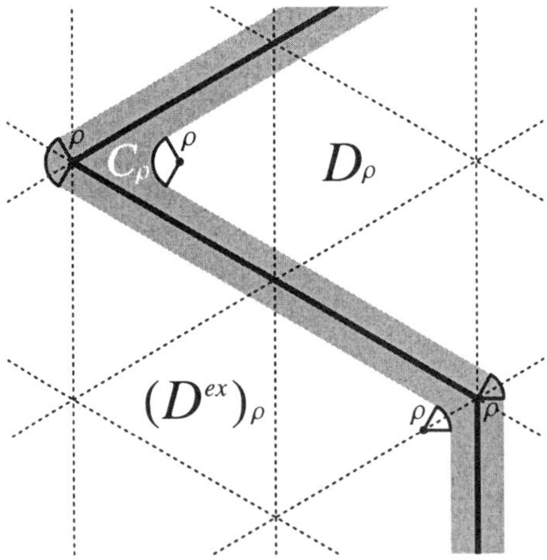

Fig. 14.2.

$w_* = \chi_D - \chi_{D^{ex}} \in \mathcal{M}_\sigma$ of (6) with a domain $D \in \mathcal{D}_\sigma$, and let us assume that $\partial D \cap \Omega \neq \emptyset$, namely that w_* is nonconstant in Ω. Let $\bar{\delta}$ and $\bar{\rho}$ be two constants such that $0 < \bar{\delta} < (1 - |\theta_*|)/3$ and $0 < \bar{\rho} < r/2$. Here, if positive numbers κ and r, as in (24), and a constant $\bar{\delta}$ satisfy:

$$0 < \frac{5\kappa}{1 - |\theta_*| - 3\bar{\delta}} < r, \tag{25}$$

if $0 \leq \delta \leq \bar{\delta}$, $0 \leq \rho \leq \bar{\rho}$, and if a (given) function $\bar{\theta}$ and a solution u of Allen-Cahn equation $(AC_\sigma; \bar{\theta})$ satisfy:

$$|\bar{\theta} - \theta_*|_{L^\infty(Q)} \leq \delta \quad \text{and} \quad |u(0) - w_*|_{L^\infty(\Omega \setminus C_\rho)} \leq \delta,$$

then:

(i) $|u(t) - w_*|_{L^\infty(\Omega \setminus C_\rho)} \leq \delta$ for any $t \geq 0$;
(ii) there exists a finite time $\bar{t}(\delta, \rho)$ such that $u(t) = w_*$ a.e. in $\Omega \setminus C_\rho$ for any $t \geq \bar{t}(\delta, \rho)$.

Remark 19: (Rough sketch of the proof) According to section 5 of [14], the above proposition is obtained by splitting the proof into discussions of stability for every connected components of the graph of the solution $w_* \in \mathcal{M}_\sigma$, and the method of such characterization has been organized as an original theory, named as "local stability".

In the framework of local stability, the behavior of each solution of the Allen-Cahn equation is estimated by comparison with the behavior of a known function, called "comparison function", which is also a solution of one Allen-Cahn equation specified by a particular kind of forcing term.

In fact, assertion (i) is directly derived from the behavior of the comparison function, while the convergence time $\bar{t}(\delta, \rho)$, as in assertion (ii) is nothing but the convergence time of the comparison function. Therefore, the comparison principle stated in Proposition 3 plays a crucial role in the argument.

Incidentally, the assumption (25) will be used to find the comparison function, in conjunction with the criterion of solutions, stated as in Proposition 4.

Remark 20: (Constant case of w_*) In the statement of Proposition 18, we have noted that the condition $\partial D \cap \Omega \neq \emptyset$ is assumed to exclude the case of the constant solution $w_* = \chi_D - \chi_{D^{ex}} \in \mathcal{M}_\sigma$ of (6). Actually, we can also obtain similar conclusions as in (i) and (ii) of Proposition 18, even when w_* is constant. Then, since the set $\Omega \setminus C_\rho$ $(0 \leq \rho < r/2)$ will just coincide with the whole domain Ω, the statement in this case will be arranged into a ρ-independent form. Moreover, when $w_* \equiv 1$ (resp. $w_* \equiv -1$) in Ω, we actually take the following solution

$$\bar{u}(t) := \begin{cases} \frac{1}{2}(1+\theta_*)e^t - \theta_* \left(\text{resp. } \frac{1}{2}(-1+\theta_*)e^t - \theta_* \right), & \text{if } 0 \leq t < \log 2, \\ 1 \text{ (resp. } -1), \text{ if } t \geq \log 2; \end{cases}$$

of equation $(AC_\sigma; \theta_*)$ as the comparison function, namely we can find the comparison function without help from condition (25). This implies that assumption (25) is unnecessary in constant case of w_*.

5. Stability for steady-state patterns

This section is devoted to the statement and the proof of the main theorem in this paper. The scope of the main theorem is to analyze the stability (stability analysis) for stationary points in the dynamical system generated by the system (S). Here, in view of Proposition 18 and Remark 20, all pairs $\{\theta_*, w_*\}$, with constant θ_* satisfying $|\theta_*| < 1$ and piecewise constant functions $w_* \in \mathcal{M}_\sigma$ as in Example 14, are supposed to be representative examples of the stationary points with some stability.

In view of this, we set a goal of the main theorem to discuss the stability for those pairs. Here, in the framework of the system (S), the range of the oscillations of parameters will be characterized by the following system of subsets in the space of solutions.

Definition 21: (Range of oscillations) Let σ be any function satisfying (24), and let $\{\theta_*, w_*\}$ be any pair with θ_* constant, satisfying $|\theta_*| < 1$, and w_* given by $w_* = \chi_D - \chi_{D^{ex}} \in \mathcal{M}_\sigma$, with domain $D \in \mathcal{D}_\sigma$. Then, for any $\lambda \geq 0$ any $0 \leq \delta \leq 1$ and any $0 \leq \rho \leq r/2$, let us define a set $U_\lambda\left(\begin{pmatrix}\theta_* \\ w_*\end{pmatrix}; \delta, \rho\right)$ by putting:

$$U_\lambda\left(\begin{pmatrix}\theta_* \\ w_*\end{pmatrix}; \delta, \rho\right) := \left\{ \begin{pmatrix}\xi \\ \eta\end{pmatrix} \in \begin{matrix} H^2(\Omega) \\ \times \\ D(V) \end{matrix} \;\middle|\; \begin{matrix} |\xi - \theta_*|_{H^2(\Omega)} \leq \lambda, \\ |\xi - \theta_*|_{H^1(\Omega)} \leq \delta, \\ |\eta - w_*|_{L^\infty(\Omega \setminus C_\rho)} \leq \delta, \\ \text{and } \Phi_\sigma(\eta) \leq \delta \end{matrix} \right\}.$$

Remark 22: Under the same notation as in Definition 21, let us take any solution $\{\theta, w\}$ of system (S) satisfying the following initial condition:

$$\begin{pmatrix}\theta_0 \\ w_0\end{pmatrix} \in U_\lambda\left(\begin{pmatrix}\theta_* \\ w_*\end{pmatrix}; \delta, \rho\right). \tag{26}$$

Then, on account of (II) of Theorem 9, we find a positive constant M_λ, depending on λ, such that

$$|\theta(t) - \theta_*|_{H^2(\Omega)} \leq M_\lambda \left(1 + \mathcal{H}^1(\partial D \cap \Omega)\right) \quad \text{for any } t \geq 0. \tag{27}$$

The above inequality will be the key point to analyze the interplay between two parameters θ and w.

Now, our main theorem is stated as follows.

Theorem 23: (Stability in an unknown setting of the temperature) Let us take any function σ satisfying (24), and any pair $\{\theta_*, w_*\}$, with θ_* constant, satisfying $|\theta_*| < 1$, and $w_* = \chi_D - \chi_{D^{ex}} \in \mathcal{M}_\sigma$ a solution of (6), with domain $D \in \mathcal{D}_\sigma$. Furthermore, only in the case $\partial D \cap \Omega \neq \emptyset$ (namely when w_* is nonconstant in Ω), let us assume that:

$$0 < \frac{5\kappa}{1 - |\theta_*|} < r \tag{28}$$

for positive numbers κ and r, as in (24). Then, for any $\lambda \geq 0$, we find (small) constants $0 < \delta_*(\lambda) \leq 1$ and $0 \leq \rho_*(\lambda) \leq r/2$, depending on λ, which characterize the stability of the pair $\{\theta_*, w_*\}$ in the following manner.

($\star$) If $0 \leq \delta \leq \delta_*(\lambda)$, $0 \leq \rho \leq \rho_*(\lambda)$ and a solution $\{\theta, w\}$ of system (S) satisfies the initial condition (26), then:

$$|w(t) - w_*|_{L^\infty(\Omega \setminus C_\rho)} \leq \delta \;\text{ for any } t \geq 0;$$

moreover, there exists a finite time $t_*(\delta,\rho)$, depending on δ and ρ, such that:

$$w(t) = w_* \text{ a.e. in } \Omega \setminus C_\rho, \quad \text{for any } t \geq t_*(\delta,\rho). \tag{29}$$

Remark 24: (ρ-independence in the case of constant w_*) In the statement of Theorem 23, let us consider the case when the function $w_* \in \mathcal{M}_\sigma$ is constant. Then, for similar reasons as in Remark 20, the statement of the main theorem actually holds independently of ρ. Especially, in this case the finite time $t_*(\delta,\rho)$ of convergence as in (29) is found in a ρ-independent form, and assertion (29) implies the uniform convergence of $w(t)$ (a.e. in Ω) in a finite time.

Proof of Theorem 23: The proof is split in two cases: w_* constant and w_* nonconstant. Since the proof in the constant case is essentially a simplified version of that in the other (nonconstant) case, here we deal with the case of w_* nonconstant, only.

Now, under the assumption (28), let us take two constants $0 < \bar{\delta} < (1-|\theta_*|)/3$ and $0 < \bar{\rho} < r/2$ satisfying (25). For any $\lambda \geq 0$, let M_λ be the constant as in (27), and let $\varepsilon_*(\lambda)$, $\delta_*(\lambda)$ and $\rho_*(\lambda)$ be positive constants given as follows:

$$\begin{cases} \varepsilon_*(\lambda) := \dfrac{\mu\,\bar{\delta}^4}{4(1+\bar{C})^4(1+M_\lambda)^4\,(1+\mathcal{H}^1(\partial D\cap\Omega))^4} \ (<1), \\ \delta_*(\lambda) := \dfrac{\varepsilon_*(\lambda)}{8\,(1+\mathcal{L}^2(\Omega))} \ (<\bar{\delta}), \quad \rho_*(\lambda) := \dfrac{\varepsilon_*(\lambda)}{16\,(1+\mathcal{H}^1(\partial D\cap\Omega))} \ (<\bar{\delta}). \end{cases}$$

Then, for any $\lambda \geq 0$, any $0 \leq \delta \leq \delta_*(\lambda)$ and any $0 \leq \rho \leq \rho_*(\lambda)$, let us take any solution $\{\theta, w\}$ satisfying the initial condition (26), and let us set

$$T_* := \sup\left\{ T \geq 0 \;\middle|\; \begin{array}{l} \mathcal{F}_{\theta_*}(w(t)) \geq \mathcal{F}_{\theta_*}(w_*) - \varepsilon_*(\lambda) \\ \text{for any } t \in [0,T] \end{array} \right\}. \tag{30}$$

We immediately see that $T_* > 0$. In fact, by (26) and (iii) of (I) in Remark 1,

$$\begin{aligned} &\mathcal{F}_{\theta_*}(w_0) \geq -\frac{1}{2}\int_\Omega |w_0+\theta_*|^2\,dx \\ &= -\frac{1}{2}\int_\Omega |w_*+\theta_*|^2\,dx - \frac{1}{2}\int_\Omega \left(|w_0+\theta_*|^2 - |w_*+\theta_*|^2\right)\,dx \\ &\geq -\frac{1}{2}\int_\Omega |w_*+\theta_*|^2\,dx - \frac{1}{2}\int_{C_\rho} 2(1+|\theta_*|)|w_0-w_*|\,dx \\ &\geq \mathcal{F}_{\theta_*}(w_*) - 8\rho\mathcal{H}^1(\partial D\cap\Omega) > \mathcal{F}_{\theta_*}(w_*) - \varepsilon_*(\lambda). \end{aligned} \tag{31}$$

This implies that $T_* > 0$, since the functional $\mathcal{F}_{\theta_*}$ is absolutely continuous on any compact interval in $[0, +\infty)$ (see (I) of Proposition 2).

Next, by (30) and (I) of Theorem 9, we have

$$\begin{aligned}
&\frac{\mu}{2}|\theta(t)-\theta_*|^2_{X_0} - \varepsilon_*(\lambda) \le \frac{1}{2}|\theta_0-\theta_*|^2_{H^1(\Omega)} + \mathcal{F}_{\theta_*}(w_0) - \mathcal{F}_{\theta_*}(w_*) \\
&\le \delta^2 + \Phi_\sigma(w_0) - \frac{1}{2}\int_\Omega \left(|w_0+\theta_*|^2 - |w_*+\theta_*|^2\right)\,dx \\
&\le 2\delta\left(1+\mathcal{L}^2(\Omega)\right) + 8\rho\mathcal{H}^1(\partial D\cap\Omega) \\
&< \varepsilon_*(\lambda), \quad \text{for any } 0\le t<T_*,
\end{aligned}$$

which is reduced to

$$|\theta(t)-\theta_*|^2_{X_0} < \frac{4}{\mu}\cdot\varepsilon_*(\lambda) = \frac{\bar{\delta}^4}{(1+\bar{C})^4(1+M_\lambda)^4\left(1+\mathcal{H}^1(\partial D\cap\Omega)\right)^4} \quad \text{for any } 0\le t<T_*. \tag{32}$$

Therefore, combining (13), (27) and (32),

$$\begin{aligned}
|\theta(t)-\theta_*|_{C(\overline{\Omega})} &\le \bar{C}|\theta(t)-\theta_*|^{1/2}_{X_0}|\theta(t)-\theta_*|^{1/2}_{H^2(\Omega)} \\
&< \frac{\bar{C}\cdot\bar{\delta}\cdot M_\lambda^{1/2}\left(1+\mathcal{H}^1(\partial D\cap\Omega)\right)^{1/2}}{(1+\bar{C})(1+M_\lambda)\left(1+\mathcal{H}^1(\partial D\cap\Omega)\right)} \\
&< \bar{\delta}, \quad \text{for any } 0\le t<T_*.
\end{aligned} \tag{33}$$

On account of (26) and (33), we can apply (i) of Proposition 18 to obtain that

$$|w(t)-w_*|_{L^\infty(\Omega\setminus C_\rho)} \le \delta \quad \text{for any } 0\le t<T_*. \tag{34}$$

This entails that $T_* = +\infty$. In fact, if $T_* < +\infty$, then by a similar computation as in (31), we would obtain that

$$\begin{aligned}
\mathcal{F}_{\theta_*}(w(T_*)) = \lim_{t\nearrow T_*}\mathcal{F}_{\theta_*}(w(t)) &\ge \mathcal{F}_{\theta_*}(w_*) - 8\rho\mathcal{H}^1(\partial D\cap\Omega) \\
&> \mathcal{F}_{\theta_*}(w_*) - \varepsilon_*(\lambda),
\end{aligned}$$

which contradicts the definition of T_*.

Furthermore, on account of (26) and (33), we can apply (ii) of Proposition 18 to find a finite time $t_*(\delta,\rho)$ satisfying (29). Thus, we conclude the main theorem. □

References

[1] L. Ambrosio, N. Fusco and D. Pallara, *Functions of Bounded Variation and Free Discontinuity Problems.* Oxford Univ. Press, New York, 2000.

[2] H. Brézis, *Opérateurs maximaux monotones et semigroupes de contractions dans les espace de Hilbert.* North-Holland, Amsterdam, 1973.

[3] L. C. Evans and R. F. Gariepy, *Measure Theory and Fine Properties of Functions.* CRC Press Inc., Boca Raton, 1992.

[4] A. Ito, N. Yamazaki and N. Kenmochi, Attractors of nonlinear evolution systems generated by time-dependent subdifferentials in Hilbert spaces. In: *Dynamical Systems and Differential Equations, Vol. I (Springfield, MO, 1996). Discrete Contin. Dynam. Systems* 1998 (Added Volume I), p. 327.

[5] N. Kenmochi, Solvability of nonlinear evolution equations with time-dependent constraints and applications. *Bull. Fac. Edu. Chiba Univ.* **30**, 1–87 (1981).

[6] N. Kenmochi, Systems of nonlinear PDEs arising from dynamical phase transitions. In: *Phase transitions and hysteresis (Montecatini Terme, 1993)* (Springer, Berlin, 1994), p. 39.

[7] N. Kenmochi and K. Shirakawa, A variational inequality for total variation functional with constraint. *Nonlinear Anal.* **46**, 435–455 (2001).

[8] N. Kenmochi and K. Shirakawa, Stability for a parabolic variational inequality associated with total variation functional. *Funkcial. Ekvac.* **44**, 119–137 (2001).

[9] N. Kenmochi and K. Shirakawa, Stability for a phase field model with the total variation functional as the interfacial energy. *Nonlinear Anal.* **53**, 425–440 (2003).

[10] J.L. Lions and E. Magenes, *Non-Homogeneous Boundary Value Problems and Applications. Vol. I.* Springer-Verlag, New York-Heidelberg, 1972.

[11] K. Shirakawa, Interfacial energies in two dimensional phase field models and related variational problems. To appear in: *Variational problems and related topics (Kyoto, 2003)* Sūrikaisekikenkyūsho Kōkyūroku (2003), p. 73.

[12] K. Shirakawa, Stability for steady-state solutions of a nonisothermal Allen-Cahn equation generated by a total variation energy. In: *Nonlinear Partial Differential Equations and Their Applications* (Gakkōtosho, Tokyo, 2004), p. 289.

[13] K. Shirakawa, Large-time behavior for a phase field system associated with total variation energy. *Adv. Math. Sci. Appl.* **15**, 1–27 (2005).

[14] K. Shirakawa, Stability analysis for Allen-Cahn equations involving indefinite diffusion coefficients. To appear in: *Mathematical Approach to Nonlinear Phenomena: Modelling, Analysis and Simulations*, Gakkōtosho, Tokyo.

[15] K. Shirakawa and M. Kimura, Stability analysis for Allen-Cahn type equation associated with the total variation energy. *Nonlinear Anal.* **60**, 257–282 (2005).

[16] A. Visintin, *Models of Phase Transitions.* Birkhäuser, Boston (1996).

GEOMETRIC FEATURES OF p-LAPLACE PHASE TRANSITIONS

Enrico Valdinoci

Dipartimento di Matematica
Università di Roma Tor Vergata
Via della Ricerca Scientifica 1, 00133 Roma, Italy
E-mail: valdinoci@mat.uniroma2.it

This note gives an informal presentation of a few recent results obtained by some (relatively) young researchers. We will not go into the full details of the (quite technical) proofs here; rather, we will try to highlight some ideas in a concrete case. The reader willing to go into the matter more thoroughly may look at [43, 39, 35, 36, 40], and [46].

1. Introduction

We are interested in the critical points of the functional

$$\mathcal{F}_\Omega(u) = \int_\Omega |\nabla u(x)|^p + W\big(u(x)\big)\, dx \,.$$

Here, $u : \Omega \subset \mathbf{R}^N \longrightarrow \mathbf{R}$ is a function defined on a domain of $\mathbf{R}^N$ and $W(u) = (1-u^2)^p$ is a *double-well potential.* The term $|\nabla u|^p$ is a *kinetic part* of p-Laplacian type, with $p \in (1, +\infty)$. Roughly speaking, the potential W tries to force the minima of $\mathcal{F}_\Omega$ to attain values close to ± 1, while the kinetic term prevents the formation of discontinuities and puts the system into a smooth phase change regime. The two phases, which are here given by the values 1 and -1, are separated by the *interface* $\{u = 0\}$.

In this framework, the case in which $W(u) = \chi_{(-1,1)}(u)$ has also been studied in [4, 35, 44] and [45], since it mimicks some features of ideal fluid jets and combustion problems, and presents interesting free boundary properties. Related models describing elastic rods have also appeared in the literature (see, e.g., [6]).

The critical points of $\mathcal{F}_\Omega$ are solutions of an elliptic PDE of the form

$$\Delta_p u = f(u), \tag{1}$$

where Δ_p is the p-Laplacian operator, i.e. $\Delta_p u = \operatorname{div}(|\nabla u|^{p-2}\nabla u)$, and $f = W'/p$. As well-known, this PDE is singular for $p \in (1,2)$ and degenerate elliptic for $p \in (2,+\infty)$, the singularity/degeneracy occurring at points of vanishing gradient. Due to the physical interpretation of the phases ± 1, f is sometimes referred to with the name of *bi-stable nonlinearity.*

For $p = 2$, the system is reduced the so-called van der Vaals (or Ginzburg-Landau, or Allen-Cahn) phase transition model, see [38, 26, 27] and [3]. Due to the microscopical interpretation of the model, an important role is also played by rescaled solutions. Namely, if u is minimal (critical) for $\mathcal{F}_\Omega$, then the rescaled function $u_\epsilon(x) := u(x/\epsilon)$ is minimal (critical) for the functional

$$\mathcal{F}^\epsilon_\Omega(u_\epsilon) = \int_{\epsilon\Omega} \epsilon^p |\nabla u_\epsilon|^p + W(u_\epsilon)\, dx\,.$$

Also, it is known[33, 41, 13] that minimizers of $\mathcal{F}^\epsilon_\Omega$ $L^1_{\rm loc}$-converge, up to subsequences, to a step function $\chi_E - \chi_{\mathbf{R}^N\setminus E}$, and that ∂E has minimal area.

2. Results

The first result we would like to outline here is a collection of **density estimates** for level sets of minimizers of $\mathcal{F}_\Omega$. Roughly speaking, this result says that level sets of minimizers behave, in measure, like "codimension one" sets. That is, in a ball of radius r the area occupied by the phases close to ± 1 is of order r^N, while the area occupied by the transition layers between the two phases is of order r^{N-1}. More precisely, we have that

Theorem 1: *There exists a universal constant $r_0 > 0$ such that, if u is a minimizer for $\mathcal{F}_\Omega$ with $|u| \le 1$ and $|u(0)| < 1/2$, then*

$$\left|B_r \cap \left\{|u| \le 1/2\right\}\right| \sim r^{N-1}\,,$$
$$\left|B_r \cap \left\{u > 1/2\right\}\right| \sim r^N\,,$$
$$\text{and} \quad \left|B_r \cap \left\{u < -1/2\right\}\right| \sim r^N\,,$$

provided $r \ge r_0$ and B_r is well-contained in the interior of Ω.

Results of this type have been proved in [14] for $p = 2$ and in [35] and [36] for any $p \in (1, +\infty)$.

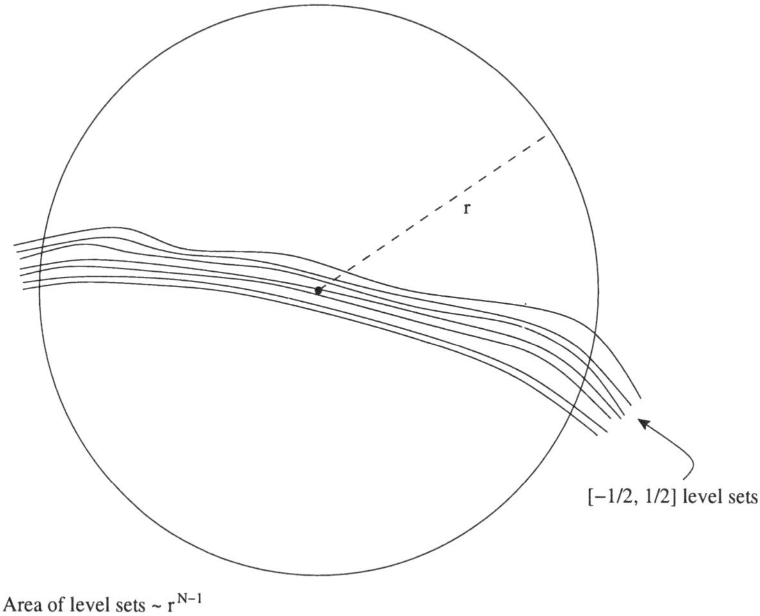

Using the above density estimates and the $L^1_{\rm loc}$-convergence of minimizers, it is not difficult to deduce that level sets of minimizers approach a minimal surface **uniformly**, see [14, 35, 36]. This also gives a more geometric interpretation of the Γ-convergence results obtained in [33] and [13].

We now discuss some results related with the **mean curvature properties** of the critical points of $\mathcal{F}_\Omega$. For this, we will disregard some technicalities and we will state the result in the following, quite heuristic, way:

Theorem 2: *Level sets of solutions of (1) possessing a certain decay property satisfy a mean curvature equation in a suitable weak viscosity sense. In particular, the level sets of the rescaled solution u_ϵ cannot be touched by below (above) by a convex (concave) paraboloid in a suitably small neighborhood.*

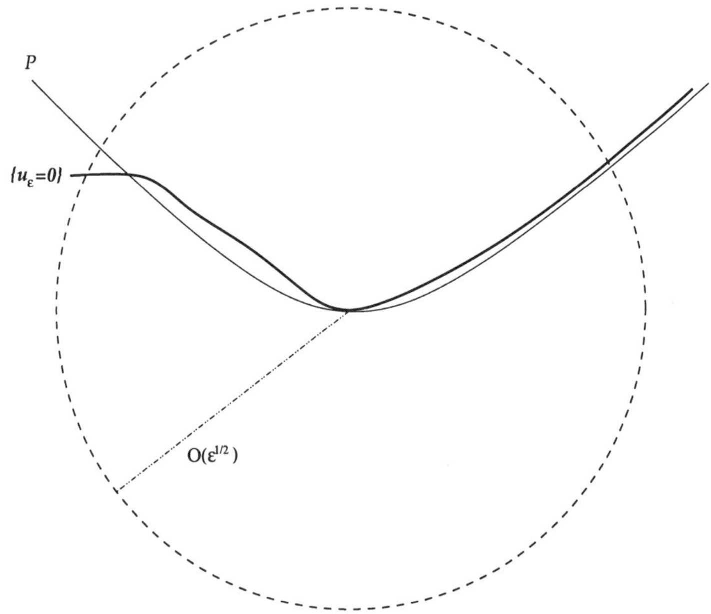

The paraboloid P does not touch the level set of the rescaled solution in a "small" neighborhood

The result above was proved in [39] for $p = 2$ (where many new fundamental ideas have been introduced), and in [40] for any $p \in (1, +\infty)$. We refer to the above papers for a more precise statement of the result. Here, we just point out that this implies that:

Corollary 3: *If the level sets of the solutions here above approach uniformly a hypersurface, then the latter has zero mean curvature.*

This somehow extends the results in [33] and [13], which hold for the minimizers of $\mathcal{F}_\Omega$, to non-minimal solutions of the equation (1): note that the area-minimality of the limit surface may get lost in this case, but the zero mean curvature property is preserved.

The above mean curvature property also plays a central role in the proof of the following **Harnack inequalities for level sets:**

Theorem 4: *Let u be a minimizer for $\mathcal{F}_\Omega$, for any $\Omega \subset \mathbf{R}^N$. Then,*

- *If the zero level set of u is trapped in a rectangle whose height is small enough, then, in a smaller neighborhood, it can be trapped in a rectangle with even smaller height.*
- *If the zero level set of u is trapped inside a flat rectangle, then, possibly changing coordinates, it is trapped in an even flatter rectangle in the interior.*

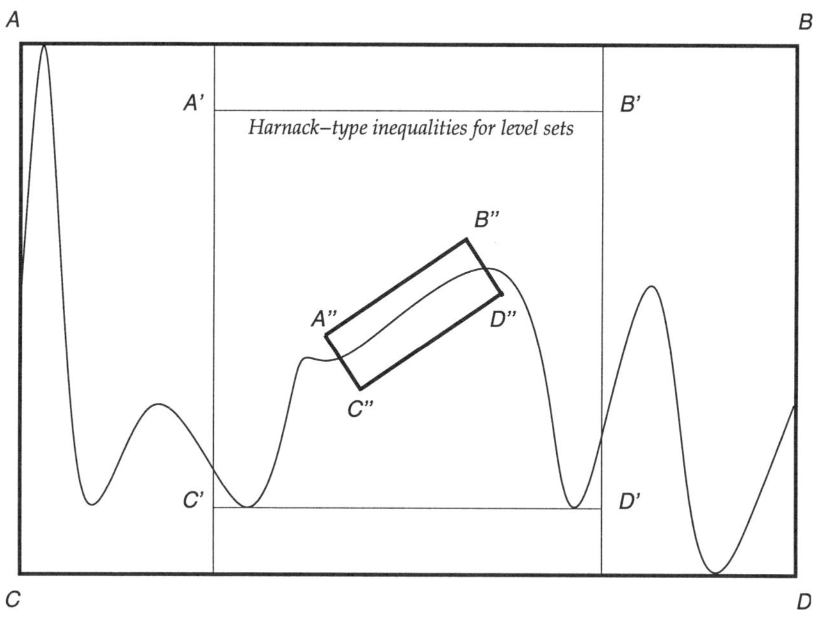

The picture here above is an attempt to describe such Harnack inequalities: indeed, if the zero level set of u, depicted as the fancy zigzag, is contained in the rectangle $ABCD$, which is assumed to be suitably short and flat, then it is contained in a shorter rectangle inside (namely, the rectangle $A'B'C'D'$) and, up to a rotation, in a flatter one (namely $A''B''C''D''$). Of course, a measure of the flatness of a rectangle is given by the height/base ratio.

The Harnack inequalities mentioned above have been proved in [39] for $p = 2$ and in [46] for any $p \in (1, +\infty)$. The proof, which is very hard, relies on the construction of suitable barriers, obtained by modifying the one-dimensional explicit heteroclinic solutions. The proof also involves some measure-theoretic estimates on the possible touching points between the barriers and the solution. In some sense, however, we may think that the above "flatness improvement" information on the interface is a consequence of the PDE structure of u and of the PDE structure of the limit surface. Further motivations and comments on the proof of such Harnack inequalities are given in [45].

By means of the Harnack inequality for level sets, it is possible to prove that, in suitably low dimension, global (i.e., defined in the whole $\mathbf{R}^N$) phase transitions do coincide with trivial (i.e., one-dimensional) ones. This rigidity property, which was first conjectured by De Giorgi[19] for $p = 2$, is related with the analogous rigidity of minimal hypersuperfaces – namely, with the fact that, in low dimension, global minimal hypersuperfaces are flat hyperplanes (see [28]).

The main result on the flatness of low dimensional level sets of global solutions for the phase transition problem may be stated as follows:

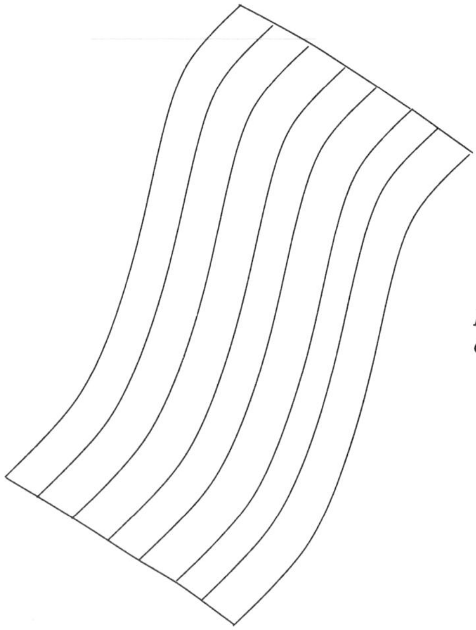

Flat level sets and one–dim. symmetry

Theorem 5: *Let $u \in W^{1,p}_{\rm loc}(\mathbf{R}^N)$ satisfy (1). Then, the level sets of u are flat hyperplanes in the following cases:*

- *if u is a minimizer and $N \le 7$;*
- *if $\partial_N u > 0$, $\lim_{x_N \to \pm\infty} u(\cdot, x_N) = \pm 1$ and $N \le 8$.*

Note that when the level sets of u are hyperplanes, then u may be thought as a function of only one variable (that is, u is "one-dimensional") and may be thence explicitly computed. The above result was proved in [24] for $p = N = 2$, in [5] and [1] for $p = 2$ and $N = 3$, in [39] for $p = 2$ and in [46] for any $p \in (1, +\infty)$. See also [18, 23, 22, 20, 8, 9, 25] and [21] for related results. Applications to the Heisenberg groups were also given in [11] and [12].

In higher dimensions, the problem remains essentially open (though flatness results can be obtained by assuming a linear growth at infinity of the level sets, see [39, 46]).

We now turn our attention to the periodic medium case, which is of obvious physical interest. In this case, we consider a modification of the above functional, in which we allow a periodic space-dependence. More precisely, we consider the functional

$$\mathcal{F}^{\rm per}_{\Omega}(u) = \int_{\Omega} \Big(a_{i,j}(x) \partial_i u \partial_j u \Big)^{p/2} + W(x, u)\, dx\,,$$

and we assume that the matrix $a_{i,j}$ is positive definite (which provides a possibly singular/degenerate elliptic operator) and that both $a_{i,j}$ and W are $\mathbf{Z}^N$-periodic in x. As expected also in the ODE setting[17], the system with a non-trivial space-dependence may exhibit some wild solutions, see [2] and [37]. Nevertheless, we have managed to construct minimizers of $\mathcal{F}^{\rm per}_{\Omega}$, for any $\Omega \subset \mathbf{R}^N$, whose level sets are at a uniform distance from a hyperplane. Hence, these transition layers form a **plane-like structure** in a periodic medium, according to the following claim:

Theorem 6: *Given any $\omega \in \mathrm{S}^{N-1}$, there exists u which is a minimizer of $\mathcal{F}^{\rm per}_{\Omega}$, for any $\Omega \subset \mathbf{R}^N$, for which the set $\{|u| < 1\}$ is contained in the strip*

$$\Big\{ x \cdot \omega \in [-M, M|] \Big\}\,,$$

with M universal.

If ω is rational, then u is periodic (with respect to the identification induced by ω); in any case, u can be approximated uniformly on compact sets by periodic minimizers.

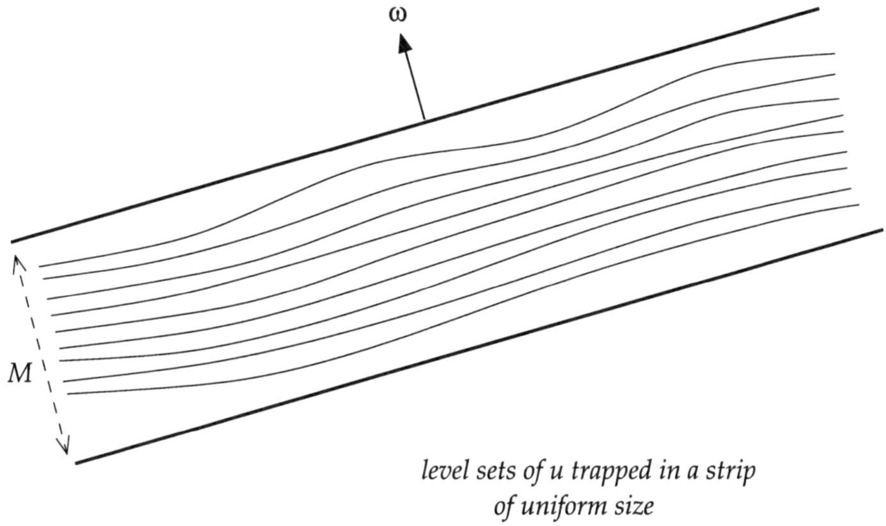

This result was proven in [42] and [43] for $p = 2$ and in [35] and [44] for any $p \in (1, +\infty)$. For related results, see [34] and [10] (in the elliptic integrand context), [15] and [7] (for minimal surfaces), and [16] (in the statistical mechanics framework). The papers [37, 10] and [32] also deal with non-minimal solutions.

For physical applications, in which the periodic medium has a microscopic modular scale, it is interesting to consider the scaled periodic case, that is the one in which the space variables are $\epsilon\mathbf{Z}^N$-periodic, for a small $\epsilon > 0$. In this case, the results in [43, 35] and [36] scale accordingly, providing the existence of minimizers in any direction whose transition layers are uniformly bounded by const ϵ.

These plane-like structures may also be seen as an attempt to extend KAM/Mather-type results from ODEs to elliptic PDEs: results of this type are indeed related with the quest of quasi-periodic structures which persist in "non-integrable" cases.

Namely, in the ODE setting, in many cases of interest, the space is laminated by minimal geodesics[29] or by quasi-periodic orbits lying on

Hamiltonian-flow-invariant tori[30]. In any case, the objects in these laminations may be seen as a deformation of flat structures. In the PDE setting, much work has been devoted to the extension of the above results to hyperbolic equations, see, e.g., [31], while the elliptic case has only very recently been addressed by quite different techniques. In the elliptic case, the plane-like structures are obtained either by solutions of the PDEs, as in [34] and [10], or by minimal hypersurfaces, as in [15], or by level sets of minimizers, see [43, 35] and [36] (the latter setting also contains the minimal hypersurfaces as a limit case). In particular, Theorem 6 asserts that, given any "frequency" ω, a minimal solution $u = u_\omega$ is constructed, whose level sets do not oscillate much far away from the flat hyperplane normal to ω.

The proof of Theorem 6 is based on a combination of ODE and PDE ingredients. The basic PDE tool is a density estimate (like in Theorem 1), which, in the course of the proof, leads to the construction of a suitably large ball outside the transition layer between the phases. Then, a convenient minimizer is selected (the so-called *minimal minimizer*), which possesses a non-self-intersection feature, known, in the dynamical system setting, with the name of *Birkhoff property.* A picture which makes this property close to intuition may be easily depicted, indeed, in the case of minimal geodesics on $\mathbf{R}^2$ with a $\mathbf{Z}^2$-periodic metric (or, analogously, in the case of minimal geodesics of $\mathbf{T}^2$). As a matter of fact, in this case the curve γ drawn here below cannot be a minimal geodesic from A to B, otherwise the curve $\gamma+e_2$ would also be minimal from $A + e_2$ and $B + e_2$. Then, the curve agreeing with γ from A to C and from D to B and with $\gamma + e_2$ from D to B would also be minimal. But the latter cannot hold, since such curve is not smooth (in C and D, otherwise $\gamma = \gamma + e_2$ by standard uniqueness ODE results), against the regularity results of geodesics.

This gives an idea of the fact that geodesics on $\mathbf{R}^2/\mathbf{Z}^2$ do not intersect their integer translations. Such property (and such proof!) also generalizes to our case and allows us to consider integer translations of the ball constructed above, which still lie outside the transition layer. This traps the transition layer in a strip of universal height, giving the desired result.

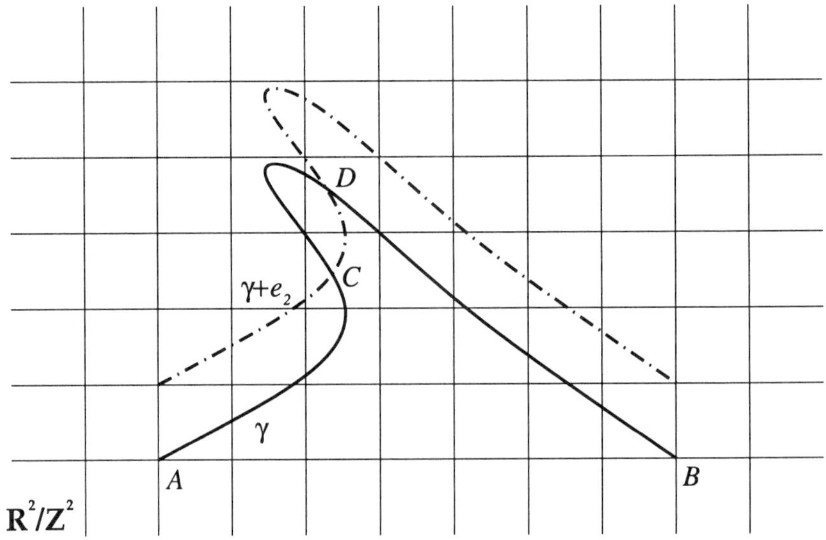

Acknowledgments

We thank the MIUR *Variational Methods and Nonlinear Differential Equations* project for partial supporting our research.

References

[1] G. Alberti, L. Ambrosio and X. Cabré, On a long-standing conjecture of E. De Giorgi: symmetry in 3D for general nonlinearities and a local minimality property. *Acta Appl. Math.* **65**, 9–33 (2001).

[2] F. Alessio, L. Jeanjean and P. Montecchiari, Existence of infinitely many stationary layered solutions in $\mathbf{R}^2$ for a class of periodic Allen-Cahn equations. *Comm. Partial Differential Equations* **27**, 1537–1574 (2002).

[3] S.M. Allen and J.W. Cahn, A microscopic theory for antiphase boundary motion and its application to antiphase domain coarsening. *Acta Metall. Mater.* **27**, 1085–1095 (1979).

[4] H.W. Alt and L.A. Caffarelli, Existence and regularity for a minimum problem with free boundary. *J. Reine Angew. Math.* **325**, 105–144 (1981).

[5] L. Ambrosio and X. Cabré, Entire solutions of semilinear elliptic equations in $\mathbf{R}^3$ and a conjecture of De Giorgi. *J. Amer. Math. Soc.* **13**, 725–739 (2000).

[6] S.S. Antman, Nonuniqueness of equilibrium states for bars in tension. *J. Math. Anal. Appl.* **44**, 333–349 (1973).

[7] F. Auer and V. Bangert, Minimising currents and the stable norm in codimension one. *C. R. Acad. Sci. Paris Sr. I Math.* **333**, 1095-1100 (2001).
[8] M.T. Barlow, R.F. Bass and C. Gui, The Liouville property and a conjecture of De Giorgi, *Comm. Pure Appl. Math.* **53**, 1007–1038 (2000).
[9] H. Berestycki, F. Hamel and R. Monneau, One-dimensional symmetry of bounded entire solutions of some elliptic equations. *Duke Math. J.* **103**, 375–396 (2000).
[10] U. Bessi, Many plane like solutions of n dimensional elliptic problems. *Preprint.*
[11] I. Birindelli and E. Lanconelli, A negative answer to a one-dimensional symmetry problem in the Heisenberg group. *Calc. Var. Partial Differential Equations* **18**, 357–372 (2003).
[12] I. Birindelli and J. Prajapat, One-dimensional symmetry in the Heisenberg group. *Ann. Scuola Norm. Sup. Pisa Cl. Sci. (4)* **30**, 269–284 (2001).
[13] G. Bouchitté, Singular perturbations of variational problems arising from a two-phase transition model. *Appl. Math. Optim.* **21**, 289–314 (1990).
[14] L.A. Caffarelli and A. Córdoba, Uniform convergence of a singular perturbation problem. *Comm. Pure Appl. Math.* **48**, 1–12 (1995).
[15] L.A. Caffarelli and R. de la Llave, *Planelike minimizers in periodic media. Comm. Pure Appl. Math.* **54**, 1403–1441 (2001).
[16] L.A. Caffarelli and R. de la Llave, Interfaces of ground states in Ising models with periodic coefficients. *J. Stat. Phys.* **118**, 687–719 (2005).
[17] V. Coti Zelati, I. Ekeland and E. Séré, *A variational approach to homoclinic orbits in Hamiltonian systems. Math. Ann.* **288**, 133–160 (1990).
[18] D. Danielli and N. Garofalo, Properties of entire solutions of non-uniformly elliptic equations arising in geometry and in phase transitions. *Calc. Var. Partial Differential Equations* **15**, 451–491 (2002).
[19] E. De Giorgi, Convergence problems for functionals and operators. In: *Proceedings of the International Meeting on Recent Methods in Nonlinear Analysis (Rome, 1978)* (Pitagora, Bologna, 1979), p. 131.
[20] J. Dolbeault and R. Monneau, On a Liouville type theorem for isotropic homogeneous fully nonlinear elliptic equations in dimension two. *Ann. Sc. Norm. Super. Pisa Cl. Sci. (5)* **2**, 181–197 (2003).
[21] Y. Du and L. Ma, Some Remarks on De Giorgi conjecture. *Proc. Amer. Math. Soc.* **131**, 2415–2422 (2002).
[22] A. Farina, One-dimensional symmetry for solutions of quasilinear equations in $\mathbf{R}^2$. *Boll. Unione Mat. Ital. Sez. B Artic. Ric. Mat. (8)* **6** 685–692 (2003).
[23] A. Farina, Rigidity and one-dimensional symmetry for semilinear elliptic equations in the whole of $\mathbf{R}^N$ and in half spaces. *Adv. Math. Sci. Appl.* **13**, 65–82 (2003).
[24] N.A. Ghoussoub and C. Gui, On a conjecture of De Giorgi and some related problems. *Math. Ann.* **311**, 481–491 (1998).
[25] N.A. Ghoussoub and C. Gui, *On the De Giorgi conjecture in dimensions* 4 *and* 5. *Ann. of Math. (2)* **157**, 313–334 (2003).
[26] V. Ginzburg and L. Landau, On the theory of superconductivity. *Zh. Èksper. Teoret. Fiz.* **20**, 1064–1082 (1950).

[27] V. Ginzburg and L.P. Pitaevskiĭ, *On the theory of superfluidity. Soviet Physics JETP* **34/7**, 858–861 (1958).
[28] E. Giusti, *Minimal Surfaces and Functions of Bounded Variation.* Birkhäuser Verlag, Basel, 1984.
[29] G.A. Hedlund, Geodesics on a two-dimensional Riemannian manifold with periodic coefficients. *Ann. of Math. (2)* **33**, 719–739 (1932).
[30] A.N. Kolmogorov, On the preservation of conditionally periodic motions under small variations of the Hamilton function. *Dokl. Akad. Nauk SSSR* **98**, 527-530 (1954). English translation in: *Selected Works of A. N. Kolmogorov. Vol. I*, Ed. V.M. Tikhomirov, Kluwer Academic Publishers, Dordrecht, 1991.
[31] S.B. Kuksin, *Nearly integrable infinite-dimensional Hamiltonian systems.* Springer-Verlag, Berlin, 1993.
[32] R. de la Llave and E. Valdinoci, Multiplicity results for interfaces of Ginzburg-Landau-Allen-Cahn equations in periodic media. *Preprint.*
[33] L. Modica and S. Mortola, Un esempio di Γ-convergenza. *Boll. Un. Mat. Ital. B (5)* **14**, 285–299 (1977).
[34] J. Moser, Minimal solutions of variational problems on a torus. *Ann. Inst. H. Poinc. Anal. Non Lin.* **3**, 229–272 (1986).
[35] A. Petrosyan and E. Valdinoci, Geometric properties of Bernoulli-type minimizers. *Interfaces Free Bound.* **7**, 55–78 (2005).
[36] A. Petrosyan and E. Valdinoci, Density estimates for a degenerate/singular phase-transition model. *SIAM J. Math. Anal.* **36**, 1057–1079 (2005).
[37] P.H. Rabinowitz and E.W. Stredulinsky, Mixed states for an Allen-Cahn type equation. *Commun. Pure Appl. Math.* **56**, 1078–1134 (2003).
[38] J.S. Rowlinson, Translation of J.D. van der Waals' "The thermodynamic theory of capillarity under the hypothesis of a continuous variation of density". *J. Statist. Phys.* **20**, 197–244 (1979).
[39] V.O. Savin, *Phase Transitions: Regularity of Flat Level Sets.* Ph.D. thesis, University of Texas at Austin, 2003.
[40] B. Sciunzi and E. Valdinoci, Mean curvature properties for p-Laplace phase transitions. *J. Eur. Math. Soc. (JEMS)* **7**, 319–359 (2005).
[41] P. Sternberg, The effect of a singular perturbation on nonconvex variational problems. *Arch. Rational Mech. Anal.* **101**, 209–260 (1988).
[42] E. Valdinoci, *Plane-like minimizers in periodic media: jet flows and Ginzburg-Landau.* Ph.D. thesis, University of Texas at Austin, 2001 (available on-line at `www.math.utexas.edu/mp_arc`).
[43] E. Valdinoci, Plane-like minimizers in periodic media: jet flows and Ginzburg-Landau-type functionals. *J. Reine Angew. Math.* **574**, 147–185 (2004).
[44] E. Valdinoci, Bernoulli jets and the zero mean curvature equation. *J. Differential Equations* (to appear).
[45] E. Valdinoci, Flatness of Bernoulli jets. *Preprint.*
[46] E. Valdinoci, B. Sciunzi and V.O. Savin. Flat level set regularity of p-Laplace phase transitions. *Preprint.*

Series on Advances in Mathematics for Applied Sciences

Aims and Scope

This Series reports on new developments in mathematical research relating to methods, qualitative and numerical analysis, mathematical modeling in the applied and the technological sciences. Contributions related to constitutive theories, fluid dynamics, kinetic and transport theories, solid mechanics, system theory and mathematical methods for the applications are welcomed.

This Series includes books, lecture notes, proceedings, collections of research papers. Monograph collections on specialized topics of current interest are particularly encouraged. Both the proceedings and monograph collections will generally be edited by a Guest editor.

High quality, novelty of the content and potential for the applications to modern problems in applied science will be the guidelines for the selection of the content of this series.

Instructions for Authors

Submission of proposals should be addressed to the editors-in-charge or to any member of the editorial board. In the latter, the authors should also notify the proposal to one of the editors-in-charge. Acceptance of books and lecture notes will generally be based on the description of the general content and scope of the book or lecture notes as well as on sample of the parts judged to be more significantly by the authors.

Acceptance of proceedings will be based on relevance of the topics and of the lecturers contributing to the volume.

Acceptance of monograph collections will be based on relevance of the subject and of the authors contributing to the volume.

Authors are urged, in order to avoid re-typing, not to begin the final preparation of the text until they received the publisher's guidelines. They will receive from World Scientific the instructions for preparing camera-ready manuscript.

SERIES ON ADVANCES IN MATHEMATICS FOR APPLIED SCIENCES

Vol. 35 Mechanics of Mixtures
by K. R. Rajagopal and L. Tao

Vol. 36 Dynamical Mechanical Systems Under Random Impulses
by R. Iwankiewicz

Vol. 37 Oscillations in Planar Dynamic Systems
by R. E. Mickens

Vol. 38 Mathematical Problems in Elasticity
ed. R. Russo

Vol. 39 On the Way to Understanding the Time Phenomenon: The Constructions of Time in Natural Science — Part 2. The "Active" Properties of Time According to N. A. Kozyrev
ed. A. P. Levich

Vol. 40 Advanced Mathematical Tools in Metrology II
eds. P. Ciarlini et al.

Vol. 41 Mathematical Methods in Electromagnetism Linear Theory and Applications
by M. Cessenat

Vol. 42 Constrained Control Problems of Discrete Processes
by V. N. Phat

Vol. 43 Motor Vehicle Dynamics: Modeling and Simulation
by G. Genta

Vol. 44 Microscopic Theory of Condensation in Gases and Plasma
by A. L. Itkin and E. G. Kolesnichenko

Vol. 45 Advanced Mathematical Tools in Metrology III
eds. P. Cianlini et al.

Vol. 46 Mathematical Topics in Neutron Transport Theory — New Aspects
by M. Mokhtar-Kharroubi

Vol. 47 Theory of the Navier–Stokes Equations
eds. J. G. Heywood et al.

Vol. 48 Advances in Nonlinear Partial Differential Equations and Stochastics
eds. S. Kawashima and T. Yanagisawa

Vol. 49 Propagation and Reflection of Shock Waves
by F. V. Shugaev and L. S. Shtemenko

Vol. 50 Homogenization
eds. V. Berdichevsky, V. Jikov and G. Papanicolaou

Vol. 51 Lecture Notes on the Mathematical Theory of Generalized Boltzmann Models
by N. Bellomo and M. Lo Schiavo

SERIES ON ADVANCES IN MATHEMATICS FOR APPLIED SCIENCES

Vol. 52 Plates, Laminates and Shells: Asymptotic Analysis and Homogenization
by T. Lewiński and J. J. Telega

Vol. 53 Advanced Mathematical and Computational Tools in Metrology IV
eds. P. Ciarlini et al.

Vol. 54 Differential Models and Neutral Systems for Controlling the Wealth of Nations
by E. N. Chukwu

Vol. 55 Mesomechanical Constitutive Modeling
by V. Kafka

Vol. 56 High-Dimensional Nonlinear Diffusion Stochastic Processes — Modelling for Engineering Applications
by Y. Mamontov and M. Willander

Vol. 57 Advanced Mathematical and Computational Tools in Metrology V
eds. P. Ciarlini et al.

Vol. 58 Mechanical and Thermodynamical Modeling of Fluid Interfaces
by R. Gatignol and R. Prud'homme

Vol. 59 Numerical Methods for Viscosity Solutions and Applications
eds. M. Falcone and Ch. Makridakis

Vol. 60 Stability and Time-Optimal Control of Hereditary Systems — With Application to the Economic Dynamics of the US (2nd Edition)
by E. N. Chukwu

Vol. 61 Evolution Equations and Approximations
by K. Ito and F. Kappel

Vol. 62 Mathematical Models and Methods for Smart Materials
eds. M. Fabrizio, B. Lazzari and A. Morro

Vol. 63 Lecture Notes on the Discretization of the Boltzmann Equation
eds. N. Bellomo and R. Gatignol

Vol. 64 Generalized Kinetic Models in Applied Sciences — Lecture Notes on Mathematical Problems
by L. Arlotti et al.

Vol. 65 Mathematical Methods for the Natural and Engineering Sciences
by R. E. Mickens

Vol. 66 Advanced Mathematical and Computational Tools in Metrology VI
eds. P. Ciarlini et al.

Vol. 67 Computational Methods for PDE in Mechanics
by B. D'Acunto

SERIES ON ADVANCES IN MATHEMATICS FOR APPLIED SCIENCES

Vol. 68 Differential Equations, Bifurcations, and Chaos in Economics
by W. B. Zhang

Vol. 69 Applied and Industrial Mathematics in Italy
eds. M. Primicerio, R. Spigler and V. Valente

Vol. 70 Multigroup Equations for the Description of the Particle Transport in Semiconductors
by M. Galler

Vol. 71 Dissipative Phase Transitions
eds. P. Colli, N. Kenmochi and J. Sprekels